PRISTIONCHUS PACIFICUS – A NEMATODE MODEL FOR COMPARATIVE AND EVOLUTIONARY BIOLOGY

PRISTIONCHUS PACIFICUS – A NEMATODE MODEL FOR COMPARATIVE AND EVOLUTIONARY BIOLOGY

Edited by
Ralf J. Sommer

David J. Hunt and Roland N. Perry (Series Editors)

NEMATOLOGY MONOGRAPHS AND PERSPECTIVES
VOLUME 11

BRILL
LEIDEN-BOSTON
2015

This book is printed on acid-free paper.

Library of Congress Cataloging-in-Publication Data

Pristionchus pacificus : a nematode model for comparative and evolutionary biology / edited by Ralf J. Sommer.
pages cm. – (Nematology monographs and perspectives ; volume 11)
Includes bibliographical references and index.
ISBN 978-90-04-26029-0 (hardback : alk. paper) – ISBN 978-90-04-26030-6 (e-book)
1. Nematodes. 2. Evolution (Biology) I. Sommer, Ralf J., 1963- editor.
QL391.N4P75 2015
592'.57–dc23

2015000351

ISBN: 978 90 04 26029 0
E-ISBN: 978 90 04 26030 6

Contents

Contributors

Laura AURILIO
Life Sciences & Bioengineering Center
Worcester Polytechnic Institute
Worcester, MA 01609, USA
E-mail: laurilio@wpi.edu

Federico D. BROWN
Departamento de Zoologia
Instituto de Biociências da Universidade de São Paulo
Rua do Matão, 14
São Paulo – SP 05508-090, Brazil
E-mail: fdbrown@usp.br

Daniel J. BUMBARGER
Allen Institute for Brain Science
Seattle, WA 98103, USA
E-mail: danb@alleninstitute.org

Christoph DIETERICH
Max-Planck Institute for the Biology of Aging
Joseph-Stelzmann Straße 9b
D-50931 Köln/Cologne, Germany
E-mail: Christoph.Dieterich@age.mpg.de

Robin M. GIBLIN-DAVIS
Fort Lauderdale Research and Education Center
University of Florida
3205 College Avenue
Davie, FL 33314, USA
E-mail: giblin@ufl.edu

Matthias HERRMANN
Department for Evolutionary Biology
Max-Planck Institute for Developmental Biology
72076 Tübingen, Germany
E-mail: matthias.herrmann@tuebingen.mpg.de

Ray L. HONG
Biology Department
California State University
Northridge, CA 91330, USA
E-mail: ray.hong@csun.edu

Natsumi KANZAKI
Forest Pathology Laboratory
Forestry and Forest Products Research Institute
Ibaraki, 305-8687, Japan
E-mail: natsumikanzaki@hotmail.com

Angela MCGAUGHRAN
Bioinformatics & Phylogenomics Team
CSIRO Ecosystem Sciences
GPO Box 1700
Canberra, ACT 2601, Australia
E-mail: ang.mcgaughran@gmail.com

Katy MORGAN
Computer Center, Room 200
University of New Orleans
2000 Lakeshore Drive
New Orleans, LA 70148, USA
E-mail: kmorgan4@uno.edu

Akira OGAWA
Laboratory for Developmental Dynamics
RIKEN Quantitative Biology Center
2-2-3 Minatojima-minamimachi
Chuo-ku, Kobe, 650-0047, Japan
E-mail: akira.ogawa@riken.jp

Robbie RAE
School of Natural Sciences & Psychology
Liverpool John Moores University
Byrom Street
Liverpool, L3 3AF, UK
E-mail: r.g.rae@ljmu.ac.uk

Erik J. RAGSDALE
Department of Biology
Indiana University
Bloomington, IN 47405, USA
E-mail: ragsdale@indiana.edu

Metta RIEBESELL
Department for Evolutionary Biology
Max-Planck Institute for Developmental Biology Tübingen
72076 Tübingen, Germany
E-mail: metta.riebesell@tuebingen.mpg.de

Christian RÖDELSPERGER
Department for Evolutionary Biology
Max-Planck Institute for Developmental Biology Tübingen
72076 Tübingen, Germany
E-mail: christian.roedelsperger@tuebingen.mpg.de

David RUDEL
Department of Biology
East Carolina University
Greenville, NC 27858, USA
E-mail: rudeld@ecu.edu

Frank C. SCHROEDER
Boyce Thompson Institute and
Department of Chemistry and Chemical Biology
Cornell University
Ithaca, NY 14853, USA
E-mail: schroeder@cornell.edu

Amit SINHA
Department of Neurobiology
University of Massachusetts Medical School
364 Plantation Street
Worcester, MA 01605, USA
E-mail: Amit.Sinha@umassmed.edu

Ralf J. SOMMER
Department for Evolutionary Biology
Max-Planck Institute for Developmental Biology Tübingen
72076 Tübingen, Germany
E-mail: ralf.sommer@tuebingen.mpg.de

Jagan SRINIVASAN
Life Sciences & Bioengineering Center
Worchester Polytechnic Institute
Worcester, MA 01609, USA
E-mail: jsrinivasan@wpi.edu

Paul W. STERNBERG
Division of Biology
Caltech
Pasadena, CA 91125, USA
E-mail: pws@caltech.edu

Foreword

Nematodes have a long history as subjects of investigation in basic and applied research. Generations of scientists have studied the influence of nematodes on major agricultural crops and, likewise, nematode parasites of humans and livestock have been investigated for more than a century. In parallel, taxonomists have catalogued the nearly endless diversity of free-living, terrestrial, marine and fresh-water nematodes. It was this subdivision into different research fields, resulting from the huge diversity of ecologies in which nematodes are found, which previously limited a comprehensive perspective of nematode biology. Two major developments over the last three decades have changed this perspective. First, one nematode has been established as a major model system for modern life sciences. *Caenorhabditis elegans* has been at the forefront of elucidating the mechanisms of development, neurobiology and behaviour. *Caenorhabditis elegans* was also the first metazoan to have its genome fully sequenced and, at the time of writing, more than a decade later, it is still the only animal for which the complete genome sequence is available. Second, unforeseeable developments in molecular biology over the last 30 years, but particularly in the last decade, have made it possible to obtain molecular insight into organisms that cannot easily be cultured in the laboratory, including many of the parasitic nematodes. With new tools and insight, novel questions can now be asked and the unfortunate divide in nematodes and nematology can be put aside. These promising research perspectives have resulted in the major paradigm shift that we are now witnessing. Research on nematodes – no matter if basic or applied – has to take an evolutionary perspective. All organisms on earth are the result of historical, evolutionary processes and therefore understanding any type of biological pattern or process will ultimately require a serious consideration of evolutionary biology. Nonetheless, establishing the basic parameters for detailed and comprehensive evolutionary studies is a demanding task.

This book summarises the attempts to take the model system approach to evolutionary biology by establishing a second nematode, *Pristionchus pacificus*, as a model for comparative and evolutionary studies. Covering the many established research fields in evolutionary biology and trying to integrate them into a holistic perspective represents an ambitious enterprise. While such a research programme never comes to an end, important milestones have been achieved in recent years making it worthwhile to provide a monographic summary of current knowledge. This book therefore aims to outline the different perspectives of mechanistic approaches in comparative and evolutionary biology and to summarise our current understanding. The nematode case study presented here will hopefully encourage others to take similar approaches in other animal and plant groups. This book will be of value to nematologists and evolutionary biologists alike. For nematologists, the *C. elegans*-*P. pacificus* juxtaposition puts mechanistic studies in a comparative framework and thereby reveals what we can learn, and what we cannot learn, from a model system approach. For evolutionary biologists, nematodes will hopefully earn a reputation as exciting study subjects, even though most are microscopic in size. But finally, I hope that this book will be of value to all those biologists still interested in a holistic perspective of organisms and life on earth.

Ralf J. SOMMER
Department for Evolutionary Biology
Max-Planck Institute for Developmental Biology,
Tübingen, Germany
Tübingen, July 2014

Acknowledgements

The Volume Editor wants to thank several groups of individuals for their support of the project. First, I want to thank several friends and colleagues who helped by reviewing the individual chapters of this book. They are James Baldwin (Riverside, CA, USA), Helge Bode (Frankfurt, Germany), David Hall (New York, NY, USA), David Hunt (Egham, UK), Roland Perry (Hertfordshire, UK) and Adrian Streit (Tübingen, Germany). I am especially grateful to the Series Editors, David Hunt and Roland Perry, for their initial encouragement in putting this project together and their helpful insight and guidance in bringing it to a conclusion.

I want to thank my many master and graduate students, postdocs and research technicians for their wonderful work over the last two decades, which was essential in placing *Pristionchus pacificus* on the map – generating tools and studying exciting research questions. Staying excited is not always easy in basic research but together we have achieved this. I hope that this research enterprise was as exciting for them as it was and still is for me. Establishing a new model sooner or later requires additional scientists to join and bring in new expertise. Among others, I am grateful to Richard Wilson and Sandra Clifton from the Genome Sequencing Center at Washington University in St Louis, MO, USA, for a wonderful and extremely professional collaboration during the *P. pacificus* Sequencing Project funded by NIH, Boris Maček from the Proteome Center of Tübingen University and Frank Schroeder from Cornell University for starting chemical biology of *P. pacificus*. Finally, I want to thank the *C. elegans* Research Community for all of the support that we have received over the years. Among them is one person without whom this project would not have been possible in the first place. My mentor, colleague and friend Paul W. Sternberg provided me with all it takes to start something novel: knowledge of state-of-the-art research in nematode developmental biology, getting excited about any fundamental question in biology and, most importantly, the vision to let me run away with it to enjoy and thrive.

Ralf J. SOMMER

Nematology Monographs & Perspectives, 2015, Vol. 11, 1-17

Chapter 1

Why *Caenorhabditis elegans* is great and *Pristionchus pacificus* might be better

Paul W. STERNBERG

Howard Hughes Medical Institute and Division of Biology & Biological Engineering, California Institute of Technology, Pasadena, CA 91125, USA
pws@caltech.edu

Introduction

Caenorhabditis elegans has been remarkably successful as an intensively studied model for human biology and, to a lesser extent, as a model for other nematodes. Worldwide, nematodes have a huge impact on human, animal and plant health. The development of an auxiliary model nematode as an experimental system is of great interest for two reasons. The first reason is that one approach to understanding evolution is to have a detailed understanding of genetic pathways in more than one species, and thus be able to compare them at a mechanistic level. The second reason is that, given the diversity of nematode taxa, any one might not be sufficiently close to taxa of importance, and we have the expectation that with two (or more) models we can extrapolate and interpolate to many relevant taxa.

In this introductory chapter, I will summarise concisely the key features of *C. elegans* that have contributed to its success as a model organism. I will also discuss a few discoveries from the *C. elegans* field that I hope are instructive. I will then briefly discuss the early history of *Pristionchus* and finish with some thoughts on prospects from my particular perspective, coloured as it is by decades of work with *C. elegans* and a variety of other nematodes.

Useful features of *C. elegans*

Almost any life scientists can recite a litany of the wonderful qualities of *C. elegans* as an experimental organism. As a young graduate student, I realised that the promise of *C. elegans* had already been made so often that it was time for my generation of *C. elegans* scientists to deliver on that promise. In practice, this meant making fundamental discoveries. We did.

EASY ON THE EYES

Under a dissecting stereomicroscope with transmitted illumination (25×), the sinuous locomotion of these worms as they undulate is mesmerising. Admittedly, incident illumination, as one would view a fruit fly, highlights the glistening cuticle and the worms look quite creepy. At higher magnification (400×-1250×) under Nomarski differential interference contrast (DIC) optics *C. elegans* is lovely to observe. (Note well that not all nematodes are as easy to observe.) Living worms can be directly observed on microscope slides, mounted on agar pads and under a sealed cover slip. The transparency is one very useful feature, but the numerical simplicity and essential invariance of its anatomy and development figure prominently in our ability rapidly to learn aspects of its anatomy and development. Invariant anatomy not only makes it more rapid to learn, but also makes experimental or genetic perturbation of that anatomy easy to study using small numbers. In principle, if an intact wild-type organism has an identical property in 100 worms, then as few as one animal could be observed to detect a variant phenotype! A researcher can thus spend less time at the microscope or, more likely, get more done for the effort.

EASY ON THE POCKETBOOK

Caenorhabditis elegans eats bacteria (*Escherichia coli* in the laboratory), which themselves eat inexpensive media. The small, 3, 6 or 10 cm diam. Petri dishes are relatively inexpensive in plastic, as is the 1.7% (w/v) agar. This low cost allows us almost to ignore the cost of genetic experiments. Microscopes are expensive, but last for decades. Liquid growth in glass flasks is significantly more economical, without having to purchase plastic or agar. Acquisition of genetic stocks from the

Caenorhabditis Genetics Center or other laboratories is as inexpensive as anything not transferable over the Internet, with light-weight mail at room temperature.

Easy on the calendar

The rapid generation time, approximately 3.5 days egg-to-egg, means genetic and molecular genetic experiments can be cycled rapidly. And if you realise that an experimental design is sub-optimal after two generations, you have only lost a week.

Easy on the researcher

Hermaphrodite genetics allows ready production of progeny homozygous for recessive alleles (Fig. 1.1A). This is delightful for genetic screens, as demonstrated by Brenner (1974) and hundreds of *C. elegans* researchers. The presence of males and thus cross-fertilisation enables many genetic experiments, notably the construction of doubly mutant strains (Fig. 1.1B). Long-term maintenance is by cryogenic storage. Mutant stocks are frozen until a future time when they can be subjected to gene therapy (that is, rescue in transgenic animals). Intermediate term storage happens by default: starved, crowded worms form dauer larvae, and almost dried agar 'chips' can be rehydrated to recover stocks after many weeks of neglect.

Cell lineage

Just by watching cells divide, migrate, differentiate or die, one learns a great deal about developmental mechanisms. These high content observations provide a treasure trove of phenotypes. The observations of Sulston & Horvitz (1977), Kimble & Hirsh (1979), Sulston *et al.* (1981, 1983) and Newman *et al.* (1996) defined the cell lineages from the zygote to the differentiated cell types in the adult. The concept of inferring a cell lineage from direct observation is illustrated in Figure 1.2. Most somatic cells that die during *C. elegans*' life do so in a reproducible manner. This reproducibility, and the fact that we know what cell will die before it is born, allowed the genetic control of programmed cell death (apoptosis) to be elucidated.

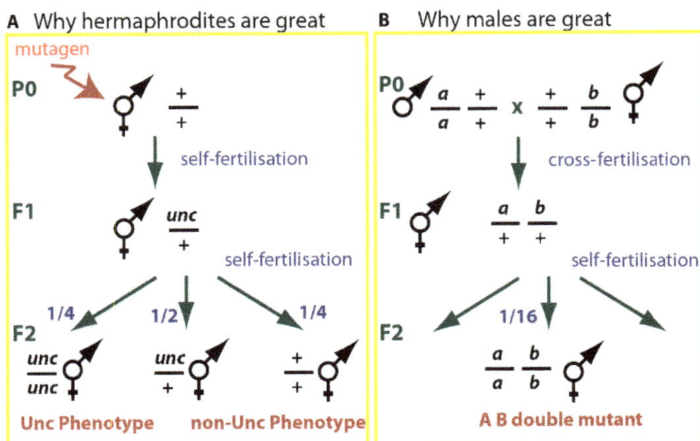

Fig. 1.1. *Useful features of hermaphrodites and males. A: Self-fertilising hermaphrodite genetics allows facile recovery of mutants after mutagenesis. Some progeny of mutagenised P0 hermaphrodites will have a recessive mutation (*unc *in this example). One-fourth of the progeny of a heterozygous F1 hermaphrodite will be homozygous for the mutation and display a phenotype (Unc in this example); B: Males allow us to construct strains that have two different mutations. If homozygous* a/a *males are mated to homozygous* b/b *hermaphrodites, cross-progeny will be heterozygous for both* a *and* b *(*a/+*;* b/+*). One sixteenth of its progeny will be the desired double mutant.*

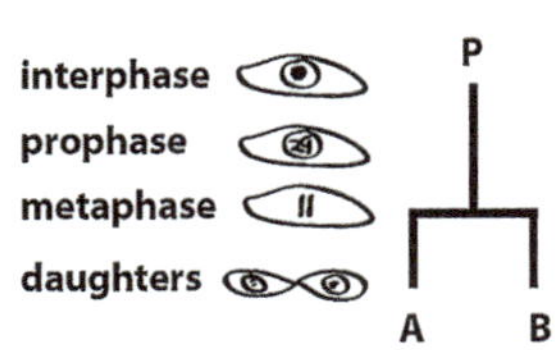

Fig. 1.2. *Inferring cell lineage from direct observation of cell division. Schematics of a post-embryonic cell division with interphase, prophase and metaphase shown on the left and the inferred lineage diagram on the right.*

Neural circuits

Having a complete physical connectome immediately suggests hypotheses about nervous system function, and helps constrain models (White *et al.*, 1986; Jarrell *et al.*, 2012). Cell ablations have defined necessity of neurons for particular behaviours, ranging from mechanosensation (Chalfie *et al.*, 1985), chemosensation (Bargmann &

Horvitz, 1991; Bargmann, 2006) to male mating (Liu & Sternberg, 1995; LeBoeuf *et al.*, 2014). The ability to activate neurons allows a gain-of-function approach; this has helped elucidate circuits underlying locomotion (Guo *et al.*, 2009; Leifer *et al.*, 2011; Stirman *et al.*, 2011) and sleep (Cho & Sternberg, 2014).

The ease of DNA-mediated transformation and transparency has allowed optogenetic tools to be applied in force to *C. elegans*. Given the rapid turnaround time for experiments, *C. elegans* is often the first choice for tests of function *in vivo*. This certainly was true for GFP (Chalfie *et al.*, 1994) and the genetically-encoded calcium sensor cameleon (Kerr *et al.*, 2000).

MICROFLUIDICS

Because of their small size and grace under pressure (they do not implode), *C. elegans* are well suited to microfluidic and optofluidic devices (San-Miguel & Lu, 2013).

Community resources

Community resources have played a huge role in making *C. elegans* research efficient. A stock centre stores and distributes strains. Some large-scale data and reagent production efforts have provided access to many genes. The genome consortium generated the first animal genome. In practice this was helped by a free-flow of information among researchers. Misplaced or contaminated clones were identified by researchers using the physical map and genome sequence as it was generated. The *C. elegans* Knockout Consortium generated thousands of deletion alleles. A resquencing effort, the Million Mutation Project, produced 2000 sequenced strains each with 300-500 random mutations (Thompson *et al.*, 2013). Information resources include, notably, WormBase (Harris *et al.*, 2014), WormAtlas (Altun *et al.*, 2002-2012) and WormBook (Girard *et al.*, 2007), together with a host of wonderful others that are highly original and well presented. These resources are not just for *C. elegans*. For example, WormBook has chapters and methods chapters for other nematodes, and WormBase handles a variety of genomes, transcriptomes, and so forth.

What *C. elegans* did for us

There have been an extraordinary number of fundamental discoveries with *C. elegans*. Many of these are historical, and you might not see many traces in the current literature, but at the time the impacts were massive.

Of particular interest to human biology include the first cloning of myosin (MacLeod *et al.*, 1981), elucidation of the EGF-RAS-RAF pathway (*e.g.*, Han *et al.*, 1993; Sundaram, 2013), defining the Notch signalling pathway (Greenwald *et al.*, 1983; Greenwald & Kovall, 2013), netrin and its receptor (Hedgecock *et al.*, 1990; Ishii *et al.*, 1992; Chan *et al.*, 1996), genetic control of apoptosis (Ellis *et al.*, 1986), the genetics of aging, RNAi, synaptic release proteins (*e.g.*, UNC-18; Hosono *et al.*, 1992) and acetylcholine receptors (Lewis *et al.*, 1980; Rand, 2007), families of transcription factors such as the POU domain (Finney *et al.*, 1988), microRNAs (Lee *et al.*, 1993), and many more.

For nematodes, studies with *C. elegans* have provided many discoveries, with notable examples of cuticle collagens, ascarosides, dafachronic acid and the targets of anti-nematode drugs. Nematode cuticle is crucial for their survival and locomotion. The unique collagen genes of nematodes were discovered in *C. elegans* (Kramer *et al.*, 1982) based on the morphological genetics first described by Brenner (1974), the Dumpy and Roller mutants. Ascarosides are secondary metabolites identified by activity-guided fractionation to play crucial roles as dauer pheromones (Jeong *et al.*, 2005; Butcher *et al.*, 2007), sex pheromones (Srinivasan *et al.*, 2008), and aggregation pheromones (Srinivasan *et al.*, 2012). There are many of these compounds produced by a variety of nematodes. The hormones that drive reproductive development, the dafachronic acids, were similarly identified by purification (Gerisch *et al.*, 2007). Lastly, the avermectin-inhibited chloride channels (Cully *et al.*, 1994) were discovered in *C. elegans*.

For all researchers, many widely used tools and methods were contributed by *C. elegans* research. These include GFP (Chalfie *et al.*, 1994), RNAi (Fire *et al.*, 1998), and integrated genome databases (AceDB; reviewed by Eeckman & Durbin, 1995; Waterston & Sulston, 1995). In addition, research helped define and refine what is now classical developmental genetics (*e.g.*, Greenwald & Horvitz, 1980; Greenwald *et al.*, 1983; Hodgkin, 1983; Ferguson *et al.*, 1987).

Limitations of *C. elegans*

One longstanding limitation has been an understanding of its natural ecology, population structure and biology and population genetics. This of course limits our ability to study evolution, as well as our ability to understand gene function, physiology and behaviour. The *C. elegans* genome has about 21 000 protein-coding genes and thousands of ncRNA genes. We are essentially clueless about the function of a majority of these genes. A typical *C. elegans* cell expresses roughly 8000 genes; again, we do not know what most of these genes are doing (Gerstein *et al.*, 2010; Schwarz *et al.*, 2012). To understand the function of these genes it is crucial to understand the life history, environment and ecology of *C. elegans* and its biotic interactions. Recently, a number of researchers have started to address this shortcoming, but this limitation sets the stage for a worm with more obvious biotic interactions.

History of *Pristionchus pacificus*

I had the pleasure of watching Dr Ralf Sommer start the *Pristionchus* field while he was a postdoctoral scholar at Caltech. In 1992, we had discussed how the time was ripe for studying the evolution of development and, when he came to Caltech in 1993, Dr Lynn Carta handed him a set of cultured nematodes to compare that were chosen based on their ability to grow and their phylogenetic distribution. Sommer analysed vulva development in these worms (Sommer & Sternberg, 1994, 1995, 1996a) and refined his interests in developing another species as a model. It needed to have useful features and be distant enough from *C. elegans* to have lots of interesting differences in its biology. *Pristionchus pacificus* strain PS312 was isolated in 1988 from soil obtained in Pasadena and brought into culture by Lynn Carta. The species was formally described in 1996 together with the isolation of mutants and the sequence of the *Ppa-let-60* gene encoding RAS (Sommer *et al.*, 1996).

Vulval development was the first aspect of *Pristionchus* developmental biology to be analysed. The focus was because this aspect of *C. elegans* was one of the most intensively studied. During early larval development, 12 ventral epithelial cells divide to produce an anterior daughter that is a neuroblast. The neuroblast generates ventral cord motor-neurons necessary for the sinusoidal locomotion; this is highly conserved. The posterior daughter of each division is epithelial, and the fate varies. In *C. elegans* hermaphrodites, of the 12 so-called Pn.p cells (P1.p, P2.p, P3.p, . . . , P12.p), the six mid-body Pn.p cells are competent to generate vulval tissue. By competent, we mean they can generate vulval cells if they receive appropriate intercellular signals, namely EGF (epidermal growth factor), notch-ligands and WNT (wingless-related integration site). The other cells differentiate as epithelial cells. In *P. pacificus*, the cells that are not part of the vulval competence group die (Sommer & Sternberg, 1996b). This is likely derived, and reproduces a theme inferred from studies of *Panagrellus redivivus* in which we had earlier found that cells of apparently arbitrary origin can be programmed to die (Sternberg & Horvitz, 1981, 1982). Indeed, in some species, for example *Turbatrix aceti*, vulval precursor cells depend on a gonadal signal for their survival (Félix & Sternberg, 1998).

However, *P. redivivus* is dioecious and thus not as amenable to genetic analysis as *C. elegans* and *P. pacificus*. *Caenorhabditis elegans* geneticists consider male-female species essentially intractable for genetic analysis, but this is patently absurd: *Drosophila melanogaster* and *Mus musculus* have certainly proven useful indeed! The finding of this first of many fascinating differences in the development of *P. pacificus* made it worth pursuing as a genetic model. Indeed, the initial genetic analysis of vulval development identified a *ced-3* loss-of-function mutant in which apoptosis fails to occur (Sommer *et al.*, 1998). In *ced-3(lf)* mutants, Pn.p cells are exhumed and can proliferate.

From these studies, in about 2 years Sommer had demonstrated that *Pristionchus* had most of the features, discussed above, that enabled *C. elegans* to become a premier model organism. We shall see in this monograph how these methods work in *P. pacificus*.

What *Pristionchus* did for me

One of the striking findings from Sommer and colleagues was the role of WNT in vulval induction in *P. pacificus*. After years of trying to demonstrate that male hook development in *C. elegans* was EGF-dependent (which it is, partially), we found that it relies on WNT signalling (Yu *et al.*, 2009). These observations raise the possibility that

WNT might have been the ancestral inducer of the vulva, but too few data points exist.

Comparative biology provides unique insight into difficult problems. For example, my laboratory staff were emboldened by the finding of apparent HOM-C dependent differences in competence along the anterior-posterior axis in other worms (Sommer & Sternberg, 1994), and revisited the issue in *C. elegans* (Clandinin *et al.*, 1997).

Science with other nematodes

In this monograph, you will see the remarkable progress that has been made with this model nematode. In each chapter, you will read of striking findings with broad biological significance. As with earthlings, *C. elegans* biologists tend to think a perfect model evolved only once. You will see that this is a myopic and narcissistic view because *Pristionchus* has clearly emerged as an experimental system in its own right. Also, we recently established the genetics of *Bursaphelenchus okinawaensis* (Shinya *et al.*, 2014). I will not comment here on the possibility of life on other planets.

Pristionchus has opened up many areas of biology

This monograph should make it clear that *Pristionchus* is contributing unique knowledge to our understanding of biology in general and nematodes in particular.

In Chapter 2, Ralf Sommer presents a cogent overview of how comparative biology benefits from the integration of mechanistic and evolutionary biological approaches.

The systematics of the group underlies any comparative analysis, and in Chapter 3, Robin Giblin-Davis and Natsumi Kanzaki discuss the phylogeny of the Diplogastridae, the family that includes *Pristionchus*. Erik Ragsdale, Matthias Herrmann and Natsumi Kanzaki describe in Chapter 4 the taxonomy and natural history of *Pristionchus* as a basis for the evolutionary biological aspect of the integration discussed by Sommer.

In Chapter 5, the genetics and molecular genetics of *Pristionchus* are summarised by Laura Aurilio and Jagan Srinivasan. These methods lay the foundation for using it as a laboratory model. As we all now fully appreciate, genomics and genome annotation enable much of what we do with an organism: sequence comparisons for proteins, non-coding RNAs, non-coding regulatory elements, transposons, and structural parts of chromosomes. The genome also allows so called functional genomic studies, largely involving genome-scale analyses, to be carried out. Christoph Dieterich and Christian Rödelsperger illuminate this crucial area of *Pristionchus* genetics in Chapter 6.

In Chapter 7, Frank Schroeder discusses the amazing diversity of nematode-derived secondary metabolites, small molecules derived from primary metabolism by presumably evolved pathways. At least some of these molecules, notably the ascarosides in *C. elegans* and the paratosides in *P. pacificus*, regulate key aspects of life cycle and style such as mating, development of mouth parts, dauers, aggregation and dispersal. One important way forward towards understanding the biosynthesis, function and evolution of these molecules is to measure their abundance in diverse strains and correlate their presence with ecological niches, life style and, ultimately, their genetic basis. *Pristionchus* is an excellent clade in which to carry out this research programme. The best-studied location for *Pristionchus* population genetics is on the island of La Réunion. In Chapter 8, Angela McGaughran and Katy Morgan summarise this case study so far.

In Chapter 9, David Rudel uses particular examples of drift in the specification process of developing organs. This line of research represents 'classic' evolution of development, popularly known as 'Evo-Devo'. Another aspect of development, physiology and behaviour is the dauer larval state of many nematodes. In Chapter 10, Akira Ogawa and Federico Brown give us the *Pristionchus* view of the dauer larva. Yet another fruitful area of *Pristionchus* research is the development and evolution of the mouth dimorphism. Erik Ragsdale describes in Chapter 11 how evolutionary novelty can be studied in this fascinating context.

Olfaction is a fundamental, widespread sensory modality. Free-living nematodes rely heavily on olfaction for many aspects of their life and, in Chapter 12, Ray Hong summarises the progress in *P. pacificus* olfaction and comparison to *C. elegans*. One of the most striking features of *Pristionchus* is the developmental plasticity of its mouth parts and associated pharyngeal nervous system. In Chapter 13, Daniel Bumbarger and Metta Riebesell describe the anatomy and connectivity of the neurons in this fascinating organ.

Lastly, animals, especially small animals that crawl around in bacteria and eat them, rely heavily on their innate immune system as a major defence against infection. Robbie Rae and Amit Sinha explore this key area of biology in Chapter 14.

These chapters describe an incredible swathe of biology. However, there are many other equally interesting aspects of *Pristionchus* that remain to be explored and exploited to generate fundamental biological knowledge. This book provides just a taste, and should serve as an enticement for those who might wish to partake. It should also demonstrate that this new model is indeed effective for exploring the integration of mechanisms of development, physiology and behaviour with population biology, natural ecology and evolutionary biology.

Prospects

Biological research will continue to be altered by technology. We are seeing the opening up to molecular analysis of new areas of study by inexpensive DNA sequencing. Enabling of new model systems or of any organism to experiments that require a genome, transcriptome, genome editing, *etc.*, is becoming routine. The technology has perhaps a greater impact on population studies (where the current affordable ten samples will become the future 100-1000) and on ecology with metagenomics. Analytical chemistry has advanced to the point where mass spectrometry and two-dimensional nuclear magnetic resonance (NMR) allow natural products to be discovered and metabolomes analysed as organismal, population or ecosystem phenotypes.

Nonetheless, the basics are still required, and this monograph demonstrates their existence and use. In addition, the success of a model organism depends greatly on the community or researchers, in particular their ability to communicate and the creativity, intensity and rigour of their science. I look forward to the next decades of *Pristionchus* research since its upward trajectory is very likely to continue.

Acknowledgements

I thank Bob Horvitz for introducing me to *P. redivivus*, *C. elegans* and their cell lineages, and Lynn Carta for getting my laboratory started on a rigorous course of nematode collection and analysis. Eric Davidson and the Marine Biology Laboratory Embryology course let me teach and meet promising young scientists. I thank the Howard Hughes Medical Institute, with which I am an Investigator, and the U.S. National Science Foundation for funding our initial comparative developmental studies.

References

ALTUN, Z.F., HERNDON, L.A., CROCKER, C., LINTS, R. & HALL, D.H. (2002-2012). *WormAtlas*. www.wormatlas.org.

BARGMANN, C.I. (2006). Chemosensation in *C. elegans*. In: The *C. elegans* Research Community (Ed.). *WormBook*. DOI:10.1895/wormbook.1.123.1.

BARGMANN, C.I. & HORVITZ, H.R. (1991). Chemosensory neurons with overlapping functions direct chemotaxis to multiple chemicals in *C. elegans*. *Neuron* 7, 729-742.

BRENNER, S. (1974). The genetics of *Caenorhabditis elegans*. *Genetics* 77, 71-94.

BUTCHER, R.A., FUJITA, M., SCHROEDER, F.C. & CLARDY, J. (2007). Small-molecule pheromones that control dauer development in *Caenorhabditis elegans*. *Nature Chemical Biology* 3, 420-422.

CHALFIE, M., SULSTON, J.E., WHITE, J.G., SOUTHGATE, E., THOMSON, J.N. & BRENNER, S. (1985). The neural circuit for touch sensitivity in *Caenorhabditis elegans*. *Journal of Neuroscience* 5, 956-964.

CHALFIE, M., TU, Y., EUSKIRCHEN, G., WARD, W.W. & PRASHER, D.C. (1994). Green fluorescent protein as a marker for gene expression. *Science* 263, 802-805.

CHAN, S.-Y., ZHENG, H., SU, M.-W., WILK, R., KILLEEN, M., HEDGECOCK, E. & CULOTTI, J.G. (1996). UNC-40, a *C. elegans* homolog of DCC (Deleted in Colorectal Cancer), is required in motile cells responding to UNC-6 netrin cues. *Cell* 87, 187-195.

CHO, J.Y. & STERNBERG, P.W. (2014). Multilevel modulation of a sensory motor circuit during *C. elegans* sleep and arousal. *Cell* 156, 249-260.

CLANDININ, T.R., KATZ, W.S. & STERNBERG, P.W. (1997). *Caenorhabditis elegans* HOM-C genes regulate the response of vulval precursor cells to inductive signal. *Developmental Biology* 182, 150-161.

CULLY, D.F., VASSILATIS, D.K., LIU, K.K., PARESS, P.S., VAN DER PLOEG, L.H., SCHAEFFER, J.M. & ARENA, J.P. (1994). Cloning of an avermectin-sensitive glutamate-gated chloride channel from *Caenorhabditis elegans*. *Nature* 371, 707-711.

EECKMAN, F.H. & DURBIN, R. (1995). ACeDB and macace. *Methods in Cell Biology* 48, 583-605.

Ellis, H.M. & Horvitz, H.R. (1986). Genetic control of programmed cell death in the nematode *C. elegans. Cell* 44, 817-829.

Felix, M.A. & Sternberg, P.W. (1998). A gonad-derived survival signal for vulval precursor cells in two nematode species. *Current Biology* 8, 287-290.

Ferguson, E.L., Sternberg, P.W. & Horvitz, H.R. (1987). A genetic pathway for the specification of the vulval cell lineages of *Caenorhabditis elegans*. *Nature* 326, 259-267.

Finney, M., Ruvkun, G. & Horvitz, H.R. (1988). The *C. elegans* cell lineage and differentiation gene *unc-86* encodes a protein with a homeodomain and extended similarity to transcription factors. *Cell* 55, 757-769.

Fire, A., Xu, S., Montgomery, M.K., Kostas, S.A., Driver, S.E. & Mello, C.C. (1998). Potent and specific genetic interference by double-stranded RNA in *Caenorhabditis elegans. Nature* 391, 806-811.

Gerisch, B., Rottiers, V., Li, D., Motola, D.L., Cummins, C.L., Lehrach, H., Mangelsdorf, D.J. & Antebi, A. (2007). A bile acid-like steroid modulates *Caenorhabditis elegans* lifespan through nuclear receptor signaling. *Proceedings of the National Academy of Sciences of the United States of America* 104, 5014-5019.

Gerstein, M.B., Lu, Z.J., Van Nostrand, E.L., Cheng, C., Arshinoff, B.I., Liu, T., Yip, K., Robilotto, R., Rechtsteiner, A., Ikegami, K. *et al.* (2010). Integrative analysis of the *Caenorhabditis elegans* genome by the modENCODE project. *Science* 330, 1775-1787.

Girard, L.R., Fiedler, T.J., Harris, T.W., Carvalho, F., Antoshechkin, I., Han, M., Sternberg, P., Stein, L.D. & Chalfie, M. (2007). WormBook: the online review of *Caenorhabditis elegans* biology. *Nucleic Acids Research* 35(Database issue), D472-D475.

Greenwald, I. & Kovall, R. (2013). Notch signaling: genetics and structure. In: The *C. elegans* Research Community (Ed.). *WormBook*. DOI:10.1895/wormbook.1.10.2.

Greenwald, I.S. & Horvitz, H.R. (1980). *unc-93(e1500)*: a behavioral mutant of *Caenorhabditis elegans* that defines a gene with a wild-type null phenotype. *Genetics* 96, 147-164.

Greenwald, I.S., Sternberg, P.W. & Horvitz, H.R. (1983). The *lin-12* locus specifies cell fates in *Caenorhabditis elegans. Cell* 34, 435-444.

Guo, Z.V., Hart, A.C. & Ramanathan, S. (2009). Optical interrogation of neural circuits in *Caenorhabditis elegans*. *Nature Methods* 6, 891-896.

Han, M., Golden, A., Han, Y. & Sternberg, P.W. (1993). *C. elegans lin-45 raf* gene participates in *let-60 ras*-stimulated vulval differentiation. *Nature* 363, 133-140.

Harris, T.W., Baran, J., Bieri, T., Cabunoc, A., Chan, J., Chen, W.J., Davis, P., Done, J., Grove, C., Howe, K. *et al.* (2014). WormBase 2014:

new views of curated biology. *Nucleic Acids Research* 42(Database issue), D789-D793.

HEDGECOCK, E.M., CULOTTI, J.G. & HALL, D.H. (1990). The *unc-5*, *unc-6*, and *unc-40* genes guide circumferential migrations of pioneer axons and mesodermal cells on the epidermis in *C. elegans*. *Neuron* 4, 61-85.

HODGKIN, J. (1983). Two types of sex determination in a nematode. *Nature* 304, 267-268.

HOSONO, R., HEKIMI, S., KAMIYA, Y., SASSA, T., MURAKAMI, S., NISHIWAKI, K., MWA, J., TAKELO, A. & KODAIRA, K.-I. (1992). The *unc-18* gene encodes a novel protein affecting the kinetics of acetylcholine metabolism in the nematode *Caenorhabditis elegans*. *Journal of Neurochemistry* 58, 1517-1525.

ISHII, N., WADSWORTH, W.G., STERN, B.D., CULOTTI, J.G. & HEDGECOCK, E.M. (1992). UNC-6, a laminin-related protein, guides cell and pioneer axon migrations in *C. elegans*. *Neuron* 9, 873-881.

JARRELL, T.A., WANG, Y., BLONIARZ, A.E., BRITTIN, C.A., XU, M., THOMSON, J.N., ALBERTSON, D.G., HALL, D.H. & EMMONS, S.W. (2012). The connectome of a decision-making neural network. *Science* 337, 437-444.

JEONG, P.-Y., JUNG, M., YIM, Y.-H., KIM, H., PARK, M., HONG, E., LEE, W., KIM, Y.H., KIM, K. & PAIK, Y.-K. (2005). Chemical structure and biological activity of the *Caenorhabditis elegans* dauer-inducing pheromone. *Nature* 433, 541-545.

KERR, R., LEV-RAM, V., BAIRD, G., VINCENT, P., TSIEN, R.Y. & SCHAFER, W.R. (2000). Optical imaging of calcium transients in neurons and pharyngeal muscle of *C. elegans*. *Neuron* 26, 583-594.

KIMBLE, J. & HIRSH, D. (1979). The postembryonic cell lineages of the hermaphrodite and male gonads in *Caenorhabditis elegans*. *Developmental Biology* 70, 396-417.

KRAMER, J.M., COX, G.N. & HIRSH, D. (1982). Comparisons of the complete sequences of two collagen genes from *Caenorhabditis elegans*. *Cell* 30, 599-606.

LEBOEUF, B., CORREA, P., JEE, C. & GARCIA, L.R. (2014). *Caenorhabditis elegans* male sensory-motor neurons and dopaminergic support cells couple ejaculation and post-ejaculatory behaviors. *eLife* 3, e02938.

LEE, R.C., FEINBAUM, R.L. & AMBROS, V. (1993). The *C. elegans* heterochronic gene *lin-4* encodes small RNAs with antisense complementarity to *lin-14*. *Cell* 75, 843-854.

LEIFER, A.M., FANG-YEN, C., GERSHOW, M., ALKEMA, M.J. & SAMUEL, A.D. (2011). Optogenetic manipulation of neural activity in freely moving *Caenorhabditis elegans*. *Nature Methods* 8, 147-152.

LEWIS, J., WU, C.-H., LEVINE, J. & BERG, H. (1980). Levamisole-resistant mutants of the nematode *Caenorhabditis elegans* appear to lack pharmacological acetylcholine receptors. *Neuroscience* 5, 967-989.

LIU, K.S. & STERNBERG, P.W. (1995). Sensory regulation of male mating behavior in *Caenorhabditis elegans*. *Neuron* 14, 79-89.

MACLEOD, A.R., KARN, J. & BRENNER, S. (1981). Molecular analysis of the *unc-54* myosin heavy-chain gene of *Caenorhabditis elegans*. *Nature* 291, 386-390.

NEWMAN, A.P., WHITE, J.G. & STERNBERG, P.W. (1996). Morphogenesis of the *C. elegans* hermaphrodite uterus. *Development* 122, 3617-3626.

RAND, J.B. (2007). Acetylcholine. In: The *C. elegans* Research Community (Ed.). *WormBook*. DOI:10.1895/wormbook.1.131.1.

SAN-MIGUEL, A. & LU, H. (2013). Microfluidics as a tool for *C. elegans* research. In: The *C. elegans* Research Community (Ed.). *WormBook*. DOI:10.1895/wormbook.1.162.1.

SCHWARZ, E.M., KATO, M. & STERNBERG, P.W. (2012). Functional transcriptomics of a migrating cell in *Caenorhabditis elegans*. *Proceedings of the National Academy of Sciences of the United States of America* 109, 16246-16251.

SHINYA, R., HASEGAWA, K., CHEN, A., KANZAKI, N. & STERNBERG, P.W. (2014). Evidence of hermaphroditism and sex ratio distortion in the fungal feeding nematode *Bursaphelenchus okinawaensis*. *G3: Genes, Genomes, Genetics* 4, 1907-1917.

SOMMER, R.J. & STERNBERG, P.W. (1994). Changes of induction and competence during the evolution of vulva development in nematodes. *Science* 265, 114-118.

SOMMER, R.J. & STERNBERG, P.W. (1995). Evolution of cell lineage and pattern formation in the vulval equivalence group of rhabditid nematodes. *Developmental Biology* 167, 61-74.

SOMMER, R.J. & STERNBERG, P.W. (1996a). Evolution of nematode vulval fate patterning. *Developmental Biology* 173, 396-407.

SOMMER, R.J. & STERNBERG, P.W. (1996b). Apoptosis limits the size of the vulval equivalence group in *Pristionchus pacificus*: a genetic analysis. *Current Biology* 6, 52-59.

SOMMER, R.J., CARTA, L.K., KIM, S.Y. & STERNBERG, P.W. (1996). Morphological, genetic and molecular description of *Pristionchus pacificus* sp. n. (Nematoda: Neodiplogastridae). *Fundamental and Applied Nematology* 19, 511-521.

SOMMER, R.J., EIZINGER, A., LEE, K.-Z., JUNGBLUT, B., BUBECK, A. & SCHLAK, I. (1998). The *Pristionchus* HOX gene *Ppa-lin-39* inhibits programmed cell death to specify the vulva equivalence group and is not required during vulva induction. *Development* 125, 3865-3873.

SRINIVASAN, J., KAPLAN, F., AJREDINI, R., ZACHARIAH, C., ALBORN, H.T., TEAL, P.E.A., MALIK, R.U., EDISON, A.C., STERNBERG, P.W. & SCHROEDER, F.C. (2008). A blend of small molecules regulates both mating and development in *Caenorhabditis elegans*. *Nature* 454, 1115-1118.

SRINIVASAN, J., VON REUSS, S.H., BOSE, N., ZASLAVER, A., MAHANTI, P., HO, M.C., O'DOCHERTY, O.G., EDISON, A.S., STERNBERG, P. & SCHROEDER, F.C. (2012). A modular library of small molecule signals regulates social behaviors in *Caenorhabditis elegans*. *PLoS Biology* 10, e1001237.

STERNBERG, P.W. & HORVITZ, H.R. (1981). Gonadal cell lineages of the nematode *Panagrellus redivivus* and implications for evolution by the modification of cell lineage. *Developmental Biology* 88, 147-166.

STERNBERG, P.W. & HORVITZ, H.R. (1982). Postembryonic nongonadal cell lineages of the nematode *Panagrellus redivivus*: description and comparison with those of *Caenorhabditis elegans*. *Developmental Biology* 93, 181-205.

STIRMAN, J.N., CRANE, M.M., HUSSON, S.J., WABNIG, S., SCHULTHEIS, C., GOTTSCHALK, A. & LU, H. (2011). Real-time multimodal optical control of neurons and muscles in freely behaving *Caenorhabditis elegans*. *Nature Methods* 8, 153-158.

SULSTON, J.E. & HORVITZ, H.R. (1977). Post-embryonic cell lineages of the nematode, *Caenorhabditis elegans*. *Developmental Biology* 56, 110-156.

SULSTON, J.E., ALBERTSON, D.G. & THOMSON, J.N. (1980). The *Caenorhabditis elegans* male: postembryonic development of nongonadal structures. *Developmental Biology* 78, 542-576.

SULSTON, J.E., SCHIERENBERG, E., WHITE, J.G. & THOMSON, J. (1983). The embryonic cell lineage of the nematode *Caenorhabditis elegans*. *Developmental Biology* 100, 64-119.

SUNDARAM, M.V. (2013). Canonical RTK-Ras-ERK signaling and related alternative pathways. In: The *C. elegans* Research Community (Ed.). *WormBook*. DOI:10.1895/wormbook.1.80.2.

THOMPSON, O., EDGLEY, M., STRASBOURGER, P., FLIBOTTE, S., EWING, B., ADAIR, R., AU, V., CHAUDHRY, I., FERNANDO, L., HUTTER, H. *ET AL*. (2013). The million mutation project: a new approach to genetics in *Caenorhabditis elegans*. *Genome Research* 23, 1749-1762.

WATERSTON, R. & SULSTON, J. (1995). The genome of *Caenorhabditis elegans*. *Proceedings of the National Academy of Sciences of the United States of America* 92, 10836-10840.

WHITE, J.G., SOUTHGATE, E., THOMSON, J.N. & BRENNER, S. (1986). The structure of the nervous system of the nematode *Caenorhabditis elegans*. *Philosophical Transactions of the Royal Society B: Biological Sciences* 314, 1-340.

YU, H., SEAH, A., HERMAN, M.A., FERGUSON, E.L., HORVITZ, H.R. & STERNBERG, P.W. (2009). Wnt and EGF pathways act together to induce *C. elegans* male hook development. *Developmental Biology* 327, 419-432.

Nematology Monographs & Perspectives, 2015, Vol. 11, 19-41

Chapter 2

Integrative evolutionary biology and mechanistic approaches in comparative biology

Ralf J. SOMMER

Department for Evolutionary Biology, Max-Planck Institute for Developmental Biology Tübingen, 72076 Tübingen, Germany
ralf.sommer@tuebingen.mpg.de

The complexity of life

> *The world of life can be studied from two points of view – that of its unity and that of its diversity. All living things, from viruses to men, have basic similarities. And yet there is an apparently endless variety of living beings. Knowledge and understanding of both the unity and the diversity are useful to man. Some biologists find the unity more inspiring, others are enthralled by the diversity.*
>
> Dobzhansky (1964). Biology, molecular and organismic.

In 1964, Theodosius Dobzhansky wrote his famous phrase that "nothing in biology makes sense except in the light of evolution" in an article entitled "Biology, molecular and organismic", published in the *American Zoologist*. In the very same article, Dobzhansky made the argument quoted above, which is of similar importance for contemporary biology, in particular evolutionary biology. Biology has two complementary facets – unity and diversity – both of which are essential for understanding the world around us. Dobzhansky's argument was phrased in the context of the 1960s, which saw the rise of molecular biology together with a growing threat to the unity of the biological sciences (Smocovitis, 1996). Molecular biologists were concerned with similarities among organisms and they discovered many general principles – from the universality of the genetic code to the principles of gene regulation – providing powerful paradigms in

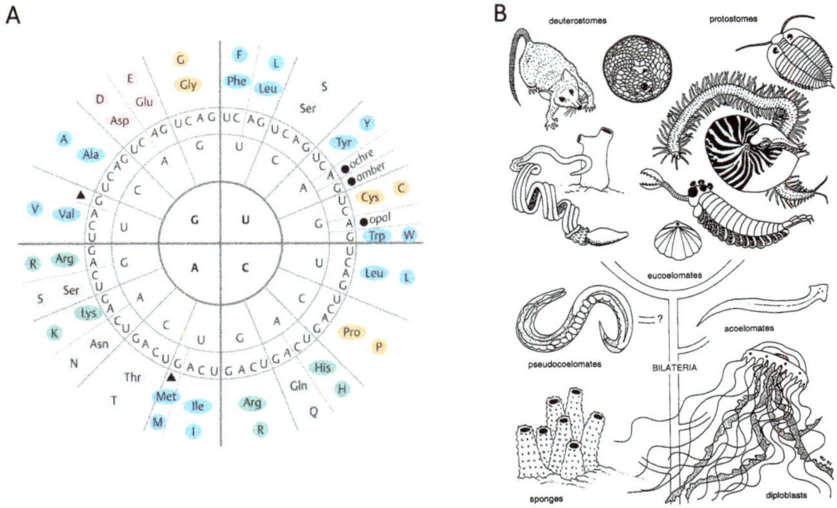

Fig. 2.1. *Unity and diversity in the biological world. A: All living beings have basic similarities. The universality of the genetic code provides a powerful testimony to the unity of biological systems and common ancestry; B: Nonetheless, there is a nearly endless diversity of biological form, here exemplified by representative animal body plans along the tree of life.*

biology, but often sidelining evolutionary biologists and their interests in diversity (Fig. 2.1). Five decades later, molecular biology represents a common and overarching methodology of all the life sciences, providing powerful examples for both the unity and the diversity of life. By now, molecular, genetic and genomic tools have produced unprecedented insights into a great number of evolutionary patterns and processes. From artificial selection of feather ornaments and colour patterns in rock pigeons (Shapiro *et al.*, 2013) to insulin signalling in the evolutionary diversification of the horns of rhinoceros beetles (Emlen *et al.*, 2012), molecular tools have helped to reveal the mechanisms of evolutionary change.

The application of molecular methodologies has provided compelling evidence that biological systems are indeed characterised by 'unity' and 'diversity'. The observed patterns, however, turned out to be very different from what was assumed in the 1960s or even earlier in the 1930s and 1940s when the Neo-Darwinian synthesis was shaped. While the discovery of the universality of the genetic code was seen as a final convincing confirmation of Darwin's concept of common ancestry of all organisms, there was overall agreement that different organisms are built by completely different molecular machineries. Although developmental biology was largely a black box at the time, there was a strong consensus that, as a result of positive and directional selection, the development of different groups of animals, such as arthropods, worms, sea urchins or vertebrates, would be controlled by unrelated and unique specification mechanisms. Developmental genetics has proven these early ideas to be totally wrong as developmental control genes are largely conserved in evolution, often throughout the animal kingdom.

But this is not the end of the story. While many developmental control genes are conserved, arguing for the unity of biological systems, they are simultaneously involved in shaping the diversity of life by being put to new uses. This completely unexpected finding stands perhaps as the most powerful testimony to Dobzhansky's unity and diversity argument. The re-use of conserved genes in different organisms and different developmental processes was termed "co-option" by Rudy Raff and represents one of the most astonishing findings of modern biology (Raff, 1996; Wilkins, 2002). Epidermal-growth factor (EGF) signalling, for example, specifies the dorso-ventral axis in *Drosophila* embryos and is later on re-used in various developmental processes, *i.e.*, eye formation. Besides co-option during the development of the same organism, signalling pathways have also been co-opted during evolution: EGF/RAS signalling specifies the *Caenorhabditis elegans* vulva, the egg-laying system of nematode females and hermaphrodites. In vertebrates, a similar signalling pathway is involved in various cell differentiation processes and in cell death, its over-activation often being involved in cancer. Another important signalling pathway with diverse functions is the Wnt pathway, which controls axis specification and head regulation in the cnidarian model *Hydra*, segmentation in *Drosophila* and other insects, and is involved in diabetes, breast and prostate cancer in humans (Wilkins, 2002). Similar examples of co-option could be listed for all signalling pathways and many transcriptional regulators, all of which lend support to the astounding notion that homologous genes control non-homologous structures (Wilkins, 2002).

The discovery of the high degree of conservation of developmental control genes and their co-option to new uses in different processes and organisms is not only a proof of the unity and diversity of biological systems. It is also the starting point for an exciting new branch of evolutionary biology with new questions: evolutionary developmental biology or in short 'evo-devo' (Raff, 1996; Gerhardt & Kirschner, 1997; Carroll *et al.*, 2001). How are genes and genetic pathways co-opted to serve such fundamentally different developmental functions as dorso-ventral patterning and eye development? How do conserved proteins perform species-specific and often cell-specific functions? Why are certain characteristics of organisms conserved, representing the unity of biology, whereas others vary to a considerable extent, giving testimony to diversity? Also, why are certain molecular machines conserved, whereas others evolve rapidly? Despite tremendous progress in developmental and evolutionary biology, we are far from having sophisticated and convincing answers to these and related questions. Those concerned with the diversity of organisms and the nearly endless variety of form in animals, plants and fungi are still searching for the mechanisms underlying these evolutionary patterns. As will be outlined below, current shortcomings result, to a large extent, from inappropriate generalisations within disciplines of biology, often unnecessarily constraining the perspectives of biological knowledge.

This book argues for the necessity of integrative studies in evolutionary biology that aim to merge comparative biology with mechanistic approaches based on molecular, genetic and genomic tools. It introduces and summarises the current state of knowledge by means of a novel model system, the nematode *Pristionchus pacificus* (Fig. 2.2), which seems well suited, among other animals and plants, to bridge the divide between molecular biology and comparative evolutionary biology. This chapter will try to lay the conceptual foundations for a comprehensive and integrative approach that aims to do justice to both the unity and the diversity of life.

The power of model system approaches

> *Now, to my amazement, I could watch the cells divide. Those Nomarski images of the worm are the most beautiful things imaginable.*
>
> Sulston & Ferry (2002). The common thread: a story of science, ethics and the human genome.

One essential factor for the success of molecular biology was its strict application to a model system and a reductionist approach as originally proposed by Delbruck, Luria and others (Judson, 1996). Many of the groundbreaking discoveries were made in a handful of selected model organisms, often viruses or bacteria. Over time research questions changed and, as a consequence, so did the type of model organism that best allowed a detailed investigation of the respective problem.

Maybe the best example for the development and establishment of new model organisms is the nematode *Caenorhabditis elegans*. Sydney Brenner searched for a simple metazoan animal that combined the complexity of multicellular organisms in development and neurobiology with the simplicity of growth and culture as found in microbes (Brenner, 1974; Brown, 2003; Friedberg, 2010). His choice turned out to be excellent. Only 1 mm in size, hundreds and thousands of *C. elegans* worms can be propagated on small Petri dishes with simple *Escherichia coli* as food source. With a 3-day generation time, two generations of worms can be grown in 1 week, thereby achieving microbial growth rates in a multicellular organism.

Free-living nematodes have a number of additional characteristics that, in combination, provide a unique entry point into developmental biology and other areas of biology. *Caenorhabditis elegans* is transparent and cell divisions can be watched under the microscope. John Sulston saw the beauty of this system (quoted above) and, together with Bob Horvitz and others, was able to determine the developmental fate of every somatic cell. This was possible because in *C. elegans* the total number of cells is small and, most importantly, the cell lineage was found to be invariant between individuals. This phenomenon, also known as eutely or cell constancy, allowed the determination of the complete cell lineage from first cleavage to maturity, giving rise to 959 cells in the adult *C. elegans* hermaphrodite and to 1031 somatic cells in the male. The elucidation of the *C. elegans* cell lineage represents the most important foundation for the establishment of this multicellular organism as a model system in developmental biology. Generations of researchers have built their *C. elegans* studies on the cell lineage diagram, allowing them to gain a detailed molecular understanding of the mechanisms underlying the unfolding of the worm (Wood, 1988; Riddle *et al.*, 1997; The *C. elegans* Research Community, 2005).

From a zoological perspective it is not surprising that a nematode was to become one of the most important model organisms in modern biology. *Caenorhabditis elegans* is a member of one of the largest animal phyla, the Nematoda. Some authors estimate the total number of nematode species to be in a range of 1 to 10 million (*e.g.*, Lambshead, 1993). As first described by Cobb, the founder of American nematology,

Fig. 2.2. *Light micrograph of* Pristionchus pacificus. *The picture shows an adult hermaphrodite, which is about 1 mm in length.* Pristionchus pacificus *has a 4-day generation time under laboratory conditions when feeding on* Escherichia coli *(20°C). This worm has a nearly cosmopolitan distribution and represents an entomophilic nematode that is found in a necromenic association with scarab beetles. © Ralf J. Sommer, MPI Developmental Biology, Tübingen, Germany.*

nematodes are found in all ecosystems and show a high degree of diversification in form, physiology and life style. They are characterised by species richness, numerical abundance and omnipresence. Millions of nematode worms can be found in a square metre of soil (Lee, 2002) and they can even be found in the most extreme ecosystems (Wharton, 2002). Parasitic species have evolved independently from free-living ancestors multiple times within the nematodes, among them some of the most devastating parasites of humans and livestock (Poulin, 2007). A molecular phylogenetic framework of nematodes, available since the late 1990s, provides a sound basis for the reconstruction of the evolution of nematode parasitism but also for the evolution of other character states, such as the mode of reproduction (Blaxter *et al.*, 1998; van Megen *et al.*, 2009).

Nematodes are mostly gonochoristic, with a male-female mode of reproduction. However, parthenogenesis and hermaphroditism have evolved multiple times independently within the nematodes (Denver *et al.*, 2011). Hermaphroditism, in particular, was one of the decisive characteristics of *C. elegans* when Sydney Brenner selected this nematode. Self-fertilisation minimises the propagation of strains, while males, when needed, can be used to transfer genetic mutations between individuals. All these characteristics and features taken together helped to establish *C. elegans* as a powerful model organism. Ever since the groundbreaking work of Brenner, Sulston and Horvitz, which won them the Nobel Prize for Medicine and Physiology in 2002, *C. elegans* has been at the forefront of modern biology. The comprehensive understanding that we now have of this small animal bears powerful testimony to the strength of the model system approach.

In addition, the tremendous knowledge about *C. elegans* can serve as a paradigm for evolutionary and comparative studies. While not all relatives of *C. elegans* are hermaphroditic, many of them also have a relatively short generation time and are often transparent. Moreover, when Sternberg and Horvitz investigated the postembryonic lineage of *Panagrellus redivivus* in the early 1980s, they could show that the entire cell lineage of other nematodes is invariant as well (Sternberg & Horvitz, 1981, 1982). Thus, some of the basic technical features that make *C. elegans* an interesting model for large-scale studies are indeed conserved among nematodes. This is also true for *P. pacificus*, which shares with *C. elegans* its short generation time, the hermaphroditic mode of reproduction and many other technical features important for large-scale experimental studies (Fig. 2.2).

Comparative biology and a need for mechanistic approaches

> *Perhaps the greatest weakness of the comparative approach used on its own is that it demonstrates an association between characters without demonstrating a causal link.*
>
> Doughty (1996) and Poulin (2007).

The model system approach as briefly summarised for *C. elegans* resulted in an unforeseeable boost to our understanding of biology. Research on *C. elegans*, yeast, *Drosophila*, *Arabidopsis* and several vertebrate species has yielded tremendous insight into biological systems. Formulated in molecular and mechanistic terms, biological patterns and processes can often be described as deducible consequences of physical and chemical laws. This is what Ernst Mayr called "proximal causation" as early as 1961, long before the full consequences of the reductionist molecular approach were apparent (Mayr, 1961). Indeed, the model system approach has not only pros but also cons. Proximal causation, Mayr argued, no matter how powerful it is in its immediate, functional context and its materialistic, mechanistic principles, is insufficient to explain fully the patterns and processes observed at the organismic, population and species level. Mayr and later Simpson (1963) forcefully argued that only evolutionary biology was able to provide an "ultimate causation", which could explain why patterns and processes observed in individual species are the way they are. They argued further that only evolutionary biology had the potential of becoming a comprehensive and unifying element in biology (Simpson, 1963; Smocovitis, 1996).

Indeed, one important principle of evolutionary biology that was largely missing from model system biology is the comparative approach. The foundations and fundamentals of comparative biology are manifold and some of them predate Darwin's Origin of Species, such as the homology concept (Owen, 1849). Rieppel, Riedl and others established a complex theoretical framework of comparative and evolutionary biology that builds on philosophical, historical and methodological issues (Riedl, 1975; Rieppel, 1988). Rieppel, in particular, highlighted the importance of the complementarity of essential concepts in comparative and evolutionary biology. He identified holism *vs* reductionism, structuralism *vs* functionalism and the hierarchical view of nature *vs* Darwinism as three essential antitheses of comparative biology. These opposite 'ways of seeing' are fundamental elements of the world that surrounds us and all serious and comprehensive attempts of understanding biology have to take these complementary views into account.

Seeking for conserved general principles, molecular research is largely unaware of these concepts and is basically built on reductionism, structuralism and Darwinism. Thus, molecular biology largely disregards the holistic and hierarchical perspectives and concepts of biology. Also, molecular biology can afford to disregard comparative biology and phylogeny because model systems can be studied independently of their phylogenetic position in the tree of life. While reductionism and structuralism were essential for the early success of molecular biology, a comprehensive understanding of biology ultimately requires inroads into the other perspectives as well. In particular, the hierarchical view of nature, as highlighted by Dobzhansky (1970), Riedl (1975) and Raff (1996), is of central importance. A look at the central finding of evo-devo, the conservation of developmental control genes throughout the animal kingdom, may serve to illustrate the point: conserved genes regulate different cell types, tissues and organs in different organisms. In many cases, homologous genes control the development of non-homologous structures. At the same time, several evo-devo studies have shown the widespread occurrence of the opposite situation: homologous structures in different organisms are regulated by non-homologous genes and signalling pathways, a phenomenon now known as 'developmental systems drift' (True & Haag, 2001). For example, vulva development in nematodes (see Rudel, Chapter 9, this volume), is just one of several mature cases that demonstrate developmental systems drift, indicating that a homologous organ built from homologous cells can be regulated genetically by distinct and unrelated signalling pathways (Sommer, 2008). Similarly, sex determination in animals (True & Haag, 2001), muscle cell specification in cnidarians (Steinmetz *et al.*, 2012) and the regulation of mating type switching in yeast (Tsong *et al.*, 2006), to name just a few, represent additional examples for developmental systems drift. Thus, conserved developmental control genes can be co-opted to regulate unrelated structures, whereas in other cases homologous structures might be controlled by unrelated regulatory networks. These opposite and complementary phenomena of co-option and developmental systems drift show the opportunistic features of nature and represent another proof for Dobzhansky's claim about the unity and diversity of life.

I argue that the time is ripe to overcome the divide between proximal and ultimate causation. Model organisms and the detailed knowledge about their biology have laid the foundations for initiating a new type of comparative and evolutionary biology. Model system approaches and laboratory studies, resulting in detailed structural analyses of biological species, should be combined with comparative biology and an integrative research paradigm. Specifically, I want to argue that evolutionary and comparative biology can open new horizons when using the knowledge and the methodology of model organisms as starting points. A comparison of related species with similar body plans can further the understanding of the mechanisms of co-option and the principles underlying modification and novelty, whereas the comparison of 'the' worm (*C. elegans*) with 'the' fly (*Drosophila*) or other unrelated 'classical' models will merely confirm the well known fact that non-homologous structures are formed by homologous genes. Only the comparison of related organisms, *i.e.*, comparisons within insects, within nematodes or within cnidarians, *etc.*, can reveal the interplay between conservation and change in the context of homology.

To be effective, comparisons require a similar level of mechanistic insight in all the species to be compared! Therefore, such studies will depend on the availability of sophisticated functional toolkits for the less well studied organisms, which should be similar, or at least related, to what is available in the respective model systems themselves (Sommer, 2009). Developing and applying functional and mechanistic approaches in comparative biology is a crucial pre-requisite that, once available, will provide a powerful tool to extend our knowledge about the evolution of multicellular organisms. Thus, the combination of comparative studies (ultimate causation) with mechanistic approaches (proximal causation) can drive the evolutionary understanding of biological systems forward. What Doughty and Poulin pointed out in the context of the evolution of parasitism in the above quotation represents a powerful paradigm for all aspects of evolutionary biology.

By now, the establishment of model systems for comparative studies, in particular in the context of evolution and development, has been underway for more than a decade. The rise of evo-devo has seen the appearance of several species as satellite models to the well established primary model organisms. The flour beetle, *Tribolium castaneum*, was the first satellite to *Drosophila*, but was soon followed by several others including the wasp *Nasonia*, which like *Tribolium* has functional genetic tools. The cnidarian model *Hydra* was complemented by *Nematostella* as an important satellite and the weed *Arabidopsis thaliana* is fruitfully complemented by other di- and monocotyledonous flowering plants, just to name a few. Several comprehensive texts, such as *Emerging model organisms* (Anon., 2009), give detailed accounts of these promising new models, providing laboratory-manual style instructions for the species mentioned above as well as others.

Integrative evolutionary biology

> *Natural selection is just one of several evolutionary mechanisms, and the failure to realize this is probably the most significant impediment to a fruitful integration of evolutionary theory with molecular, cellular and developmental biology.*
>
> Lynch (2008). The origins of genome architecture.

Evo-devo and comparative developmental biology are most fruitful in delineating patterns in evolutionary biology but they represent only one of several important branches of contemporary evolutionary biology. Also, many evo-devo studies work along a purely selectionist line emphasising the adaptedness of structures (see above quote). By contrast, population genetics has long elucidated additional evolutionary forces, from genetic drift in population bottlenecks to neutral evolution (Wright, 1932; Kimura, 1983). Variation, as a result of three distinct mechanisms, selection, drift and neutral evolution, is crucial to the understanding of evolution, but is largely missing from many research programmes in cell and developmental biology or evo-devo. Therefore, evolutionary studies in developmental biology and evo-devo have to incorporate evolutionary theory and its different facets (Fig. 2.3).

In addition, evolutionary biologists have become aware of the influence of ecology on evolution and development. It was Van Valen (1973) who first stated that evolution is the influence of ecology on development. The last two decades have seen a growing awareness of the principal importance of the environment on development. One concept that

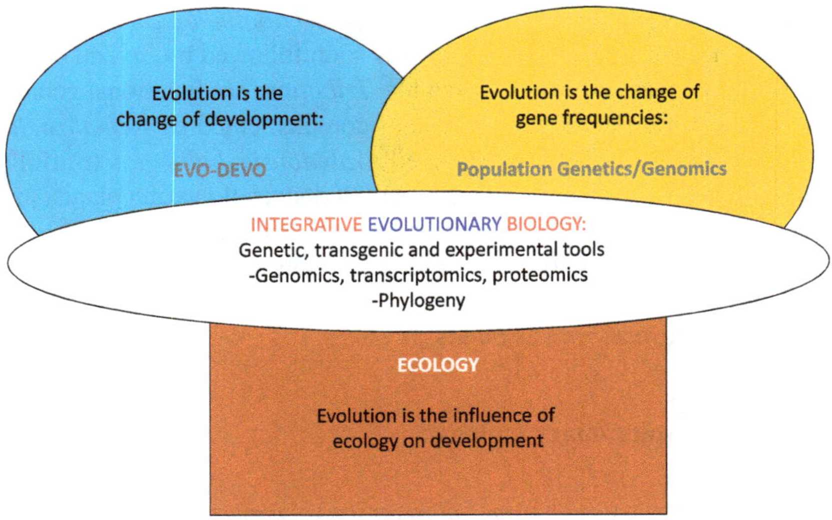

Fig. 2.3. *Integrative evolutionary biology and the need for mechanistic studies. Contemporary evolutionary biology that is concerned with the processes underlying the diversity of form and shape has to be integrative by linking:* i*) developmental biology and evo-devo; with* ii*) population genetics; and* iii*) ecology. Integration can be achieved by developing a platform of molecular tools that allow mechanistic insight into developmental and evolutionary patterns and processes.*

aims for the integration of development and ecology is 'phenotypic plasticity', the ability of an organism to generate distinct phenotypes under the influence of variable environments. Multiple authors have proposed phenotypic plasticity as a facilitator of phenotypic novelty (West-Eberhard, 2003; Moczek *et al.*, 2011). For example, butterfly species can develop different morphs with distinct wing patterns (Beldade & Brakefield, 2002), and rhinoceros beetles, as well as other beetles, form distinct horns (Emlen *et al.*, 2012) depending on their growth conditions. In addition to insects, flowering plants and nematodes are known to show several examples of phenotypic plasticity (Schlichting & Piggliucci, 1998; Sommer & Ogawa, 2011). Case studies from these taxa indicate that, indeed, phenotypic plasticity is often correlated with diversity and the evolutionary appearance of novel structures. Therefore, both variation and the environment have ultimately to be considered in the context of evolution and development (Fig. 2.3).

Integrative evolutionary biology needs comparative approaches

Natural systems are too complex to be reducible to a unique description.
Loreau (2010). From populations to ecosystems.

The last two paragraphs put forward related, but distinct, arguments. First, I argued that comparative studies have to incorporate molecular tools, thereby aiming for a mechanistic understanding of evolutionary patterns and processes. Second, rather than seeking generalisations within one discipline, one should aim for integrative studies that link the largely disconnected areas of evolutionary biology, such as evo-devo, population genetics and ecology (Fig. 2.3).

This is not to argue against the many sophisticated purely evo-devo or purely population genetics studies. But I strongly advocate that through integrative studies of mechanistic approaches with comparative biology, due consideration can be given to the hierarchical view of nature, building bridges between disciplines and thereby revealing new principles that neither discipline is able to find on its own. Neither population genetics nor ecology or evo-devo alone can explain the many complementary facets of the diversity of life. Nor can cell biology or developmental biology prove the unity of life. Similarly, no single animal or plant species is sufficient to represent the full complement of the complexity of biological systems. No single model can describe all facets of the unity and diversity of life. Instead, several inroads have to be taken in parallel in the attempt to extract from the powerful paradigm available in the 'classical' model organisms a new integrative and comparative research programme, which will be better suited to represent the complexity of evolution and the biological world.

I have argued previously that evolutionary research in developmental biology and evo-devo has to be integrated with studies in the areas of population genetics and ecology (Sommer, 2009). Such an integrated approach, backed by molecular, genetic and genomic tools, will be most powerful because it can generate novel interdisciplinary principles and perspectives (Fig. 2.3). Working across the borders of disciplines will ultimately extend our knowledge about the complexity of biological systems. What Loreau (2010), quoted above, pointed out for the sub-disciplines of ecology, is similarly true for development and evolution; natural systems are simply too complex to be reducible to unique descriptions. Integration will establish new interfaces, which are often missing when research programmes are limited to well established disciplines. The remainder of this chapter, and, basically, this book, will describe the research endeavour to establish the nematode *Pristionchus pacificus* as one such model and to summarise the current achievements in this direction (Figs 2.2, 2.3).

Pristionchus pacificus – a nematode model to combine integrative evolutionary biology and mechanistic approaches in comparative biology

This book focuses on *P. pacificus* and the progress and prospects of developing this nematode as a model system to combine integrative evolutionary biology with mechanistic approaches in comparative biology (Fig. 2.3). Five research concepts provide anchor points for integrative and mechanistic studies:

1. *Functional tools in laboratory studies*: the *C. elegans* paradigm as a comparative framework in the context of homology.
2. *'Holistic' mechanistic tools*: from genomics to small molecule chemistry.
3. *Phylogeny*: a conceptual paradigm to identify reference points for comparison.
4. *Ecology*: the interaction with scarabaeid beetles and the consequences for the biology of the organism.
5. *Population genetics*: from cosmopolitan sampling to Island biology – the La Réunion case study.

Mechanistic understanding requires a reductionist approach and laboratory studies (Fig. 2.3). In *P. pacificus*, genetic and molecular tools were developed in parallel to the description of this new species in 1996 (Sommer & Sternberg, 1996; Sommer *et al.*, 1996). With self-fertilisation as mode of reproduction, the generation of spontaneous males and a 4-day life cycle with *E. coli* as food source, *P. pacificus* can be grown as easily as *C. elegans*. In line with typical nematode development, *P. pacificus* proceeds through four juvenile stages, called J1-J4, before becoming adult (Fig. 2.4). In contrast to *C. elegans*, however, *P. pacificus* and other diplogastrid nematodes have an embryonic moult, which means that the J1 stays in the egg and hatching occurs when the J2 stage emerges from the egg (Fürst von Lieven, 2005).

Since the description of *P. pacificus* in 1996, many functional tools have been developed in this species: forward genetics (Sommer *et al.*, 1996), genetic and physical maps (Srinivasan *et al.*, 2002, 2003), reverse genetics (Tian *et al.*, 2008; Witte *et al.*, 2015), whole genome sequencing (Diederich *et al.*, 2008), DNA-mediated transformation (Schlager *et al.*, 2009) and various -omics technologies (Borchert *et al.*, 2010; Sinha *et al.*, 2012a, b) (see Aurilio & Srinivasan, Chapter 5, this volume). The similarity of the technological platforms in *P. pacificus* and *C. elegans* allows detailed comparisons in the context of homology: *i*) are the six chromosomes of the two species homologous and is there micro- or even macrosynteny?; *ii*) does recombination follow the same rules, *i.e.*, is there interference in *P. pacificus*?; and *iii*) are genome size and gene content, transcriptome and proteome related or do they differ given the long evolutionary separation time of more than 200 million years (Dieterich *et al.*, 2008)? Starting from these more technical questions, other biological features, such as development, the nervous system or cell biological and higher structural features, can be compared. One important insight that can be gained from this type of comparative study is to what extent the findings obtained in *C. elegans* represent general phenomena in nematode worms, invertebrates or even animals. Is what has been found in 'the' worm true for all nematodes or are there different solutions to the same problem? Comparative studies can tell.

All comparisons have to be seen in a phylogenetic context; *P. pacificus* and *C. elegans* are clade V nematodes, but are members of different nematode families, the Diplogastridae and Rhabditidae, respectively (Blaxter *et al.*, 1998). Genome-based studies suggest that they were separated more than 200 million years ago (Dieterich *et al.*, 2008). It is crucial to recognise that different comparative and evolutionary studies need different reference points. While evo-devo studies depend heavily on the comparison between two rather distant, but still clearly related, genetically tractable organisms, other aspects of the *P. pacificus* research agenda need more closely related reference points. Research in the last few years has been able to provide such reference points with the description of the presumptive sister species *P. exspectatus* (Kanzaki *et al.*, 2012a), the sister genus *Parapristionchus* (Kanzaki *et al.*, 2012b), as well as many additional *Pristionchus* species and Diplogastridae genera (see Kanzaki & Giblin-Davis, Chapter 3, and Ragsdale *et al.*, Chapter 4, this volume). With the help of modern molecular tools the phylogeny of animals can be reconstructed at very high resolution. The

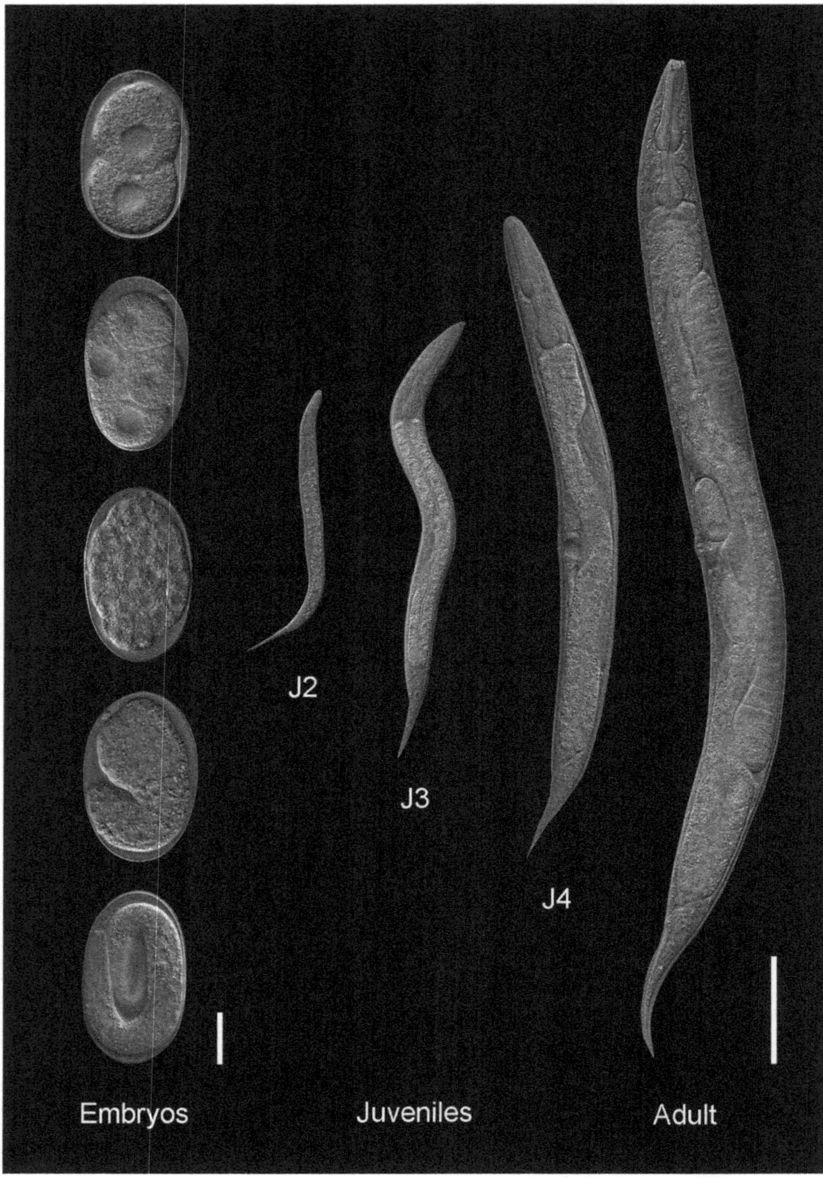
J2
J3
J4
Embryos
Juveniles
Adult

application of these tools to: *i*) *P. pacificus* strains; *ii*) *Pristionchus* species; *iii*) Diplogastridae genera; and *iv*) higher-level Rhabditomorpha taxa has produced a phylogenetic framework that serves as an important backbone for comparative studies and character polarisation.

There is often a negative correlation between the ability to develop a species as a model organism for laboratory studies and understanding the ecology of the species. *Pristionchus* nematodes live in close association with scarabaeid beetles, *i.e.*, cockchafers, dung beetles and stag beetles (Herrmann *et al.*, 2006); *P. pacificus* itself was found to be associated with the Oriental beetle *Exomala orientalis* (Herrmann *et al.*, 2007). This entomophilic association is best described as necromeny because the *Pristionchus* nematodes found on the living beetle are all in the non-feeding dauer stage waiting for the beetle to die before feeding on the developing microbes. These findings about *Pristionchus* ecological aspects have been a starting point for many studies at the interface between genetics, development and ecology and have provided valuable insight into the functional specification of ecologically relevant traits. Several of these traits will be discussed in this book.

Building on the understanding of the *Pristionchus*-scarab beetle ecosystem and the well documented invasion of *P. pacificus* from Japan to the USA with the *E. orientalis* vector (Herrmann *et al.*, 2007), population genetics soon started to concentrate on Island systems. The Mascareine Island, La Réunion, was identified as a hot spot for *P. pacificus* biodiversity and is the focus of intense studies (Herrmann *et al.*, 2010; Morgan *et al.*, 2012) (see McGaughran & Morgan, Chapter 8, this volume). Such accomplishments over the last few years allow integrative studies of developmental genetics and evo-devo with ecology and population genetics. Building on a resource platform, with phylogeny, genetic and genomic tools, as well as small molecule chemistry (Fig. 2.3), *Pristionchus* is well equipped to become an important addition to other organisms such as *Drosophila* that

Fig. 2.4. *Life cycle of* Pristionchus pacificus. Pristionchus pacificus *has a simple life cycle that can be completed in 4 days under laboratory conditions at 20°C if sufficient bacterial food is provided. The self-fertilising hermaphrodite lays fertilised eggs, which develop into adults through four juvenile stages (J1-J4) separated by moults. The first moult into a J1 takes place within the eggshell; hatching occurs at the J1-J2 moult. © Ralf J. Sommer, MPI Developmental Biology, Tübingen, Germany.*

allow integrative insight into the patterns and processes of evolutionary biology (Powell, 1997).

Conclusion

A comprehensive understanding of the biological world requires the simultaneous consideration of complementary perspectives from different fields of biology. Biological systems – ecosystems, species, individuals and cells – are hierarchical structures and depend on the interactions of their units that can often be described in molecular, physico-chemical or mathematical terms. At the same time, biological systems result from historical evolutionary processes. Ultimately, approaches and knowledge that provide insight into the molecular mechanisms of model organisms and the genomic toolkit of biological species have to be united with the population genetic and ecosystem functions of species. In this chapter, I put forward two important requirements for a more comprehensive understanding of biological complexity. First, comparative studies – an important and central element of evolutionary biology – have to incorporate molecular tools aiming for molecular understanding. Second, evolutionary biology has to seek generalisations by crossing borders and disciplines. Integrative evolutionary biology has to make use of laboratory studies searching for molecular and mechanistic insight and, simultaneously, must try to integrate development, population genetics and ecology. This book will describe the recent work on the nematode *P. pacificus* as a new model system to achieve just that: to apply comparative biology with mechanistic tools that combine laboratory studies and fieldwork. The last decade has seen several such developments in selected organisms that aim to study biological complexity from different angles. Hopefully, *P. pacificus* will be a useful addition.

Acknowledgements

I thank Metta Riebesell for carefully reading this manuscript and an external expert for review.

References

ANON. (2009). *Emerging model organisms*, Vols 1 & 2. Cold Spring Harbor, NY, USA, Cold Spring Harbor Laboratory Press.

BELDADE, P. & BRAKEFIELD, P.M. (2002). The genetics and evo-devo of butterfly wing patterns. *Nature Reviews Genetics* 3, 442-452.

BLAXTER, M.L., DELEY, P., GAREY, J.R., LIU, L.X., SCHELDEMAN, P., VIERSTRAETE, A., VANFLETEREN, J.R., MACKEY, L.Y., DORRIS, M., FRISSE, L.M. *ET AL.* (1998). A molecular evolutionary framework for the phylum nematode. *Nature* 392, 71-75.

BORCHERT, N., DIETERICH, C., KRUG, K., SCHÜTZ, W., JUNG, S., NORDHEIM, A., SOMMER, R.J. & MACEK, B. (2010). Proteogenomics of *Pristionchus pacificus* reveals distinct proteome structure of nematode models. *Genome Research* 20, 837-846.

BRENNER, S. (1974). The genetics of *Caenorhabditis elegans*. *Genetics* 77, 1-94.

BROWN, A. (2003). *In the beginning was the worm*. New York, NY, USA, Columbia University Press.

CARROLL, S.B., GRENIER, J.K. & WEATHERBEE, S.D. (2001). *From DNA to diversity*. Edinburgh, UK, Blackwell Science.

DENVER, D.R., CLARK, K.A. & RABOIN, M.J. (2011). Reproductive mode evolution in nematodes: insights from molecular phylogenies and recently discovered species. *Molecular Phylogenetics and Evolution* 61, 584-592.

DIETERICH, C., CLIFTON, S.W., SCHUSTER, L., CHINWALLA, A., DELEHAUNTY, K., DINKELACKER, I., FULTON, L., FULTON, R., GODFREY, J., MINX, P. *ET AL.* (2008). The *Pristionchus pacificus* genome provides a unique perspective on nematode lifestyle and parasitism. *Nature Genetics* 40, 1193-1198.

DOBZHANSKY, T. (1964). Biology, molecular and organismic. *American Zoologist* 4, 443-452.

DOBZHANSKY, T. (1970). *Genetics of the evolutionary process*. New York, NY, USA, Columbia University Press.

DOUGHTY, P. (1996). Statistical analysis of natural experiments in evolutionary biology: comments on recent criticism of the use of comparative methods to study adaptation. *American Naturalist* 148, 943-956.

EMLEN, D.J., WARREN, I.A., JOHNS, A., DWORKIN, A. & LAVINE, L.C. (2012). A mechanism of extreme growth and reliable signaling in sexually selected ornaments and weapons. *Science* 337, 860-864.

FRIEDBERG, E.C. (2010). *Sidney Brenner, a biography*. Cold Spring Harbor, NY, USA, Cold Spring Harbor Laboratory Press.

FÜRST VON LIEVEN, A. (2005). The embryonic molt in diplogastrids (Nematoda) – homology of developmental stages and heterochrony as a prerequisite for morphological diversity. *Zoologischer Anzeiger* 244, 79-91.

GERHARDT, J. & KIRSCHNER, M. (1997). *Cells, embryos and evolution*. Edinburgh, UK, Blackwell Science.

HERRMANN, M., MAYER, E.W. & SOMMER, R.J. (2006). Nematodes of the genus *Pristionchus* are closely associated with scarab beetles and the Colorado potato beetle in western Europe. *Zoology* 109, 96-108.

HERRMANN, M., MAYER, W., HONG, R., KIENLE, S., MINASAKI, R. & SOMMER, R.J. (2007). The nematode *Pristionchus pacificus* (Nematoda: Diplogastridae) is associated with the Oriental beetle *Exomala orientalis* (Coleoptera: Scarabaeidae) in Japan. *Zoological Science* 24, 883-889.

HERRMANN, M., KIENLE, S., ROCHAT, J., MAYER, W. & SOMMER, R.J. (2010). Haplotype diversity of the nematode *Pristionchus pacificus* on Réunion in the Indian Ocean suggests multiple independent invasions. *Biological Journal of the Linnean Society* 100, 170-179.

JUDSON, H.F. (1996). *The eighth day of creation*. Cold Spring Harbor, NY, USA, Cold Spring Harbor Laboratory Press.

KANZAKI, N., RAGSDALE, E., HERRMANN, M., MAYER, W.E. & SOMMER, R.J. (2012a). Description of three *Pristionchus* species (Nematoda: Diplogastridae) from Japan that form a cryptic species complex with the model organism *P. pacificus*. *Zoological Science* 29, 403-417.

KANZAKI, N., RAGSDALE, E., HERRMANN, M., MAYER, W.E., TANAKA, R. & SOMMER, R.J. (2012b). *Parapristionchus giblindavisi* n. gen., n. sp. (Rhabditida: Diplogastridae) isolated from stag beetles (Coleoptera: Lucanidae) in Japan. *Nematology* 14, 933-947.

KIMURA, M. (1983). *The neutral theory of molecular evolution*. Cambridge, UK, Cambridge University Press.

LAMBSHEAD, P.J. (1993). Recent developments in benthic biodiversity research. *Océanis* 19, 5-24.

LEE, D.L. (Ed.) (2002). *The biology of nematodes*. London, UK, Taylor & Francis.

LOREAU, M. (2010). *From populations to ecosystems*. Princeton, NJ, USA, Princeton University Press.

LYNCH, M. (2008). *The origins of genome architecture*. Sunderland, MA, USA, Sinauer Associates Inc.

MAYR, E. (1961). Cause and effect in biology. *Science* 134, 1501-1506.

MOCZEK, A.P., SULTAN, S., FOSTER, S., LEDON-RETTIG, C., DWORKIN, I., NIJHOUT, F., ABOUHEIF, E. & PFENNIG, D.W. (2011). The role of developmental plasticity in evolutionary innovations. *Proceedings of the Royal Society B: Biological Sciences* 278, 2705-2713.

MORGAN, K., MCGAUGHRAN, A., WITTE, H., BARTELMES, G., VILLATE, L., HERRMANN, M., ROCHAT, J. & SOMMER, R.J. (2012). Multilocus analysis of *Pristionchus pacificus* on La Réunion Island reveals an evolutionary history shaped by multiple introductions, constrained dispersal events, and rare out-crossing. *Molecular Ecology* 21, 250-266.

OWEN, R. (1849). *On the nature of limbs*. (Re-printed in 2007.) Chicago, IL, USA, Chicago University Press.

POULIN, R. (2007). *Evolutionary ecology of parasites*. Princeton, NJ, USA, Princeton University Press.

POWELL, J.R. (1997). *Progress and prospects in evolutionary biology. The* Drosophila *model*. Oxford, UK, Oxford University Press.

RAFF, R. (1996). *The shape of life*. Chicago, IL, USA, Chicago University Press.

RIDDLE, D.L., BLUMENTHAL, T., MEYER, B.J. & PRIESS, J.R. (Eds) (1997). C. elegans *II*. Cold Spring Harbor, NY, USA, Cold Spring Harbor Laboratory Press.

RIEDL, R. (1975). *Die Ordnung des Lebendigen*. Hamburg, Germany, Paul Parey.

RIEPPEL, O.C. (1988). *Fundamentals of comparative biology*. Zürich, Switzerland, Birkhäuser.

SCHLAGER, B., WANG, X., BRAACH, G. & SOMMER, R.J. (2009). Molecular cloning of a dominant Roller mutant and establishment of DNA-mediated transformation in the nematode model *Pristionchus pacificus*. *Genesis* 47, 300-304.

SCHLICHTING, C.D. & PIGLIUCCI, M. (1998). *Phenotypic evolution: a reaction norm perspective*. Sunderland, MA, USA, Sinauer Associates Inc.

SHAPIRO, M.D., KRONENBERG, Z., LI, C., DOMYAN, E.T., PAN, H., CAMPBELL, M., TAN, H., HUFF, C.D., HU, H., VICKREY, A.I. *ET AL.* (2013). Genomic diversity and evolution of the head crest in rock pigeon. *Science* 339, 1063-1067.

SIMPSON, G.G. (1963). Biology and the nature of science. *Science* 139, 81-88.

SINHA, A., RAE, R., IATSENKO, I. & SOMMER, R.J. (2012). System wide analysis of the evolution of innate immunity in the nematode model species *Caenorhabditis elegans* and *Pristionchus pacificus*. *PLoS ONE* 7, e44255.

SINHA, A., SOMMER, R.J. & DIETERICH, C. (2012). Divergent gene expression in the conserved dauer stage of the nematodes *Pristionchus pacificus* and *Caenorhabditis elegans*. *BMC Genomics* 13, 254.

SMOCOVITIS, V.B. (1996). *Unifying biology*. Princeton, NJ, USA, Princeton University Press.

SOMMER, R.J. (2008). Homology and the hierarchy of biological systems. *BioEssays* 30, 653-658.

SOMMER R.J. (2009). The future of evo-devo: model systems and evolutionary theory. *Nature Reviews Genetics* 10, 416-422.

SOMMER, R.J. & OGAWA, A. (2011). Hormone signaling and phenotypic plasticity in nematode development and evolution. *Current Biology* 21, R758-R766.

SOMMER, R.J. & STERNBERG, P.W. (1996). Apoptosis limits the size of the vulval equivalence group in *Pristionchus pacificus*: a genetic analysis. *Current Biology* 6, 52-59.

SOMMER, R.J., CARTA, L.K., KIM, S.-Y. & STERNBERG, P.W. (1996). Morphological, genetic and molecular description of *Pristionchus pacificus* sp. n. (Nematoda, Diplogastridae). *Fundamental and Applied Nematology* 19, 511-521.

SRINIVASAN, J., SINZ, W., LANZ, C., BRAND, A., NANDAKUMAR, R., RADDATZ, G., WITTE, H., KELLER, H., KIPPING, I., PIRES-DASILVA, A. *ET AL.* (2002). A BAC-based genetic linkage map of the nematode *Pristionchus pacificus*. *Genetics* 162, 129-134.

SRINIVASAN, J., SINZ, W., JESSE, T., WIGGERS-PEREBOLTE, L., JANSEN, K., BUNTJER, J., VAN DER MEULEN, M. & SOMMER, R.J. (2003). An integrated physical and genetic map of the nematode *Pristionchus pacificus*. *Molecular Genetics and Genomics* 269, 715-722.

STEINMETZ, P.R.H., KRAUS, J.E., LARROUX, C., HAMMEL, J.U., AMON-HASSENZAHL, A., HOULISTON, E., WÖRHEIDE, G., NICKEL, M., DEGNAN, B.M. & TECHNAU, U. (2012). Independent evolution of striated muscles in cnidarians and bilaterians. *Nature* 487, 231-234.

STERNBERG, P.W. & HORVITZ, H.R. (1981). Gonadal cell lineages of the nematode *Panagrellus redivivus* and implications for evolution by modification of cell lineage. *Developmental Biology* 88, 147-166.

STERNBERG, P.W. & HORVITZ, H.R. (1982). Postembryonic non-gonadal cell lineages of the nematode *Panagrellus redivivus*: description and comparison with those of *Caenorhabditis elegans*. *Developmental Biology* 93, 181-205.

SULSTON, J. & FERRY, G. (2002). *The common thread: a story of science, ethics and the human genome*. London, UK, Bantam Press.

THE *C. ELEGANS* RESEARCH COMMUNITY (2005). *Wormbook, the online review of* C. elegans *biology*. www.wormbook.org.

TIAN, H., SCHLAGER, B., XIAO, H. & SOMMER, R.J. (2008). Wnt signaling by differentially expressed Wnt ligands induces vulva development in *Pristionchus pacificus*. *Current Biology* 18, 142-146.

TRUE, J.R. & HAAG, E.S. (2001). Developmental systems drift and flexibility in evolutionary trajectories. *Evolution & Development* 3, 109-119.

TSONG, A.E., TUCH, B.B., LI, H. & JOHNSON, A.D. (2006). Evolution of alternative transcriptional circuits with identical logic. *Nature* 443, 415-420.

VAN MEGEN, H., VAN DEN ELSEN, S., HOLTERMAN, M., KARSSEN, G., MOOYMAN, P., BONGERS, T., HOLOVACHOV, O., BAKKER, J. & HELDER, J. (2009). A phylogenetic tree of nematodes based on about 1200 full-length small subunit ribosomal DNA sequences. *Nematology* 11, 927-950.

Van Valen, L. (1973). Festschrift. *Science* 180, 488.

West-Eberhard, M.-J. (2003). *Developmental plasticity and evolution*. Oxford, UK, Oxford University Press.

Wharton, D.A. (2002). *Life at the limits*. Cambridge, UK, Cambridge University Press.

Wilkins, A.S. (2002). *The evolution of developmental pathways*. Sunderland, MA, USA, Sinauer Associates Inc.

Wood, W.B. (1988). *The nematode* Caenorhabditis elegans. Cold Spring Harbor, NY, USA, Cold Spring Harbor Laboratory Press.

Witte, H., Moreno, E., Rödelsperger, C., Kim, J., Kim, J.-S., Streit A. & Sommer, R.J. (2015). Gene inactivation using the CRISPR/Cas9 system in the nematode *Pristionchus pacificus*. *Development Genes and Evolution* 225, 55-62.

Wright, S. (1932). The roles of mutation, inbreeding, crossbreeding and selection in evolution. In: *Proceedings of the 6th International Congress of Genetics*. Ithaca, NY, USA.

Nematology Monographs & Perspectives, 2015, Vol. 11, 43-76

Chapter 3

Diplogastrid systematics and phylogeny

Natsumi KANZAKI [1,2] and Robin M. GIBLIN-DAVIS [1]

[1] *Fort Lauderdale Research and Education Center, University of Florida/IFAS, 3205 College Avenue, Davie, FL 33314, USA*
giblin@ufl.edu

[2] *Forestry and Forest Products Research Institute, 1 Matsunosato, Tsukuba, Ibaraki, 305-8687, Japan*
nkanzaki@affrc.go.jp

Systematics and phylogeny

A phylogenetic systematist and an evolutionary systematist may make very different classifications, while inferring much the same phylogeny. If it is the phylogeny that gets used by other biologists, their differences about how to classify may not be important. I have consequently announced that I have founded the fourth great school of classification, the It-Doesn't-Matter-Very-Much school.

Felsenstein (2004). Inferring phylogenies.

Classification is a separate activity in science from theory and history. By all means we should try to reconstruct the past, but we do this not by subordinating classification to phylogeny, but by doing phylogeny on the basis of classificatory information. I do think that we can, to some degree of confidence, reconstruct past sequences. But this is always hypothetical and requires that we have empirical foundations for our reconstructions independent of our prior assumptions about how biological history unfolds, because biology is a bitch, and she won't be tamed by simplistic schemes, not even of common descent and models of speciation.

Wilkins (2011). What is systematics and what is taxonomy?

Introduction

This chapter will provide an overview of our current knowledge of the systematics and phylogeny of *Pristionchus pacificus* within the infraorder Diplogastromorpha. We will also discuss the morphological and life history traits, as well as the molecular phylogenetic inferences, that are helping us to generate hypotheses about the evolutionary history of this remarkable nematode and its relatives.

Systematics involves the description (taxonomy), classification (arrangement), morphology (terminology of parts), and naming (nomenclature) of organisms relative to their natural groupings (Wilkins, 2011). Setting aside Felsenstein's "*It-Doesn't-Matter-Very-Much school*", quoted above, for the moment, there are three basic approaches to classification. Pre-Darwinian or Linnaean classification creates lists of named species that are classified upon the levels of morphological similarity or dissimilarity. This is considered an artificial classification system because it is based upon potentially arbitrary characters of interest. The problem with an artificial classification system is its inherent lack of predictability because the easily measured, but arbitrarily selected, characters may not correlate well with each other or be useful for understanding evolutionary relationships. Darwinian or evolutionary classification allows for groups of species to give rise to new groups, and creates order using phylogenetic inferences and the degree of evolutionary change. This is considered a natural classification system because it groups organisms based upon shared ancestral characters (plesiomorphies) that can be used predictively. One problem with natural classification schemes is that they change as new information becomes available. Monophyly in the evolutionary systematics sense means that a group is derived from a single common ancestor. This allows for derived taxa to be excluded from their parent taxa in discussions of nomenclature and common ancestry. Lastly, phylogenetic systematics is derived from cladistics and defines monophyly or holophyly as a lineage containing the ancestral species and all of its descendants (clades). This is also a natural classification system that creates paraphyly out of the parental groupings in some evolutionary classification schemes. For example, birds and dinosaurs can be considered separate monophyletic lineages in evolutionary systematics, whereas they are paraphyletic according to phylogenetic systematics, which goes to the crux of the Felsenstein quote.

If we were strictly to employ Felsenstein's approach to classification, summarised in the opening quote (= "statistical phylogenetics" as renamed by Wilkins, 2011), then classification would be reduced to assigning names or codes to terminal clades and systematics would be subsumed under the operation of phylogenetic reconstruction, which in the case of nematodes currently entails mostly molecular phylogenetics. However, Wilkins (2011) rejects this idea on the grounds that even though we can reconstruct past evolutionary history with some level of confidence it is always hypothetical, and classification is an important empirical tool for independent reconstructions concerning "how biological history unfolds." He instead favours a modernised version of Asa Gray's (1879) scheme of classification where it is a separate operation that is not subsumed by phylogeny but helps support it. Systematics in this sense is the ability to identify and communicate about a specimen and its possible evolutionary history and includes taxonomy (species descriptions), classification (hierarchical arrangement of groups of species), nomenclature (Linnaean naming conventions, binomial nomenclature, *etc.*), and morphology (study and terminology of anatomical structures) (Wilkins, 2011). Phylogeny is defined as our attempt to reconstruct the evolutionary history of species (a species here is defined as coalescent populations of organisms that are on independent and non-coalescing lineage trajectories), but is done with the aid of named entities and groupings that can be arranged in practical ways based upon myriad types of morphological, developmental, molecular, genetic and ecological data at different scales. We will be combining De Ley & Blaxter's (2002, 2004) phylogenetic classification of the phylum Nematoda and the infraorder Diplogastromorpha into higher ranks and incorporating the catalogue of named and described species of Sudhaus & Fürst von Lieven (2003) and others (Kanzaki *et al.*, 2009a, 2012b; Fürst von Lieven *et al.*, 2011; Susoy & Herrmann, 2012; Herrmann *et al.*, 2013).

De Ley & Blaxter (2002, 2004) produced a phylogenetic classification of the phylum Nematoda by combining developmental and morphological characters with a molecular phylogenetic analysis of the phylum using the small subunit ribosomal rRNA gene (Blaxter *et al.*, 1998; Meldal *et al.*, 2007; van Megen *et al.*, 2009) and assigning ranks and taxa to clades. There are still issues with resolution of the deep branches of the nematode tree of life (De Ley & Blaxter, 2004) that create problems for hierarchical hypotheses but the clade that includes *P. pacificus* is clearly demarcated. According to their classification, *P. pacificus* belongs to the order Rhabditida, suborder Rhabditina, infraorder Diplogastromorpha, superfamily Diplogastroidea, and family Diplogastridae (De Ley & Blaxter, 2002, 2004). Simultaneously, Sudhaus & Fürst von Lieven

(2003) summarised a long-standing argument about taxonomic epithets, and changed the family name Diplogasteridae to Diplogastridae, because they argued that the correct construction of the name Diplogastr- should have been done from the root -gastros rather than -gaster. Over the past decade, many taxonomists, including us, have used this familial designation and spelling. We have therefore standardised the spellings of the higher classification as Diplogastromorpha, Diplogastroidea and Diplogastridae throughout this chapter.

Sudhaus & Fürst von Lieven (2003) used a morphological phylogenetic approach and recognised the clade classified by De Ley & Blaxter (2002) as Diplogastromorpha, but chose the family name Diplogastridae (= Diplogasteridae) Micoletzky 1922 for their 'rankless' higher classification of the equivalent natural grouping of 'diplogastrids' that is delimited by two morphological plesiomorphies described by Fürst von Lieven & Sudhaus (2000) as: *i*) loss of valves in terminal bulb; and *ii*) dorsal tooth present. They rejected the use of almost all higher category ranks for what we are calling Diplogastromorpha for anything above the species level on the grounds of subjectivity in describing degrees of distinctiveness between ranks. However, because we are historically bound by the International Code of Zoological Nomenclature (ICZN, 1999) to naming species with two epithets (generic and species names), the genus remains a valid ranking and, in the spirit of Wilkins (2011), will be discussed in greater detail below. Thus, the major classificatory operation at this time in the Diplogastromorpha is the designation of genera and generic groupings.

Members of the Diplogastromorpha typically possess variable stomas with variably-shaped gymnostoms armed with a large dorsal stegostomatal tooth, a pharynx with a muscular pro- and metacorpus, as for the other two infraorders (see below), but with a glandular isthmus and postcorpus, and, in males, papilliform or setiform genital sensilla without a bursa (except in the clade containing *Rhabditolaimus*). Members of the Diplogastromorpha, as the highly malleable stomatal morphology might suggest, are found in diverse habitats and include bacteriovores, fungivores, omnivores, and/or predators and parasites that are commonly found in saprobic environments, often with specialised and synchronised associations with insects, typically as phoretics, but also as necromenic or parasitic symbionts. This natural grouping may also exhibit the synapomorphy of possessing only three juvenile stages after hatching (this needs further verification) (De Ley & Blaxter, 2004) and currently includes 37 genera and over 300 valid species (Sudhaus & Fürst von Lieven, 2003; Kanzaki *et al.*, 2009a, 2012b, 2014a; Fürst von Lieven *et al.*, 2011; Susoy & Herrmann, 2012; Herrmann *et al.*, 2013; Susoy *et al.*, 2015; this chapter) with many more predicted (Giblin-Davis *et al.*, 2013).

According to De Ley & Blaxter (2002, 2004), there are two additional infraorders within the suborder Rhabditina; the Bunonematomorpha, which includes an interesting group of occasional insect-associated bacterivorous nematodes that share the characters of body asymmetry, the right side being ornamented with warts or papillae in one or two longitudinal rows and the left side with relatively deep longitudinal ridges, setiose lip appendages that have biradial symmetry from the right or left lateral aspects, a tubular stoma comprised mostly of gymnostom, a muscular pharynx comprising procorpus, expanded metacorpus, isthmus, and valvate and expanded postcorpus (terminal bulb), and the Rhabditomorpha, which includes the now famous rotting plant- and invertebrate-associated bacterivorous nematode model, *Caenorhabditis elegans* (Kiontke *et al.*, 2011), and possesses an annulated cuticle, six lips in the labial region with papilliform labial sensilla, a cylindrical stoma comprised mostly of gymnostom, muscular pharynx as described for the Bunonematomorpha, and in males, typically possessing setiform genital sensilla enclosed within expansions of the caudal alae forming a bursa.

Systematics

Overview

The taxonomy and classification of the Diplogastromorpha has a confusing history, but has recently been made clearer by the efforts of Sudhaus & Fürst von Lieven (2003) using a phylogenetic classification approach based upon morphology with subsequent refinements following incorporation of molecular and morphological phylogenetic data (Mayer *et al.*, 2007, 2009; Kanzaki *et al.*, 2011; Susoy & Herrmann, 2012). Much of the confusion derives from the relative plasticity of the stoma under evolutionary pressure and time, rampant homoplasy, and the lack of fossils to corroborate ancestral *vs* derived states when looking at contemporary lineages. Sudhaus & Fürst von Lieven (2003) tried to solve this problem by creating a candidate stem species of the Diplogastromorpha, and then applied cladistic rules for generic designations as they toiled through inadequate species descriptions, the lack of type specimens, and nomenclatural tangles. This resulted in a new 'line in the sand' relative to this natural grouping and a working hypothesis for all future generic and species descriptions. Since this time, nematode taxonomy has moved into the 'digital age' with most taxonomic work requiring digital photographs of voucher specimens, molecular barcodes, and type cultures (when possible) for future comparative work and mating studies, in addition to the traditionally accepted line drawings and morphometrics of type material. This is all moving us steadily to the desired point when nematode diversity can be adequately catalogued and easily and accurately identified and named for further study. We are definitely not there yet for the Diplogastromorpha because of the predicted number of species, many of which are associated with myriad insect species and probably involve cryptic species that are difficult to discern using traditional morphological methods (Kanzaki *et al.*, 2009b, 2012a; McFrederick & Taylor, 2012; Giblin-Davis *et al.*, 2013).

Classification at genus and higher levels

Superfamily and family designations

For the purpose of this chapter, we agree with the simplified 'rankless' higher classification solution proposed by Sudhaus & Fürst von Lieven (2003) and, because we agree with De Ley & Blaxter's pragmatic attempt at simplifying nematode classification while reflecting the expanding molecular phylogenetic data and inferences, we will use the infraorder as the only rank higher than genus. The reader is referred to De Ley & Blaxter (2002) if they are interested in their higher level classification and the authors of their chosen higher rankings (superfamilies and families) relative to the Diplogastromorpha. However, because of recent progress with the molecular and morphological phylogeny of the Diplogastromorpha (Susoy *et al.*, 2015), we have performed some classificatory housekeeping for the group. For example, *Odontopharynx* de Man, 1912b (Odontopharyngoidea, Odontopharyngidae) is transferred out of the Diplogastromorpha because molecular phylogenetic sequence inferences suggests that the 'diplogastrid' morphological similarities in stomatal and pharyngeal morphology are due to convergence and not shared ancestry (van Megen *et al.*, 2009; Susoy *et al.*, 2015).

Generic designations

Sudhaus & Fürst von Lieven (2003) argued that the genus (and by association all higher rankings) is a man-made convention because organisms classified as such cannot be defended as biologically equivalent in rank to taxa in any other genus (or equivalent higher ranking). However the genus is a rank of convention and must be applied with a level of practicality in Diplogastromorpha, for which they offered the following suggestions: *i*) genera should be groups of taxa for which "generalised statements" can be made, information stored, and testable hypotheses generated about understudied characters; *ii*) monotypic genera are to be avoided; and *iii*) the basis for their practical groupings is derived from a phylogenetic tree modified from Fürst von Lieven & Sudhaus (2000) and is informed by possible derivations from a hypothetical stem species or the 'ground pattern' allowing hypotheses about character polarities (apomorphic or uniquely derived *vs* plesiomorphic or ancestral). This typological and conventional operation for creating genera fits Wilkins' (2011) suggestion of a modernised version of Gray's (1879) scheme of classification that can be informed by molecular phylogeny to help create a natural classification, especially when clear monophyletic clades can be inferred that corroborate or challenge the 'understudied characters' used in proposing genera or species. Recent molecular and morphological trait analyses (Kanzaki *et al.*, 2014b; Susoy *et al.*, 2015) are currently focused on this goal for the Diplogastromorpha.

General morphology

The morphology of the infraorder Diplogastromorpha (= family Diplogastridae of Sudhaus & Fürst von Lieven, 2003) can be very challenging. The infraorder contains many cryptic species groups, and some species have species-specific apomorphies and secondary loss of characters. Thus, the common characters to define the infraorder are very limited. The general morphological characters of the infraorder are schematically illustrated in Figure 3.1.

Body surface

The thickness of the cuticle is variable among genera, and surface structure, *i.e.*, with/without striation and annulations, is also variable among genera. The lip region is not clearly distinguishable from the

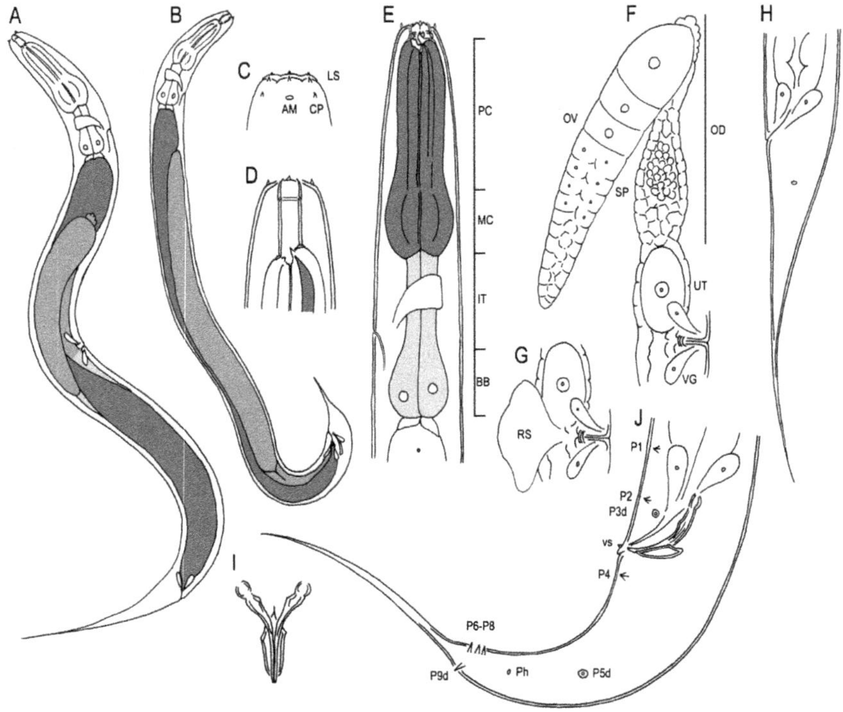

Fig. 3.1. *Schematic drawings for general morphology of diplogastrid nematodes. A: Adult female; B: Adult male; C: Anterior end (AM: amphid; CP: cephalic papillae; LS: labial sensilla); D: Stomatal region; E: Pharynx (PC: procorpus; MC: metacorpus or median bulb; IT: isthmus; BB: basal bulb or postcorpus); F: Female gonad (OV: ovary; OD: oviduct; SP: sperm stored in oviduct; UT: uterus; VG: vaginal gland or vulval gland); G: Vulval region of female (RS:* receptaculum seminis*); H: Female tail; I: Ventral view of paired male spicules and gubernaculum; J: Lateral view of male tail (P + number: paired genital papillae; vs: ventral single papilla; Ph: phasmid).*

rest of the body, *i.e.*, it is connected smoothly with the rest of body. The anterior end of the labial region is usually separated into six lip sectors but sometimes some sectors are fused to form one element. Six labial sensilla are present at the anterior end. Usually, each lip sector has a sensillum. Males have an extra four cephalic papillae occurring subdorsally and subventrally slightly outside the ring of labial sensilla. A pair of amphids are present laterally slightly outside the ring of labial sensilla, *i.e.*, mostly at the same level with male cephalic papillae. An excretory pore is present ventrally in the region of the mid-pharynx.

A pair of deirids is present on the lateral field, which is usually difficult to observe with light microscopy. Post-deirids and deirid-like pores have been reported in several genera/species. A pair of phasmids is present at the tail region, usually located posterior to the anal/cloacal opening.

Stoma

Stomatal morphology is highly variable among genera. Typically, the stoma is separated into three elements, cheilostom, gymnostom and stegostom, and the stegostom is further separated into three (four) sections, pro-/mesostegostom, metastegostom and telostegostom. Detailed morphology of each stomatal element will be introduced below. The schematic drawings of stomatal elements are provided in Figures 3.2 and 3.3.

Digestive tract

The digestive tract is relatively similar among the genera. The pharynx is separated into two parts, muscular anterior pharynx and glandular posterior pharynx. The anterior pharynx is composed of a procorpus and metacorpus. The procorpus is a muscular tube connecting the stoma (telostegostom) and metacorpus. The metacorpus comprises a muscular median bulb. The posterior pharynx consists of an isthmus and basal bulb. Both are glandular and lack the valve apparatus, except for *Pseudodiplogasteroides*, which has a heavily sclerotised lumen forming a valve apparatus. The isthmus is surrounded by a nerve

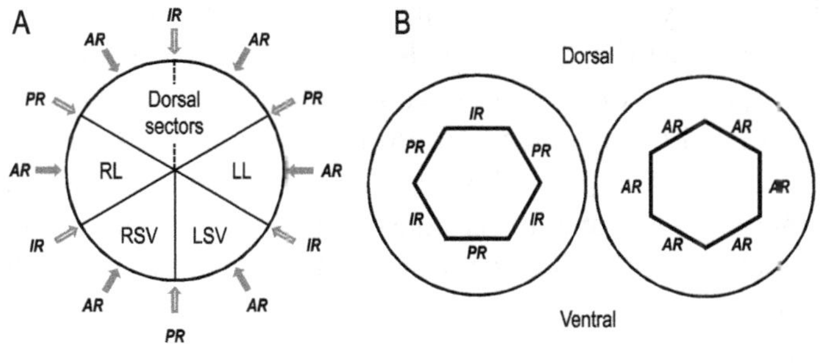

Fig. 3.2. *Schematic drawing of the cross section of stomatal region. A: Radial positions of sectors (AR: adradial; IR: interradial; PR: perradial; LL: left lateral sector; RL: right lateral sector; LSV: left subventral sector; RSV: right subventral sector); B: Cross sections showing cheilostomatal positions (left: per and interradial plates; right: adradial plates).*

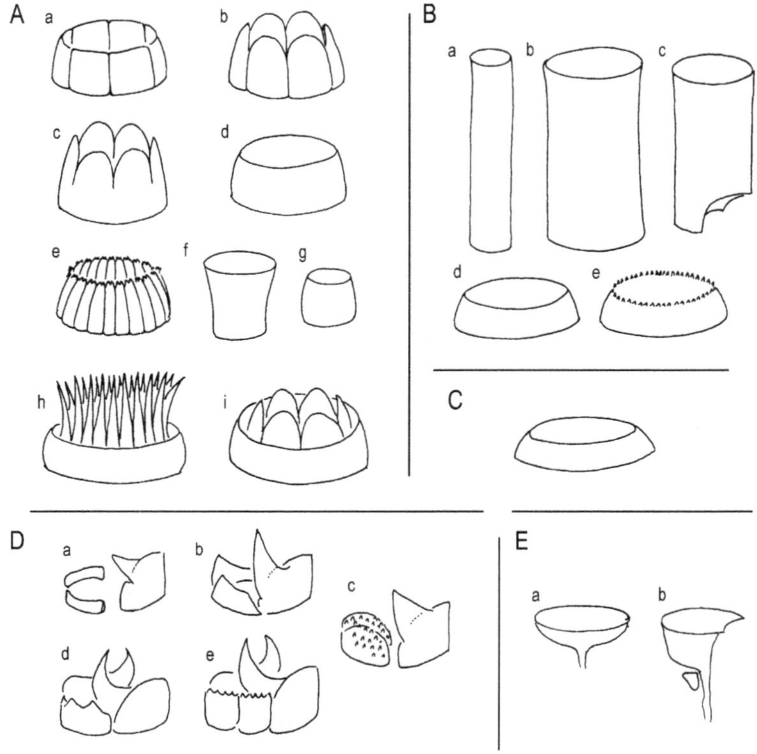

Fig. 3.3. *Variation of stomatal elements. A (a-i): Cheilostom; B (a-e): Gymnostom; C: Pro-/mesostegostom; D (a-e): Metastegostom; E (a, b): Telostegostom.*

ring (circumpharyngeal commissure). The basal bulb and intestine are connected by a cardia (pharyngeal-intestinal valve), which is usually well-developed and distinguishable. The intestine is a relatively simple tube constructed by flattened and tile-like cells. Three (one dorsal and two subventral) rectal (anal/cloacal) glands are present at the intestinal-rectal junction. The junction is constricted by a sphincter muscle but this is sometimes difficult to confirm because it is masked by the rectal glands. The rectum ends with an anal opening which is usually a dome-shaped slit.

Female reproductive tract

The number of gonad(s) is often variable among species of the same genus. Typically two gonadal branches are present, the anterior gonad being on the right side of the intestine and the posterior gonad on the left. In a species with a single gonad, the posterior gonad is always vestigial, becoming a post-uterine branch, or a simple sac-like structure. From the distal end to the vulval opening, the organs are arranged as ovary, oviduct, uterus and vagina. The ovary is reflexed for its entire length (antidromously reflexed). The ovary and oviduct are connected by a distinguishable tissue type which is typically composed of small and rounded cells. The oviduct is a simple tube composed of large and flattened cells. The spermatheca is absent, with part of the oviduct serving as the spermatheca. The uterus has a relatively thick wall, and in some genera (species) a *receptaculum seminis* is present on the dorsal side of the uterus. The vagina is distinctively sclerotised and usually forms a simple tube with a sphincter muscle at the uterus-vaginal junction. The vulval opening is usually pore-like, but forms a dome-shaped slit in some species. Four vaginal (vulval) glands are present around the vagina, and their size is variable among genera.

MALE SECONDARY SEXUAL CHARACTERS

A single testis is present on the right side and/or ventral to the intestine. The posterior part of the testis functions as a *vas deferens* but is not easily distinguished from the testis. The posterior end of the *vas deferens* is fused with the rectum to form a narrow cloacal tube. The cloacal opening is a dome-shaped slit in ventral view. Nineteen genital papillae (a ventral single papilla and nine pairs of subventral/lateral/subdorsal papillae) are present in the tail region, but the number of papillae is sometimes reduced and the arrangement is variable among genera/species. Typically, the ventral single papilla is located immediately anterior to the cloacal opening (on the anterior cloacal lip) and paired papillae include three pre- and six post-cloacal pairs. Within the three precloacal pairs, the second or third pair are usually located latero-ventrally, and the other two are subventral. In the postcloacal pairs, the fourth pair is subventrally situated, the fifth pair is lateral, the sixth to eighth pairs form a triplet-like arrangement at the ventral side near the tail tip, and the ninth pair occurs on the dorsal side at the same level as the sixth-eighth pairs. A pair of spicules and a gubernaculum are present, the size and shape being variable among genera/species.

FEMALE TAIL CHARACTERS

The female tail is typically an elongated cone with or without a filiform terminus. However, some genera have short and conical tails.

List of genera and their morphological characters

Currently, we recognise 37 genera in Diplogastromorpha. In the present classification system, the diplogastrid genera are mostly defined by the typological character of stomatal morphology (*e.g.*, Sudhaus & Fürst von Lieven, 2003). Thus, the variations of each stomatal element are introduced first followed by the current diplogastrid genera and their generic characters.

STOMATAL MORPHOLOGY AS A GENERIC DIAGNOSTIC CHARACTER

The stoma of diplogastrid nematodes is basically separated into three sections, cheilostom, gymnostom and stegostom, and the stegostom is further separated into three subsections, pro-/mesostegostom, metastegostom and telostegostom. Each section (subsection) is schematically illustrated in Figure 3.3.

Cheilostom

The cheilostom is the most anterior element of the stoma, *i.e.*, the stomatal opening, and is produced by the epidermis. The element is often manifested as a short tube or ring, and sometimes comprises plates, rugae, or a corona (Fig. 3.3A a-i).

Gymnostom

The gymnostom is the intermediate element, connecting the cheilostom and stegostom. It is associated with two rings of arcade syncytial cells that extend anteriad from the procorpus. This element is relatively simple compared with the other two elements. It usually comprises a thick or thin short tube or ring. Dorsal and ventral walls have the same length (isotopic) (Fig. 3.3B a, b, d, e) or, in many species, the dorsal wall is shorter than the ventral wall (anisotopic) (Fig. 3.3B c). In some species the anterior end of the tube is serrated (Fig. 3.3B e).

The anterior end often inserts into the posterior end of the cheilostom internally.

Stegostom

The stegostom is separated into three (four) sections, pro-/mesostegostom, metastegostom and telostegostom from the anterior. The stegostom is produced by and associated with succeeding layers of epidermal cells in the anterior procorpus (pharynx). Prostegostom and mesostegostom are fused to form the pro-/mesostegostom, and this element usually forms a ring connecting the posterior end of the gymnostom with the metastegostom (Fig. 3.3C). The metastegostom is usually separated into three sectors, the right and left subventral and dorsal sectors, with each bearing teeth, ridges, serrated plates or denticles (Fig. 3.3D a-e). The telostegostom is relatively simple, and often forms a shallow plate or funnel shape to connect the metastegostom with the procorpus (Fig. 3.3E a). In some genera, small denticles or subventral apodemes are observed in this region (Fig. 3.3E b).

LIST OF GENERA

The currently recognised genera and their stomatal morphology are listed below. The stomatal characters were simplified based upon Sudhaus & Fürst von Lieven (2003), Kanzaki *et al.* (2009a, 2012b, 2014b), Fürst von Lieven *et al.* (2011), Susoy & Herrmann (2012) and Herrmann *et al.* (2013). The genera for which molecular sequence information is available and used for phylogenetic inferences are highlighted with an asterisk after the genus name.

***Acrostichus** Rahm, 1928**

Cheilostom is separated into six adradial plates, with the anterior end of each plate being elongate and forming a short flap. Gymnostom is a relatively simple tube. Metastegostom bears dagger-like tooth on dorsal sector and triangular ridges on both right and left subventral sectors. Stomatal polymorphism has been reported in several species, where the stoma is widened, cheilostomatal elements divided (usually in multiples of three), gymnostom widened and each tooth and ridge becomes enlarged. In addition to stomatal morphology, females have an oval-shaped *receptaculum seminis*, and males have P1 and P2 papillae arranged in tandem (forming a doublet) and a massive gubernaculum, which are considered good generic characters.

_Allodiplogaster_* Paramonov & Sobolev _in_ Skrjabin, Shikobalova, Sobolev, Paramonov & Sudarikov, 1954

Cheilostom is separated into six per- and interradial vertically striated plates or rugae. Gymnostom is a short tube or ring-like. A dorsal claw-like tooth, right subventral claw-like tooth and left subventral serrated plates are present on metastogostomatal sectors. Telostegostom has two subventral apodemes. The genus is separated into two ecological groups, the *henrichae* group and the *striata* group. Stomatal dimorphism has been reported in several species in the *henrichae* group. In the eurystomatous form, the cheilostom and gymnostom become wider, and each tooth or serrated plate in the metastegostom becomes enlarged and more pronounced. Most species in the *henrichae* group have been recovered from insects, *e.g.*, Hymenoptera and Coleoptera. The *striata* group contains aquatic or semi-aquatic species. Morphologically, the group is distinguished from the *henrichae* group by its long and setiform labial sensilla and male genital papillae and long tail of males and females. None of the *striata* group has been molecularly analysed so far. If members are subsequently discovered to be phylogenetically separated from the *henrichae* group, a genus *Gobindonema* Khera, 1970 will be resurrected for the home of the species in the *striata* group.

Anchidiplogaster Paramonov, 1952

This monotypic genus had been synonymised with *Koerneria*, mostly because of its unclear description (Sudhaus & Fürst von Lieven, 2003). However, because of the absence of stegostomatal apodemes (Hnatewytsch, 1929), a requisite apomorphy of *Koerneria* and *Allodiplogaster*, the monotypic type species, *A. dubia* was returned to the resurrected genus, *Anchidiplogaster* (Kanzaki *et al.*, 2014a). The genus is morphologically defined with its miniscule, undivided stoma with two small, similarly sized, pyramidal teeth (one dorsal and one right subventral), and absence of male genital papillae and testis flexure. The type species was isolated from wood in a mine in Germany.

_Butlerius_* Goodey, 1929

Cheilostom forms a wide tube with anterior flaps (adradial or per- and interradial was not specified). Gymnostom is a wide ring or short tube. The metastegostom possesses a dorsal flap-like or thorn-like tooth, and a ridge or a denticular plate is present on subventral sectors. In addition to stomatal morphology, long and bristle-shaped labial sensilla are considered a generic character. Most species have been isolated from nutritionally rich soil, *e.g.*, rhizosphere and humus.

***Cephalobium* Cobb, 1920**

Stomatal morphology has not been examined closely using modern standards. There are two clear stomatal elements, anterior and posterior cuticular tubes. Because of the presence of a dorsal tooth, the posterior part is considered as being stegostom. Thus, the anterior tube is either cheilostom or gymnostom and, judging from its position in drawings from previous descriptions, the tube is hypothesised to be gymnostom. Therefore, the cheilostomatal element is assumed to be degenerate or composed of a thin membrane-like structure. All species in the genus are known to be parasites of crickets (Grylloidea: Orthoptera).

***Cutidiplogaster* Fürst von Lieven, Uni, Ueda, Barbuto & Bain, 2011**

The heavily cuticularised cheilostom forms a short tube or collar, and the gymnostomatal cuticular tube inserts into the overlapping posterior edge of the cheilostom. The metastegostom has a dorsal tooth and the telostegostom is an unusually long tube. In addition to stomatal characters, the unusually long and coiled tail is reported as a generic character. This monotypic genus was described from skin lesions of a manatee.

***Demaniella** * **Steiner, 1914**

Short tube-like cheilostom is not strongly sclerotised and its anterior end is connected to the cuticularised lip edge. Gymnostom is a simple and anisotopic tube. A flap-like process (tooth) present on the dorsal metastegostom. The species in the genus are isolated from nutritionally rich environments, *e.g.*, sewage and compost.

***Diplogaster** * **Schultze *in* Carus, 1857**

The cheilostom forms a short and wide tube, and its anterior part is separated and elongated to form six per- and interradial flaps. Gymnostom is a short and cuticular ring-like element. Metastegostom possesses a dagger-like tooth on the dorsal sector, and a small ridge on the subventral sectors. Two species are known, both aquatic.

***Diplogasteriana** * **Meyl, 1960**

Cheilostom is heavily cuticularised and anteriorly separated into inner and outer rings. Thus, it seems that two short tubes are connected at the posterior end. The outer ring is heavily sclerotised, with the inner ring being thinner and the anterior edge being split into six per- and interradial flaps. Gymnostom is a short and thick cuticular ring. Large triangular tooth and a triangular subventral ridge present on dorsal and subventral sectors, respectively. In addition to the stomatal morphology, the presence of male bursal flap is suggested as a generic character. This monotypic genus was isolated from slime flux of deciduous trees associated with *Nosodendron fasciculare* (Coleoptera).

***Diplogasteroides** de Man, 1912a**

Cheilostom is a short cuticular tube and the gymnostom is a rather simple anisotopic cuticular tube. The anterior end of the cheilostom bears short flaps in some species. Metastegostom bears three small rod-like teeth on its dorsal sector, and both subventral sectors have a sclerotised surface. Females are didelphic. Many species are associated with insects, especially wood-boring beetles. According to recent molecular phylogenetic analyses (Kanzaki *et al.*, 2013; Susoy *et al.*, 2015), the genus is clearly paraphyletic as envisioned by Sudhaus & Fürst von Lieven (2003) (discussed below in *Fuchsnema*).

***Diplogastrellus** Paramonov, 1952**

Cheilostom a cuticular tube, its anterior end split into six per- and interradial flaps. Gymnostom a simple anisotopic cuticular tube. Cheilostom and gymnostom roughly similar in length. Stegostom bears a flap-like or dagger-like tooth and small denticles on the dorsal and subventral sectors, respectively. Because all known species in the genus are monodelphic, it is considered a generic character. The isolation source varies from insects to nutritionally rich environments. Although more material needs to be examined to confirm morphology and molecular phylogenetic relationships, Susoy *et al.* (2015) suggested that the genus might be separated into two subgenera, *Diplogastrellus* and *Metadiplogaster*.

***Eudiplogasterium** Meyl, 1960**

Cheilostom wide and shallow and separated into 12 plates. Gymnostom forms deep barrel-shaped cuticular tube, with dorsal wall being much shorter than ventral wall. An anteriorly directed claw-like dorsal tooth and small subventral denticles present on metastegostomatal sectors. The genus is currently monotypic, and the type species was isolated from cow dung. *Eudiplogasterium* was considered a junior synonym of *Fictor* by Sudhaus & Fürst von Lieven (2003), but was distinct from the genus in recent molecular phylogenetic analysis (Susoy *et al.*, 2015); thus, the genus is listed here as valid.

***Fictor** Paramonov, 1952**

Cheilostom composed of narrow stick-like plates (rugae). Gymnostom forms a cuticular ring. Claw-like dorsal and right subventral teeth, and serrated left subventral plates present on metastegostomatal sectors. The isolation source varies from insects to nutritionally rich environments.

***Fuchsnema** Andrássy, 1984**

Cheilostom a short cuticular tube, gymnostom a rather simple anisotopic cuticular tube. The anterior end of the cheilostom bears short flaps in some species. Metastegostom bears three small rod-like teeth on its dorsal sector, and both subventral sectors have a sclerotised surface. Many species are associated with insects, especially wood-boring beetles. Stomatal typological characters overlap with those of *Diplogasteroides*, but the synonymy with *Diplogasteroides* by Sudhaus & Fürst von Lieven (2003) is not accepted (see Andrássy, 2005 and Susoy *et al.*, 2015) and *Fuchsnema* can be distinguished from *Diplogasteroides* by molecular phylogenetic distance, and the possession of a monodelphic and prodelphic female gonad.

***Goffartia* Hirshmann, 1952**

Cheilostom a cuticular tube which narrows anteriorly. Gymnostom is barrel-shaped, and no tooth, ridge or denticles are present on stegostomatal sectors. The species are aquatic and sometimes associated with riparian beetles.

***Heteropleuronema* Andrássy, 1970**

The stomatal morphology of this monotypic genus has not been reported in detail. The genus is mostly characterised by body surface ornamentation. The body surface of the type species is asymmetrical, *i.e.*, left side has annulations and striations typical of diplogastrids, but the right side has three deep longitudinal ridges. The presence of a bursa in males is also suggested as a diagnostic generic character. The type species has been isolated from wood infected by fungi.

***Hugotdiplogaster* Morand & Baker, 1995**

The stomatal morphology of this monotypic genus has not been reported in detail. According to the original description, the stoma is tube-like and the separation of elements is not clear. The type species was isolated as a parasite from the genital tract of a slug.

***Koerneria** Meyl, 1960**

Cheilostom a tube with fine longitudinal striations. Gymnostom a short tube or ring-like. A dorsal claw-like tooth, right subventral claw-like tooth and left subventral serrated plates are present on metastogostomatal sectors. Telostegostom has two subventral apodemes. Stomatal dimorphism has been reported in several species. In the eurystomatous form, the cheilostom and gymnostom become wider, and each tooth or serrated plate in the metastegostom becomes enlarged and more pronounced. The species of the genus with known bionomics are often associated with beetles, especially stag beetles (Lucanidae).

***Leptojacobus** Kanzaki, Ragsdale, Susoy & Sommer, 2014b**

Stoma small with minute armature. Cheilostom divided into six adradial plates, but the plates are unclear and difficult to observe with light microscopy. The subventral stegostomatal sectors are symmetrical and the dorsal tooth is thin and conical. Adults possess very thin and delicate bodies. This monotypic genus is associated with stag beetles (Lucanidae) imported from Indonesia to Japan.

***Levipalatum** Ragsdale, Kanzaki & Sommer, 2014**

The dorsal metastegostomatal tooth is long and hooked and connected to a 'palate' that projects anteriad and mediad. Stoma possesses telostegostomatal ridges of denticles. Anterior region of pharynx bulges. Males possess ten pairs of genital papillae. This monotypic genus is associated with scarab beetles in Texas, USA.

***Longibucca* Chitwood, 1933**

Detailed stomatal morphology has not been reported, but is generally characterised as being extremely long, with the gymnostom occupying most of its length. Further, females have a single gonad in all three known species. The genus is characterised more clearly by its life history feature, *i.e.*, parasites of several different vertebrates. Currently, the molecular profile has not been determined for any of the species in this genus. Considering their unique morphology and biological characters, the placement of this genus in Diplogastromorpha should be examined and confirmed by molecular analyses.

***Mehdinema** Farooqui, 1967**

Cheilostom short and tube-like and gymnostom a long and narrow tube, occupying most of the stoma. Stegostom short and does not have a clear tooth, but both dorsal and subventral sectors are sclerotised. This genus is characterised by its body surface, *i.e.*, covered by posteriorly directed spines, and the biological character of being specialised parasites of crickets (Orthoptera: Gryllidae).

***Micoletzkya** Weingärtner, 1955**

Cheilostom separated into six per- and interradial plates, with the anterior end of each plate elongated to form a rounded flap. Gymnostom cuticularised and ring-like. A claw-like tooth present on the dorsal metastegostomatal sector and a ridge and two small denticles are on the right and left subventral sectors, respectively. Stomatal dimorphism has been reported in several species. In the eurystomatous form, the entire stoma becomes wider and each tooth, ridge or denticle becomes larger, especially the right subventral ridge, which becomes a large claw-like tooth. In addition to the stomatal morphology, the position of the labial sensilla, *i.e.*, distance from the stomatal opening, is longer in the two lateral sensilla compared with the dorsal and subventral ones and male genital papillae arrangement, *i.e.*, the P1 and P2 are closely aligned, are all listed as generic characters. Almost all species are isolated from wood-boring beetles, especially, the bark beetles (Scolytidae), and their associated environments.

***Mononchoides** Rahm, 1928**

Cheilostom separated into narrow and stick-like plates (rugae). Gymnostom a short tube or ring-like. In the metastegostom, a large claw-like tooth occurs on the right subventral and dorsal sectors and the left subventral sector has two serrated plates. The telostegostomatal wall is heavily sclerotised and its ventral side is clearly deeper than the dorsal. The members of the genus have been isolated from various environments including aquatic situations, *e.g.*, river sediment, insects, manure and soil.

***Neodiplogaster** Cobb, 1924**

Stoma narrow and deep, sometimes confused with the stylet of tylenchid nematodes under lower magnification because of the long and cuticularised stegostom. Cheilostom separated into rugae, gymnostom a short tube or ring-like. Small claw-like tooth can be observed on the right subventral and dorsal sectors of metastegostom but tooth or plate is not observed on the left subventral sector. Telostegostom is well-cuticularised and deep. Dorsal wing-like apodemes present at the posterior end of telostegostom. Stomatal dimorphism has been reported in several species and the eurystomatous form has an obviously wider stoma, and well-developed right subventral and dorsal teeth. In addition to the stomatal morphology, the relatively short male tail is considered diagnostic at generic level. The species in the genus are associated with wood-boring beetles, *i.e.*, weevils, bark and ambrosia beetles and cerambycids.

***Odontopharynx* de Man, 1912b (this genus, which was treated by Sudhaus & Fürst von Lieven, 2003 as a member of the "Diplogastridae", is transferred out of the Diplogastromorpha)**

Cheilostom a short tube. Gymnostom barrel-shaped, heavily cuticularised and possesses several small ridges. A large dagger-shaped dorsal tooth appears to be fixed (non-moveable) and mostly present in the gymnostom (see Figs 20A, B; 21A in Fürst von Lieven, 2000). However, its developmental origin needs to be verified because it might not be produced by pharyngeal cells but could originate from arcade syncytial or epithelial cells. This contrasts with the moveable and stegostomatally-derived diplogastrid dorsal tooth and supports the molecular phylogenetic inferences of van Megen *et al.* (2009) and Susoy *et al.* (2015), which place this genus in a clade outside of the Diplogastromorpha. The well-developed procorpus and asymmetry of the paired spicules (left spicule is larger than the right spicule; see Figure 21B, C, E in Fürst von Lieven, 2000) might be good diagnostic characters for delineating the genus and clade when further work is done.

***Oigolaimella** Paramonov, 1952**

Cheilostom separated anteriorly into two sections. The outer part forms a ring-like short tube and the inner ring is separated into many triangular plates (coprona). Gymnostom is a wide and very short tube, or ring-like. In metastegostom, claw-like tooth and triangular tooth are present on the dorsal and right subventral sectors, respectively, and left subventral sector does not have a tooth or denticles, but has a sclerotised surface.

Parapristionchus *** Kanzaki, Ragsdale, Herrmann, Mayer, Tanaka & Sommer, 2012b**

Cheilostom separated into 12 plates. Gymnostom a short tube, with its dorsal side thicker than the ventral. Metastegostom possesses a claw-like dorsal tooth, a right subventral ridge and three denticles on the left subventral sector. Stomatal dimorphism has been reported and the eurystomatous form has a wider stoma with large teeth and ridges. Notably, the right subventral ridge becomes a claw-like tooth and the left subventral denticles become wide-ridged plates. The genus is monotypic, and the type species is associated with stag beetles (Lucanidae).

Parasitodiplogaster *** Poinar, 1979**

The stomatal morphology is variable among species and the genus could conceivably be separated into several genera. The cheilostom is not sclerotised, forming a vestigial or weakly cuticularised ring. Gymnostom a short tube or ring, internally inserted into the cheilostom. Stegostom possesses claw-like or diamond-shaped dorsal and right subventral teeth, but left subventral sector does not have any armature. Telostegostom has vestigial apodeme. Stomatal dimorphism has been reported in the *P. maxinema* clade, where the stenostomatous form has a narrow and tube-like stoma with long and stick-like right subventral and dorsal teeth, and the eurystomatous form has a wide and shallow stoma with large and triangular subventral and dorsal teeth. In addition to the stomatal structures, the very posteriorly located phasmid is considered as a typologically diagnostic generic character. All species in the genus are necromenic associates/parasites of fig wasps, and live in the fig syconia. Because of stomatal dimorphism, some species could be predators in the syconia.

Paroigolaimella *** Paramonov, 1952**

Cheilostom separated into six adradial plates, the anterior end of each plate elongated to form a rounded flap. Gymnostom a short anisotopic tube. Metastegostom bears a triangular dorsal tooth and subventral serrated bulges. The species in the genus are isolated from nutritionally rich environments, *e.g.*, dung and sewage water. Molecular data from two isolates of this genus suggest that it is paraphyletic (Susoy *et al.*, 2015) and in need of further work.

Pristionchus * **Kreis, 1932**

Cheilostom separated into six (or 12 in some species) per- and interradial plates, the anterior end of each plate elongated to form a rounded flap. Gymnostom a short tube or ring-like. Metastegostom bears a triangular dorsal tooth, a right subventral ridge and three left subventral denticles. Stomatal dimorphism has been reported in many species. In the eurystomatous form, metastegostomatal teeth, ridge and denticles become enlarged, *i.e.*, a large claw-like dorsal tooth, a claw-like right subventral tooth and three left subventral ridged plates. Although the members of the genus have been isolated from various environments, *i.e.*, soil, wood and humus, most species are considered as associates of various insects and arthropods. Detailed morphology, life history and carrier associations of the genus are discussed by Ragsdale *et al.* (Chapter 4, this volume).

Pseudodiplogasteroides * **Körner, 1954**

Cheilostom a short tube. Gymnostom an anisotopic tube with a vertical ridge on the inner surface of its dorsal side. Metastegostom has a dorsal flap-like tooth and sclerotised subventral sectors. In addition to the stomatal morphology, the structure of the basal bulb, which has a sclerotised inner lining that manifests as a valve-like apparatus, is an important generic character. There are two nominal species, both associated with wood-boring beetles (Lucanidae and Cerambycidae).

Rhabditidoides * **Rahm, 1928**

Cheilostom a short tube and gymnostom an anisotopic tube. Metastegostom has a dorsal flap-like tooth and sclerotised subventral sectors. In addition to the stomatal morphology, the arrangement of the male genital papillae is considered a generic apomorphy, *i.e.*, ventral triplet papillae (P6-P8) are clearly separated into (P6, P7) P8. The species in the genus are isolated from nutritionally rich environments, *e.g.*, rotting plants, compost and slime flux. An insect (stag beetle) association has also been reported.

Rhabditolaimus * **Fuchs, 1914**

Cheilostom forms a short tube with crown-like anterior end, *i.e.*, anterior end of the stomatal tube has four (dorsal, two lateral and ventral) rounded expansions. Gymnostom a very long and slender tube which occupies most of the stoma. Stegostom with sclerotised dorsal and subventral sectors, but does not possess any teeth, denticles or ridges. The pharynx morphology is also considered as a generic apomorphy, *i.e.*, the procorpus and metacorpus are fused to form a sausage-like muscular tube.

***Sachsia** Meyl, 1960**

The stomatal morphology of this monotypic genus is not sufficiently described. According to the information about the type species, both the cheilostom and gymnostom are equally long short tubes, with a thorn-like tooth on the dorsal sector of the metastegostom. The type species was recovered from cow dung.

***Sudhausia** Herrmann, Ragsdale, Kanzaki & Sommer, 2013**

Cheilostom triradiate, its anterior rim punctuated by grooves. Gymnostom tube-shaped, divided into two distinct regions, *i.e.*, an offset and short anterior region and a longer posterior section. A cylindrical metastegostom bears a stick-like dorsal tooth and a pair of narrow, conical, equally sized and axially oriented denticles on both ventral sides. Telostegostom plate-like and bears minute, conical, axially oriented denticles arranged in three pairs (one dorsal, two subventral). Biologically, the genus is characterised by its viviparous reproductive mode. Both nominal species were isolated from dung beetles.

***Teratodiplogaster** Kanzaki, Giblin-Davis, Davies, Ye, Center & Thomas, 2009a**

Cheilostom a short tube, gymnostom has an isotopic cuticular tube-like shape. Metastegostom has a large and triangular right subventral tooth, and the dorsal metastegostom forms a cuticular plate that is concave with the middle part integrating with tip of right subventral tooth. The left subventral sector is variable. Telostegostom sclerotised. Dorsal pharyngeal gland observed in some individuals but position of its opening has not been confirmed. The most characteristic morphology of the genus is the lip shape. The lip region of the genus is hypothesised to be highly specialised for their feeding, *i.e.*, lip expanded to form a large scoop-like structure that is assumed to be used for scooping liquid material inside the fig syconia. Both right and left lateral lips are thin, semicircular-shaped membrane. Two (right and left) subventral lips fused to form a large scoop-like membrane consisting of a large spoon-like central region and two small fin-like triangular flaps on either side. Dorsal lips form a mirrored structure to ventral lips. The genus contains two nominal species both of which were described from the syconia of subgenus *Sycomorus* figs.

***Tylopharynx** de Man, 1876**

Stoma narrow and deep, sometimes confused with the stylet of tylenchid nematodes under lower magnification because of the long and cuticularised stegostom. Cheilostom forms a short tube, gymnostom a short tube or ring-like. Small claw-like tooth can be observed on the right subventral and dorsal sectors of metastegostom, but no teeth or plates have been observed on the left subventral sector. Telostegostom is well-cuticularised and deep. Dorsal spherical apodemes present at the posterior end of telostegostom. There are two nominal species isolated from soil (mud) and manure.

Phylogeny or reconstructing evolutionary history of diplogastrids

Overview

The phylogenetic relationships among diplogastrid genera have not been adequately addressed in detail. For example, several superfamilies harbouring families and genera were simply listed under the order Diplogastrida in most previous taxonomic systems (*e.g.*, Maggenti, 1991). Recently, a comprehensive evolutionary hypothesis was proposed by Sudhaus & Fürst von Lieven (2003) based on the concepts of a 'stem species' and 'genus specific apomorphy'. They hypothesised an ancestral form that possessed a tube-like simple stoma, based upon comparisons with diplogastrids and their close relatives, *e.g.*, Bunonematomorpha and *Rhabditoides* spp. According to this stem species concept, they hypothesised that the diplogastrid nematodes evolved to have more complex stomatal morphology, *e.g.*, *Pristionchus*, *Neodiplogaster* and *Koerneria*, which led to complex metastegostomatal elements and stomatal dimorphism.

Thereafter, Mayer *et al.* (2009) conducted molecular phylogenetic analyses based upon ribosomal RNA gene sequences and 12 ribosomal protein gene sequences to infer the relationships among 14 divergent diplogastrid genera. The phylogenetic relationships inferred by the molecular sequence analyses were not clearly correlated with the previous hypotheses based upon morphology, *i.e.*, *Koerneria*, which has one of the most complex stomatal morphologies, was considered as the basal clade of the diplogastrids, and the resolution of the other generic groupings was not clear.

An updated molecular phylogeny based upon ribosomal RNA and multiple ribosomal protein gene sequences is introduced below with the caveat that we really are just at the beginning of our understanding of the possible evolutionary history of this interesting group of nematodes.

SSU-BASED PHYLOGENETIC RELATIONSHIPS

To infer a robust phylogeny, analyses employing multiple genetic loci (*e.g.*, Mayer *et al.*, 2009) are desirable. Until recently, the level of molecular sequence information about ribosomal genes was not sufficient to estimate infraorder-wide phylogeny. Thus, most inferences were done based on near-full-length SSU. However, the most comprehensive phylogenetic analysis using disparate 28 diplogastrid genera was just completed by Susoy *et al.* (2015). Thus, the phylogenetic tree (Fig. 3.4) and the relationship between phylogeny and stomatal morphology (Fig. 3.5) are introduced here. Molecular phylogenetic analysis suggests that six well-supported phylogenetic clades can be recognised. *Leptojacobus* (1) and *Koerneria* (2) are clearly separate from other diplogastrids and form independent clades as the sisters of the other genera; *Allodiplogaster* and two other fig and fig wasp-associated genera (*Parasitodiplogaster* and *Teratodiplogaster*) form a well-supported clade (3); *Fictor*, *Sudhausia*, *Mononchoides*, *Neodiplogaster*, *Tylopharynx*, *Paroigolaimella*, *Eudiplogasterium* and *Sachsia* form a clade (4); *Acrostichus*, *Diplogasteriana*, *Micoletzkya*, *Parapristionchus* and *Pristionchus* form a clade (5); and the remaining genera, *i.e.*, *Oigolaimella*, *Rhabditolaimus*, *Levipalatum*, *Pseudodiplogasteroides*, *Rhabditidoides*, *Diplogasteroides*, *Mehdinema*, *Butlerius*, *Diplogastrellus* and *Fuchsnema* belong to the same clade (6). These groups seem to have some morphological similarities that may be grounded in homology. For example, *Tylopharynx*, *Neodiplogaster* and *Mononchoides*, which have dorsal and right subventral claw-like teeth and a deep telostegostom, form a well-supported clade and genera with

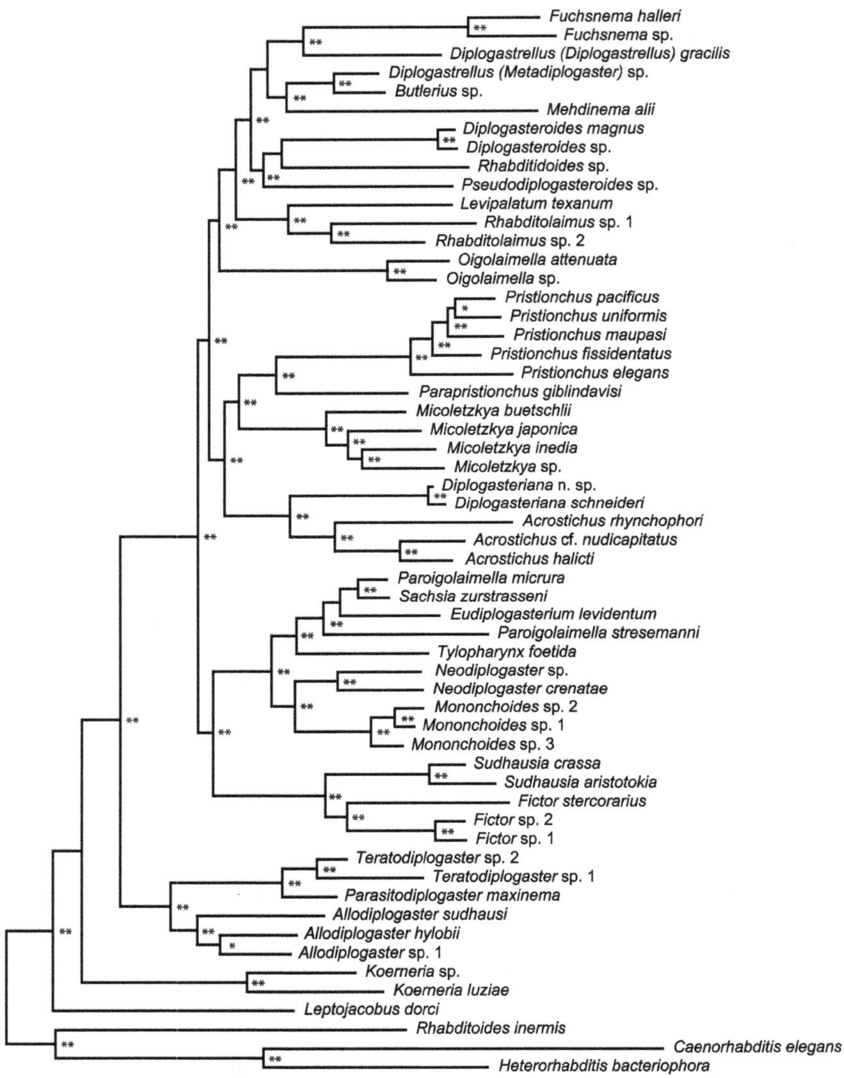

Fig. 3.4. *Phylogenetic relationships inferred for nematodes of Diplogastridae and outgroups from an alignment including SSU rRNA, LSU rRNA, 11 ribosomal protein genes, and RNA polymerase II. Aligned sequences of Diplogastridae and outgroups used to infer this phylogeny contained 667 kb excluding missing data and 6354 parsimony informative sites. **, 100% posterior probability (PP); *, 99% PP. Modified from Susoy* et al. *(2015).*

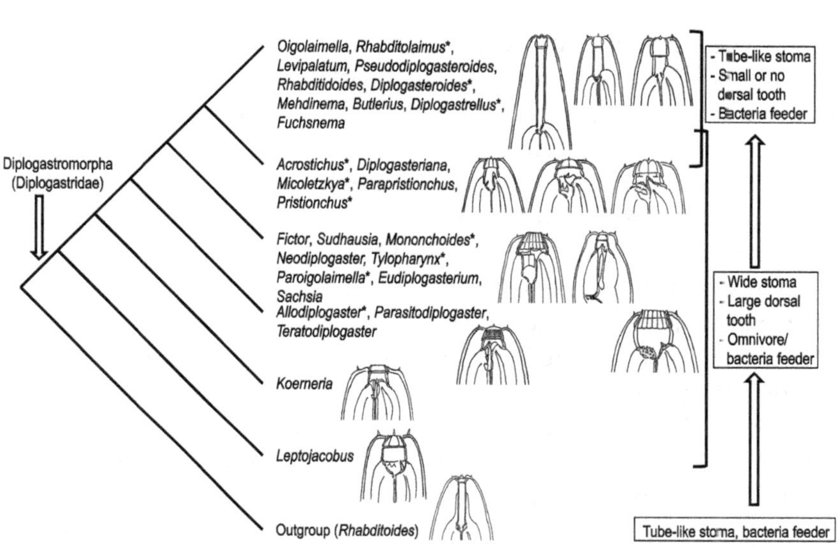

Fig. 3.5. *General trends in stomatal morphology relative to molecular phylogenetic inference. * indicates that the genus was drawn to represent each lineage involving more than one genus. In some groupings,* e.g., *(*Allodiplogaster, Parasitodiplogaster *and* Teratodiplogaster*), there exists significant morphological variation that can involve a tube-like stoma and significant modifications of the teeth.*

an elongated gymnostom that manifests as a tube-like stoma, *Rhabditidoides*, *Diplogasteroides*, *Diplogastrellus*, *Rhabditolaimus*, *Mehdinema*, *Fuchsnema* and *Pseudodiplogasteroides* belong to the same clade. Furthermore, three other genera, *Pristionchus*, *Acrostichus* and *Micoletzkya*, each form a well-supported clade that agrees with their generic morphological apomorphies (Sudhaus & Fürst von Lieven, 2003). Comparisons of *Koerneria* and *Leptojacobus* with outgroup taxa (*Rhabditoides*, *Caenorhabditis* and *Heterorhabditis*) suggest that there are significant differences in the stomatal morphology, *i.e.*, wide and asymmetric stoma with complex stegostom (*Koerneria*), wide and symmentric stoma with plated cheilostom and toothed stegostom (*Leptojacobus*) and simple tube-shaped stoma (outgroup taxa). There could be many 'missing links' that have an intermediate form between these two, or diplogastrid nematodes might have extremely high plasticity to allow for dramatic morphological alterations under evolutionary pressure. The latter possibility seems most plausible given the importance of feeding in niche utilisation and specialisation for nematodes. Thus, many new characters need to be examined and vetted for their potential phylogenetic signal in terms of reconstructing the evolutionary history of the Diplogastromorpha. The molecular phylogenetic analyses also revealed the presence of paraphyletic genera, or characteristic morphological traits shared with different clades. As stated above, *Koerneria* and *Allodiplogaster* are characterised by subventral stegostomatal apodemes, with some variation in cheilostomatal structure, *i.e.*, separated into plates or forming a short tube without plates. In this case, the apodeme could be a convergent character (or ancestral character shared by basal clades) and cheilostomatal morphology may reflect their phylogenetic status. Although this is not reflected in the current phylogenetic tree, the genus *Diplogasteroides* is also paraphyletic and is clearly separated into two different clades, *D. magnus* + *Diplogasteroides* sp. *vs D. andrassyi* (Kanzaki *et al.*, 2013). Interestingly, these two clades share some morphological characters besides the stomatal morphology. Kiontke *et al.* (2001) and Kanzaki *et al.* (2013) reported that *D. magnus* and *D. andrassyi* each have a *receptaculum seminis* containing a spermatophore-like structure and P7 genital papillae possessing three tips, which have not been reported in the other genera. Therefore, these characteristic structures are considered to have occurred at least twice independently in the different clades.

Integrated systematics based on morphology and molecular phylogeny

As in other organisms, robust molecular phylogeny, detailed morphological characterisation and understanding of development and homology, and bidirectional feedback between phylogeny and morphology are necessary to create a comprehensive and integrated systematics in diplogastrids.

Currently, the biggest problem is material collection and availability. To obtain molecular sequence information and comprehensive morphological characters, authentic DNA samples and well-preserved morphological materials are necessary. Cultured materials will fulfill these conditions. However, at present, most of the previously described species lack even the rudimentary information needed for modern classification, *i.e.*, even the authentic type materials have been lost in many species. Therefore, obtaining cultured or fresh specimens of previously described species is a first step for building a comprehensive systematics for the Diplogastromorpha.

Further, as mentioned above, seeking the 'missing link' clade will be important. Presently in morphological and molecular datasets there is a big gap between *Koerneria* + *Leptojacobus*, the most basal genera of diplogastrid, and outgroup clades. There could be more intermediate species that help to fill this gap and clarify the evolutionary history of the diplogastrids.

Recent methodological progress enables us to create video-capture-based photo-documentation, obtain whole (or partial) genome sequences from small numbers of individuals and study morphogenesis at the ultrastructural level. Much of this work has been pioneered with model organisms such as *P. pacificus*. These current and future technological advances, together with much deeper sampling of nematode associates of invertebrates, should help move us forward in elucidating the phylogeny and systematics of the Diplogastromorpha.

Acknowledgements

Special thanks to Erik Ragsdale, Matthias Herrmann, Vladislav Susoy, David Hunt and Ralf Sommer for suggestions and discussions that improved the chapter.

References

ANDRÁSSY, I. (1970). Freilebende Nematoden aus Vietnam. *Opuscula Zoologica Budapest* 10, 5-31.

ANDRÁSSY, I. (2005). *The free-living nematode fauna of Hungary I. (Nematoda errantia)*. In: Csuzdi, C. & Mahunka, S. (Eds). *Pedozoologia Hungarica*, No. 3. Budapest, Hungary, Hungarian Natural History Museum and Systematic Zoology Research Group of the Hungarian Academy of Sciences.

BLAXTER, M.L., DE LEY, P., GAREY, J.R., LIU, L.X., SCHELDEMANN, P., VIERSTRAETE, A., VANFLETEREN, J.R., MACKEY, L.Y., DORRIS, M., FRISSE, L.M. *ET AL.* (1998). A molecular evolutionary framework for the phylum Nematoda. *Nature* 392, 71-75.

CARUS, J.V. (1857). *Icones zootomica. Erste Hälfte: Die wirvellosen Thiere*. Leipzig, Germany, Engelmann.

CHITWOOD, B.G. (1933). On some nematodes of the superfamily Rhabditoidea and their status as parasites of reptiles and amphibians. *Journal of the Washington Academy of Sciences* 23, 508-520.

COBB, N.A. (1920). One hundred new nemas (type species of 100 new genera). *Contribution to a Science of Nematology* 9, 217-343.

COBB, N.A. (1924). *Neodiplogaster tropica* n. g. (?) n. sp. *Journal of Parasitology* 11, 105.

DE LEY, P. & BLAXTER, M. (2002). Systematic position and phylogeny. In: Lee, D. (Ed.). *The biology of nematodes*. London, UK, Taylor & Francis, pp. 1-30.

DE LEY, P. & BLAXTER, M. (2004). A new system for Nematoda: combining morphological characters with molecular trees, and translating clades into ranks and taxa. In: Cook, R. & Hunt, D.J. (Eds). *Proceedings of the Fourth International Congress of Nematology, Tenerife, Spain, 8-13 June 2002. Nematology Monographs and Perspectives* 2. Leiden, The Netherlands, Brill, pp. 633-653.

DE MAN, J.G. (1876). Onderzoekigen over vrij in de aarde levende Nematoden. *Tijdschrift der Nederlandsche Dierkundige Vereenigning* 2, 78-196 + plates 3-13.

DE MAN, J.G. (1912a). Helminthologische Beiträge. *Zoologische Jahrbücher (Systematik), Supplement* 15, 1-104.

DE MAN, J.G. (1912b). *Odontopharynx longicaudata* n. g., n. sp. Eine neue Form von Anguilluliden. *Zoologische Jahrbücher (Systematik), Supplement* 15, 439-464.

FAROOQUI, M.N. (1967). On a new family Mehdinematidae fam. nov. from *Gryllus domesticus*. *Zoologischer Anzeiger* 178, 322-326.

FELSENSTEIN, J. (2004). *Inferring phylogenies*. Sunderland, MA, USA, Sinauer Associates Inc.

FUCHS, G. (1914). Über Parasiten und andere biologische an die Borkenkäfer gebundene Nematoden. *85. Verhandlung der Gesellschaft Deutscher Naturforscher und Ärzte* 2, 688-692.

FÜRST VON LIEVEN, A. (2000). *Vergleichende und funktionelle Morphologie der Mundhöhle der Diplogastrina (Nematoda) mit einem ersten Entwurf der Phylogenic dieses Taxons*. Ph.D. Dissertation, Freie Universität Berlin.

FÜRST VON LIEVEN, A. & SUDHAUS, W. (2000). Comparative and functional morphology of the buccal cavity of Diplogastrina (Nematoda) and a first outline of the phylogeny of this taxon. *Journal of Zoological Systematics and Evolutionary Research* 38, 37-63.

FÜRST VON LIEVEN, A., UNI, S., UEDA, K., BARBUTO, M. & BAIN, O. (2011). *Cutidiplogaster manati* n. gen., n. sp. (Nematoda: Diplogastridae) from skin lesions of a West Indian manatee (Sirenia) from the Okinawa Churaumi Aquarium. *Nematology* 13, 51-59.

GIBLIN-DAVIS, R.M., KANZAKI, N. & DAVIES, K.A. (2013). Nematodes that ride insects: unforeseen consequences of arriving species. *Florida Entomologist* 96, 770-780.

GOODEY, T. (1929). On some new and little-known free-living nematodes. *Journal of Helminthology* 5, 25-32.

GRANT, V. (2003). Incongruence between cladistic and taxonomic systems. *American Journal of Botany* 90, 1263. DOI:10.3732/ajb.90.9.1263.

GRAY, A. (1879). *Structural botany*. New York, NY, and Chicago, IL, USA, Ivison, Blakeman, Taylor & Co.

HERRMANN, M., RAGSDALE, E.J., KANZAKI, N. & SOMMER, R.J. (2013). *Sudhausia aristotokia* n. gen., n. sp. and *S. crassa* n. gen., n. sp. (Nematoda: Diplogastridae): viviparous new species with precocious gonad development. *Nematology* 15, 1001-1020

HIRSCHMANN, H. (1952). Die Nematodender Wassergrenze mittelfränkishcer Gewässer. *Zoologische Jahrbücher (Systematik)* 81, 303-407.

KANZAKI, N. & GIBLIN-DAVIS, R.M. (2014). Phylogenetic status and morphological characters of *Rhabditolaimus anoplophorae* (Rhabditida: Diplogastridae). *Journal of Nematology* 46, 44-49.

KANZAKI, N., GIBLIN-DAVIS, R.M., DAVIES, K.A., YE, W., CENTER, B.J. & THOMAS, W.K. (2009a). *Teratodiplogaster fignewmani* gen. nov., sp. nov. (Nematoda: Diplogastridae) from the syconia of *Ficus racemosa* in Australia. *Zoological Science* 26, 569-578.

KANZAKI, N., GIBLIN-DAVIS, R.M., ZENG, Y., YE, W. & CENTER, B.J. (2009b). *Acrostichus rhynchophori* n. sp. (Rhabditida: Diplogastridae): a phoretic associate of *Rhynchophorus cruentatus* Fabricius and *R. palmarum* L. (Coleoptera: Curculionidae) in the Americas. *Nematology* 11, 669-688.

KANZAKI, N., TAKI, H., MASUYA, H., OKABE, K., TANAKA, R. & ABE, F. (2011). Diversity of stag beetle-associated nematodes in Japan. *Environmental Entomology* 40, 281-288.

KANZAKI, N., RAGSDALE, E.J., HERRMANN, M., MAYER, W.E. & SOMMER, R.J. (2012a). Description of three nematode species from Japan, *Pristionchus exspectatus* n. sp., *P. arcanus* n. sp., and *P. japonicus* n. sp., which form a cryptic species complex including the model organism *P. pacificus* (Diplogastridae). *Zoological Science* 29, 403-417.

KANZAKI, N., RAGSDALE, E.J., HERRMANN, M., MAYER, W.E., TANAKA, R. & SOMMER, R.J. (2012b). *Parapristionchus giblindavisi* n. gen., n. sp. (Rhabditida: Diplogastridae) isolated from stag beetles (Coleoptera: Lucanidae) in Japan. *Nematology* 14, 933-947.

KANZAKI, N., TANAKA, R., HIROOKA, Y. & MAEHARA, N. (2013). Description of *Diplogasteroides andrassyi* n. sp., associated with *Monochamus grandis* and Pinaceae trees in Japan. *Journal of Nematode Morphology and Systematics* 16, 35-47.

KANZAKI, N., RAGSDALE, E.J. & GIBLIN-DAVIS, R.M. (2014a). Revision of the paraphyletic genus *Koerneria* Meyl, 1960 and resurrection of two other genera of Diplogastridae (Nematoda). *ZooKeys* 442, 17-30.

KANZAKI, N., RAGSDALE, E.J., SUSOY, V. & SOMMER, R.J. (2014b). *Leptojacobus dorci* n. gen., n. sp. (Nematoda: Diplogastridae), an associate of *Dorcus* stag beetles (Coleoptera: Lucanidae). *Journal of Nematology* 46, 50-59.

KHERA, S. (1970). Nematodes from the banks of still and running waters. IX. Two new genera belonging to subfamily Diplogasterinae Micoletzky from India. *Revista Brasileira de Biologia (Rio de Janeiro)* 30, 405-409.

KIONTKE, K., MANEGOLD, A. & SUDHAUS, W. (2001). Redescription of *Diplogasteroides nasuensis* Takaki, 1941 and *D. magnus* Völk, 1950 (Nematoda: Diplogastrina) associated with Scarabaeidae (Coleoptera). *Nematology* 3, 817-832.

KIONTKE, K.C., FÉLIX, M.-A., AILION, M., ROCKMAN, M.V., BRAENDLE, C., PÉNIGAULT, J.-B. & FITCH, D.H.A. (2011). A phylogeny and molecular barcodes for *Caenorhabditis*, with numerous new species from rotting fruits. *BMC Evolutionary Biology* 11, 339.

KÖRNER, H. (1954). Die Nematodenfauna des vergehenden Holzes und ihre Beziehungen zu den Insekten. *Zoologische Jahrbücher (Systematik)* 82, 245-353.

KREIS, H.A. (1932). Beiträge zur Kenntnis pflanzenparasitischer Nematoden. *Zeitschrift für Parasitenkunde* 5, 184-194.

MAGGENTI, A.R. (1991). Nemata: higher classification. In: Nickle, W.R. (Ed.). *Manual of agricultural nematology.* New York, NY, USA, Marcel Dekker, pp. 147-187.

MAYER, W.E., HERRMANN, M. & SOMMER, R.J. (2007). Phylogeny of the nematode genus *Pristionchus* and implications for biodiversity, biogeography and the evolution of hermaphroditism. *BMC Evolutionary Biology* 7, 104.

MAYER, W.E., HERRMANN, M. & SOMMER, R.J. (2009) Molecular phylogeny of beetle associated diplogastrid nematodes suggests host switching rather than nematode-beetle coevolution. *BMC Evolutionary Biology* 9, 212.

MCFREDERICK, Q.S. & TAYLOR, D.R. (2012). Evolutionary history of nematodes associated with sweat bees. *Molecular Phylogenetics and Evolution* 66, 847-866.

MELDAL, B.H.M., DEBENHAM, N.J., DE LEY, P., TANDINGAN DE LEY, I., VANFLETEREN, J., VIERSTRAETE, A., BERT, W., BORGONIE, G., MOENS, T., TYLER, P.A. *ET AL.* (2007). An improved molecular phylogeny of the Nematoda with special emphasis on marine taxa. *Molecular Phylogenetics and Evolution* 42, 622-636.

MEYL, A.H. (1960). Die freilebenen Erd- und Süßwassernematoden (Fadenwürwer). In: Brohmer, P., Ehrmann, P. & Ulmer, G. (Eds). *Die Tierwelt Mitteleuropas*, Bd. 1 (5a). Leipzig, Germany, Quelle & Meyer.

MICOLETZKY, H. (1922). Die freilebenden Erd-Nematoden. *Archiv für Naturgeschichte* 87A(1921), 1-650.

MORAND, S. & BAKER, G.M. (1995). *Hugotdiplogaster neozelanaia* n. gen., n. sp. (Nematoda: Diplogasteridae), a parasite of the New Zealand endemic slug, *Athoracophorus bitentaculatus* (Quey & Gaimard, 1832) (Gastropoda: Athoracophoridae). *New Zealand Journal of Zoology* 22, 109-113

PARAMONOV, A.A. (1952). Opyt ekologicheskoi klassifikatsii fitonematod. *Trudy Gelmintologicheskoi Laboratorii, Akademia Nauk SSSR (Moskva)* 6, 338-369.

POINAR JR, G.O. (1979). *Parasitodiplogaster sycophilon* gen. n., sp. n. (Diplogasteridae: Nematoda), a parasite of *Elisabethiella stuckenbergi* Grandi (Agaonidae: Hymenoptera) in Rhodesia. *Proceedings of the Koninklijke Nederlandse Akademie van Wetenschappen, Series C: Biological and Medical Sciences* 82, 375-381.

RAGSDALE, E.J., KANZAKI, N. & SOMMER, R.J. (2014). *Levipalatum texanum* n. gen., n. sp. (Nematoda: Diplogastridae), an androdioecious species from the south-eastern USA. *Nematology* 16, 695-709.

RAHM, G. (1928). Alguns nematodes parasitas e semiparasitas das plantas culturaes do Brasil. *Archivos do Instituto Biologico de Defesa Agricola e Animal (São Paulo)* 1, 239-251.

SKRJABIN, K.I., SHIKOBALOVA, N.P., SOBOLEV, A.A., PARAMONOV, A.A. & SUDARIKOV, A.A. (1954). Camellanata, Rhabditata, Tylenchata, Trichocephalata and Dioctophymata and the distribution of parasitic nematodes by hosts. *Izdatel'stvo Akademii Nauk SSSR (Moskva)* 4, 1-927.

STEINER, G. (1914). Freilebende Nematoden aus der Schweiz. 1. + 2. Teil. *Archiv für Hydrobiologie und Planktonkunde* 9, 259-276, 420-438.

SUDHAUS, W. & FÜRST VON LIEVEN, A. (2003). A phylogenetic classification and catalogue of the Diplogastridae (Secernentea, Nematoda). *Journal of Nematode Morphology and Systematics* 6, 43-89.

SUSOY, V. & HERRMANN, M. (2012). Validation of the genus *Rhabditolaimus* Fuchs, 1914 (Nematoda: Diplogastridae) supported by integrative taxonomic evidence. *Nematology* 14, 595-604.

SUSOY, V., RAGSDALE, E.J., KANZAKI, N. & SOMMER, R.J. (2015). Rapid diversification associated with a macroevolutionary pulse of developmental plasticity. *eLife*. DOI:10.7554/eLife.05463

VAN MEGEN, H., VAN DEN ELSEN, S., HOLTERMAN, M., KARSSEN, G., MOOYMAN, P., BONGERS, T., HOLOVACHOV, O., BAKKER, J. & HELDER, J. (2009). A phylogenetic tree of nematodes based on about 1200 full length small subunit ribosomal DNA sequences. *Nematology* 11, 927-950.

WEINGÄRTNER, I. (1955). Berichtigung zur Arbeit Versuch einer Neuordung der Gattung "*Diplogaster* Schulze 1857 (Nematoda)". *Zoologische Jahrbücher (Systematik)* 83, 638.

WILKINS, J.S. (2011). *What is systematics and what is taxonomy*? Evolvingthoughts.net. http://evolvingthoughts.net/2011/02/what-is-systematics-and-what-is-taxonomy/.

Nematology Monographs & Perspectives, 2015, Vol. 11, 77-120

Chapter 4

Taxonomy and natural history: the genus *Pristionchus*

Erik J. RAGSDALE [1], Natsumi KANZAKI [2] and Matthias HERRMANN [3]

[1] *Department of Biology, Indiana University, 915 E. 3rd Street, Bloomington, IN 47405, USA*
ragsdale@indiana.edu
[2] *Forest Pathology Laboratory, Forestry and Forest Products Research Institute, 1 Matsunosato, Tsukuba, Ibarakim, 305-8687, Japan*
nkanzaki@affrc.go.jp
[3] *Department for Evolutionary Biology, Max-Planck Institute for Developmental Biology, 72076 Tübingen, Germany*
matthias.herrmann@tuebingen.mpg.de

Introduction

> *I enumerated the chief objections which might be justly urged against the views maintained in this volume... One, namely the distinctness of specific forms, and their not being blended together by innumerable transitional links, is a very obvious difficulty.*
>
> Charles Darwin (1859).

The programme to discover and catalogue the diversity of life provides the "general reference system" (Hennig, 1966) needed for comparative analysis of morphological, developmental or ecological characters. Inferences of process from macroevolutionary pattern require many more taxa than the minimum three necessary for evolutionary hypotheses. Phylogenetic intermediates, when discovered, can bridge the otherwise inexplicable gaps between forms that confound our understanding of evolution, such as how apparently novel structures arise. Although some nematode fossils are known (Poinar, 2011), their relative paucity with respect to animals with hard parts makes a truly historical study of nematodes difficult. However, thorough sampling of extant species and detailed character (*i.e.*, phylogenetic) analysis – which we consider the foremost missions of taxonomy – can recover lost history by inference.

In addition to yielding a historical context for comparative analysis, the taxonomy of a model system provides and distinguishes subjects for experimental biology. Alternative models give strength to the evo-devo of living organisms, in which molecular mechanisms are identifiable and testable. The discovery of new strains and attention to their life history allow new models to be brought into the laboratory. Molecular taxonomy expedites this work, both by revealing cryptic species and by providing a wealth of characters for testing phylogenetic relationships. Although descriptions of more than sequence characters may be intractable for nematode diversity at large (Blaxter, 2004; Félix *et al.*, 2014), knowledge of morphological traits and their ecological relevance is essential for comparative biology, particularly of model organisms. We have therefore endeavoured to refine the taxonomic system of *Pristionchus* through multiple molecular markers as well as detectable, if subtle, morphological differences.

The recent discovery of many new species of *Pristionchus*, which currently includes 48 valid species, has given the model of *P. pacificus* a solid comparative context. What has brought most of these recently reported species to light is the exploration of their natural history: collecting efforts have targeted their association with insects, specifically beetles (Herrmann *et al.*, 2006a). In this chapter, we review the biological associations and the morphological traits that distinguish and delimit *Pristionchus* species. By giving a current synthesis of *Pristionchus* taxonomy and bionomics, we aim to provide a foundation for macroevolutionary studies in the genus as well as a stepping-off point for the future discovery of forms.

Natural history

PRISTIONCHUS AND BEETLES

When *P. pacificus* was originally described (Sommer *et al.*, 1996; Figs 4.1, 4.2), next to nothing was known about the ecology of the

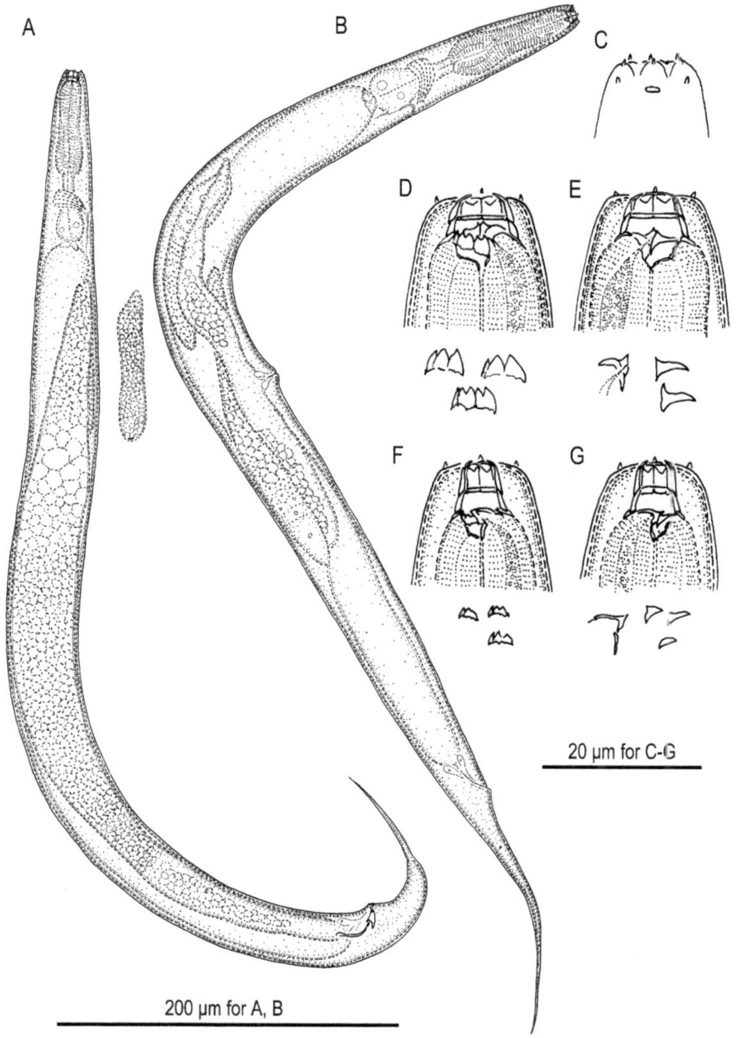

Fig. 4.1. *General morphology of the model organism* Pristionchus pacificus. *A: Entire body of male, right lateral aspect. Testis flexure shown to right of body; B: Entire body of hermaphrodite, right lateral aspect; C: Lip region of male, lateral aspect; D: Stomatal region of eurystomatous (Eu) hermaphrodite, left lateral aspect. Three variations of left subventral denticles shown below; E: Eu hermaphrodite, right lateral aspect. Dorsal tooth and two variations of subventral tooth shown below; F: Stenostomatous (St) hermaphrodite, left lateral aspect. Three variations of left subventral denticles shown below; G: St hermaphrodite, right lateral aspect. Variation of dorsal tooth and three variations of right subventral ridge shown below.*

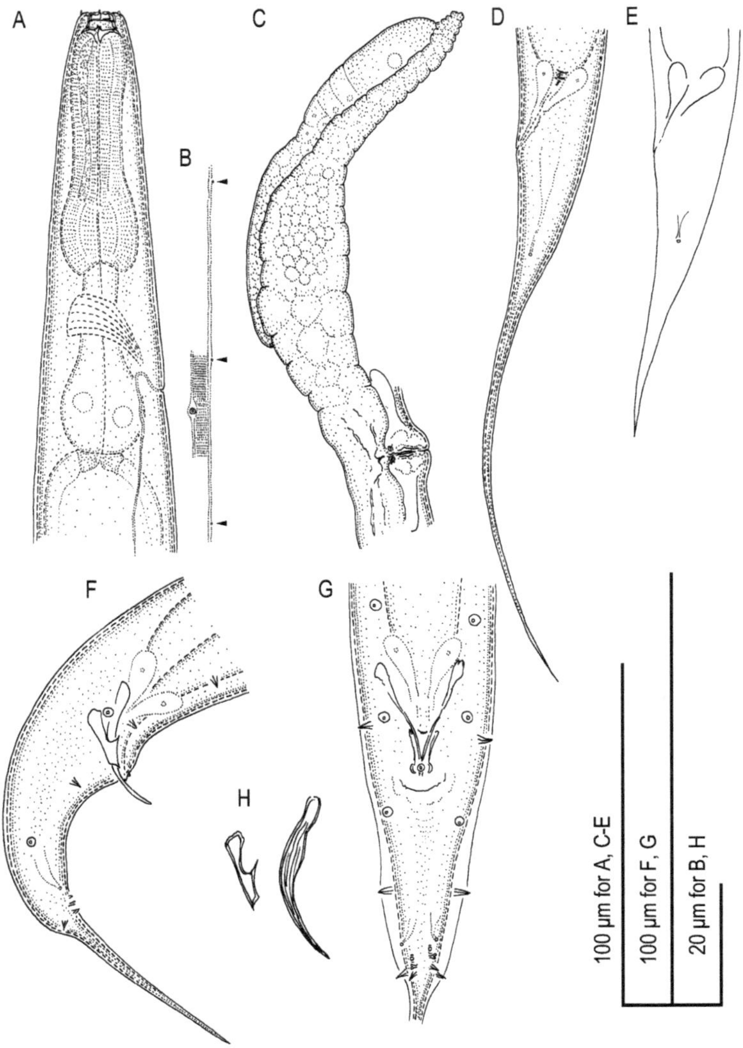

Fig. 4.2. *General morphology of the model organism* Pristionchus pacificus. *A: Neck region of Eu hermaphrodite, right lateral aspect; B: Body surface structure showing lateral field, deirid and secretory pores (arrowhead); C: Anterior gonad branch of young hermaphrodite; D: Tail region of hermaphrodite, left lateral aspect; E: Variation of hermaphrodite tail; F: Male tail, right lateral aspect; G: Male tail, ventral aspect; H: Gubernaculum and spicule.*

species. The first strain, which has since become the reference strain for genetic and genomic studies (Eizinger & Sommer, 1997; Srinivasan *et al.*, 2002, 2003; Dieterich *et al.*, 2008), had been isolated from a soil sample taken by students in Pasadena, CA, USA, in 1988. As *P. pacificus* became increasingly used as a satellite model organism for comparison with *Caenorhabditis elegans*, so the desire grew for more strains of *P. pacificus* and other *Pristionchus* species. The first attempts to collect more isolates were by sampling soil, following conventional approaches for isolating free-living Rhabditina *sensu* De Ley & Blaxter (2002), including Diplogastridae such as *Pristionchus*. However, soil samples had recovered only five *Pristionchus* species and four strains of *P. pacificus* between 1994 and 2004 (Sommer *et al.*, 2001; Herrmann *et al.*, 2006a; R. Sommer, pers. comm.).

Anecdotal observations had previously suggested that *Pristionchus* species were associated with insects, often cadavers (Völk, 1950; Sudhaus & Fürst von Lieven, 2003). Furthermore, field collections and inoculation experiments had shown one *Pristionchus* species, *P. uniformis*, to be associated with the Colorado potato beetle *Leptinotarsa decemlineata* Say (Coleoptera: Chrysomelidae) and the European cockchafer *Melolontha melolontha* L. (Coleoptera: Scarabaeidae) (Fedorko & Stanuszek, 1971). Following these indications, Herrmann *et al.* (2006a) conducted a systematic screen for *Pristionchus* associated with beetles, particularly of Scarabaeidae. Beetles were indeed a reliable source for *Pristionchus*, as a screen of some 4500 beetles collected in Europe produced 371 isolates of the genus. Mating tests and molecular markers, particularly a fragment of the small subunit (SSU) rRNA gene, revealed the isolates belonged to seven biological species. Four of these species were androdioecious (*i.e.*, consisting of males and self-fertile hermaphrodites) and thus a boon for genetic studies. Furthermore, with the discovery of hundreds of strains, microevolutionary studies by population genetics instantly became possible (see McGaughran & Morgan, Chapter 8, this volume). The techniques established for isolating *Pristionchus* strains have since become a standard for isolating new strains of *Pristionchus*. Collections targeting beetles in other geographic regions soon revealed a complex of new and previously described species from North America (Herrmann *et al.*, 2006b), as well as a consistent association of *P. pacificus* with the Oriental beetle, *Exomala orientalis* (Scarabaeoidea), in Japan (Herrmann *et al.*, 2007). Expanding this search strategy to other parts of the world, especially East Asia, has now led to the discovery of six previously described and 18 new insect-associated species of *Pristionchus* (Kanzaki *et al.*, 2011, 2012a, b, 2013a, b, c, 2014a; Ragsdale *et al.*, 2013; Fig. 4.3).

Nature of the host association

The association of *Pristionchus* nematodes with their insect carriers is considered *necromenic*, a term coined to mean "waiting for the cadaver" (Sudhaus & Schulte, 1989). On the live hosts, nematodes exist as dauers, a metabolically dormant juvenile stage that is dispersed by the adult insect (see Ogawa & Brown, Chapter 10, this volume). When the insect dies, the nematodes resume development and proliferate on the host cadaver, which is colonised by other organisms that the nematodes use as food. Early reports had suggested an association of some *Pristionchus* species, for example *P. brevicauda* (Kotlán, 1928) and the so-named *P. entomophagus* (Steiner, 1929), with the dead insects on which they were collected. However, it was the empirical observations and experiments with *P. uniformis* (Fedorko & Stanuszek, 1971) and other *Pristionchus* species (Herrmann *et al.*, 2006a, b, 2007; Rae *et al.*, 2008; Weller *et al.*, 2010; D'Anna & Sommer, 2011) that determined the beetle association to be a necromenic one.

The simple technique that has brought hundreds of *Pristionchus* strains into culture reflects the suitability of a dead host as a habitat for these nematodes (Herrmann *et al.*, 2006a). In this technique, living beetles are captured and brought into the laboratory, after which they are sacrificed and placed onto standard nematode growth medium (NGM) agar. Populations of microorganisms, fungi and nematodes already associated with the host then increase in an otherwise sterile environment (see Rae & Sinha, Chapter 14, this volume). Several species of nematodes can emerge in this microhabitat. The succession of organisms on the decomposing cadaver can be complex and *Pristionchus* nematodes are often not the first nematodes to appear. For example, Rhabditidae such as *Pelodera* and *Oscheius* emerge rapidly and in large numbers from the same beetles as *Pristionchus* (Weller *et al.*, 2010). Other Diplogastridae such as *Diplogasteroides* can also appear earlier than *Pristionchus* (Herrmann, unpubl. data). After 4-7 days, *Pristionchus* nematodes appear in appreciable numbers, where they can be observed feeding on cohabiting species, including other nematodes. By isolating

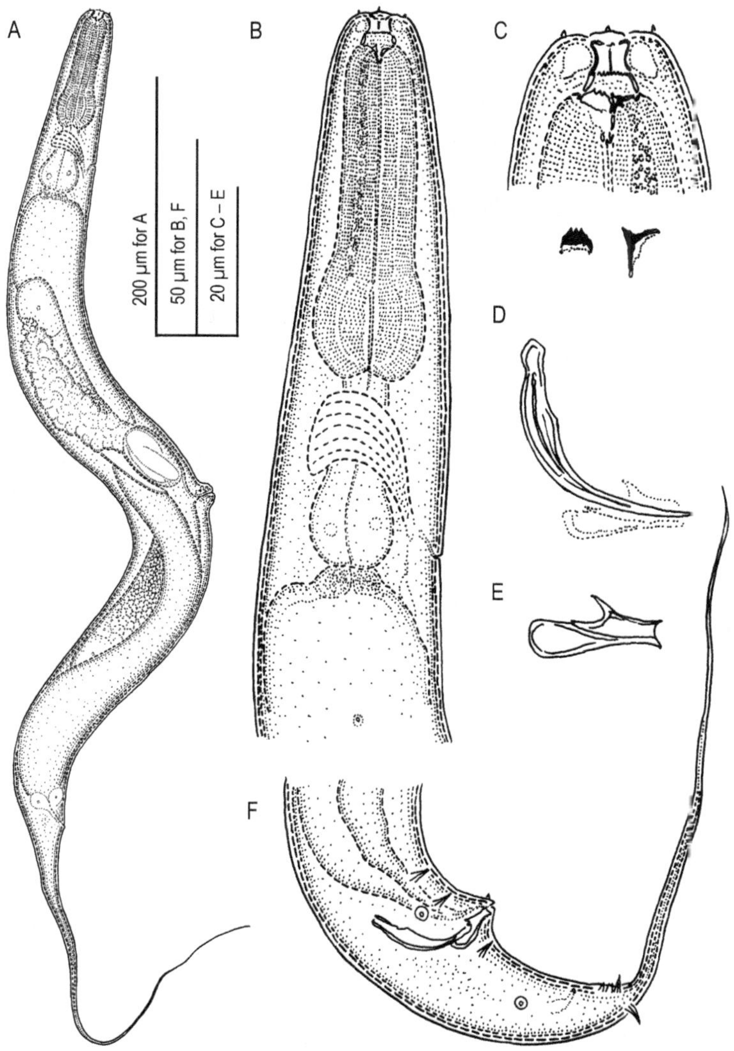

Fig. 4.3. Pristionchus elegans, *one of many recently discovered* Pristionchus *species from East Asia. This species was found in associated with scarab beetles of the genus* Phleotrupes. *Illustrations are modified from Kanzaki* et al. *(2012b). A: Entire body of female, right lateral aspect; B: Neck of female, right lateral aspect; C: Stomatal region of female, left lateral aspect. Left subventral ridge and dorsal tooth are additionally shown below; D: Spicule, right lateral aspect; E: Gubernaculum, right lateral aspect; F: Tail of male, right lateral aspect.*

nematodes with this technique, *Pristionchus* strains can then often be brought into monoxenic culture on a diet of *Escherichia coli* OP50, on which the nematodes can be kept indefinitely.

The ability to induce nematode development and propagation by sacrificing their carrier insects has demonstrated necromeny in *Pristionchus*. Whether a necromenic lifestyle is facultative or obligatory has not been tested for most *Pristionchus* species, although it is apparent that the death of the beetle is not required for *P. pacificus* to resume development. For example, a technique to isolate *P. pacificus* from protected beetle species without sacrificing the host has achieved preliminary success. Providing beetles with wet, *E. coli* OP50-laden tissue in a closed container can bait dauer individuals, which can then be coaxed to exit the dauer stage on bacteria-rich NGM plates (Herrmann *et al.*, unpubl. data). The association observed in *P. pacificus* is thus similar to general phoresy, a phenomenon whereby nematodes disperse with vector animals and, once a suitable habitat is reached, disembark from the living or dead carrier to resume development and proliferate (Bovien, 1937). Because necromeny is only a small step beyond phoresy (Poulin, 2007; Sudhaus, 2010), phoretic and necromenic lifestyles may constitute complementary strategies for a given species. The particular resilience of dauers in *P. pacificus*, which in contrast to *C. elegans* can last almost a year in the absence of potential hosts (Mayer & Sommer, 2011), may lend flexibility to its life history, especially to include associations with long-lived hosts.

In addition to observational evidence, functional genetic and anatomical experiments support an intricate relationship between *Pristionchus* nematodes and their insect hosts, presumably due to refined interactions in evolution. In *P. pacificus*, adults are attracted to the sex pheromone of the Oriental beetle, specifically the compound *z*-7-tetradece-2-one (ZDTO) (Herrmann *et al.*, 2007; Hong *et al.*, 2008a; Hong, Chapter 12, this volume). However, the same molecule inhibits embryonic and early juvenile development as well as dauer exit in *P. pacificus*, although ZDTO is neither attractive nor harmful to *C. elegans* (Cinkornpumin *et al.*, 2014). Furthermore, the nematicidal activity of ZDTO against *P. pacificus* embryos and young juveniles is neutralised through species-specific expression of the predicted lipid-binding protein OBI-1 in chemosensory (amphid) support cells. The complex interaction between *P. pacificus* and the Oriental beetle may thus be the result of attenuated antagonism, supporting a co-evolutionary history of the nematodes and their hosts (Cinkornpumin *et al.*, 2014).

INSECT HOSTS AND OTHER SOURCES OF *PRISTIONCHUS*

Of the groups of insects that have yielded *Pristionchus* nematodes (Table 4.1), the most common hosts are scarab beetles (Coleoptera: Scarabaeidae). These beetles include most of the sampled carriers of *P. pacificus* (Herrmann *et al.*, 2006a, 2007, 2010; Kanzaki *et al.*, 2011; Morgan *et al.*, 2012). Preference for the Oriental beetle in particular has been shown by the attraction of *P. pacificus* to the sex pheromone of that species (Herrmann *et al.*, 2007). Scarabs are also the known hosts for 18 other species of *Pristionchus* (Table 4.1). Host preferences for several of these species have likewise been inferred from chemoattraction profiles (Hong & Sommer, 2006). For example, whereas *P. pacificus* is attracted to the Oriental beetle pheromone, phenol produced by *Melolontha* sp. was found to synergise with plant volatiles to make up a potent attractant for *P. maupasi* (Hong *et al.*, 2008b).

In addition to scarabs, stag beetles (Lucanidae), which are also scarabaeoids, are common hosts of *Pristionchus*. These hosts have yielded *P. exspectatus*, *P. maxplancki*, *P. lucani* and the closest known outgroup to *Pristionchus*, *Parapristionchus giblindavisi* (Mayer *et al.*, 2007; Kanzaki *et al.*, 2011, 2012c). Consistent with an association with stag beetles, which live most of their lives in rotting wood, are reports of *Pristionchus* from termites. In particular, *P. aerivorus*, *P. arcanus* and even *P. pacificus* have all been collected from these insects (Poinar, 1990; Poinar *et al.*, 2006; Kanzaki *et al.*, 2012a). It is therefore likely that rotting wood features often in the life histories of *Pristionchus* species. Congruent with this idea is the isolation of other species, for example, *P. macrospiculum* and *P. micoletzkyi*, from rotting wood, even in the absence of an apparent vector (Hnatewytsch, 1929; Altherr, 1938).

Besides scarab and stag beetles, hosts of *Pristionchus* also fall into several other beetle families, including carrion beetles (Silphidae), shining fungus beetles (Scaphidiidae), pleasing fungus beetles (Erotylidae) (Kanzaki *et al.*, 2014b) and, as already mentioned for *P. uniformis*, leaf beetles (Chrysomelidae). Other potential hosts of *Pristionchus* are Lepidoptera and Hymenoptera, from which *P. brevicauda* and *P. entomophagus*, respectively, have been described (Kotlán, 1928; Steiner,

Table 4.1. *Known hosts or habitats and type localities for* Pristionchus *species.*

Species epithet	Host or habitat	Type locality
aequidentatus	Soil	Tshamugussa, Congo
aerivorus	*Leucotermes lucifugus*	Kansas, USA
americanus	*Polyphylla* sp.	Centerville, MA, USA
arcanus	*Odontotermes formosanus*	Iriomote, Japan
atlanticus	Soil	Cold Spring Harbor, NY, USA
biformis	Laboratory crossing experiments within *P. lheritieri*	Erlangen, Germany
boliviae	*Cyclocephala amazonica*	Near Buena Vista, Bolivia
brachycephalus	Dead stem of papaya	La Serena, Chile
brevicauda	*Ostrinia nubialis* Soil	Peremarton, Hungary
breviflagellum	*Coffea* berries	Costermansville, Congo
bucculentus	*Episcapha gorhami* *Encaustes praenobilis*	Sapporo, Hokkaido, Japan
bulgaricus	*Cetonia aurata*	Tabachka, Bulgaria
clausii	Fungi	Germany
clavus	*Allium vineale* Soil	Göttingen, Germany
dentatus	Not given	Germany
dubius	Rotten wood	Schneeberg, Germany
elegans	*Phleotrupes auratus*	Kutsuki, Japan
entomophagus	*Pamphilius stellatus* Various Scarabaeioidea	Eberswalde, Germany
exspectatus	*Prismognathus angularis*	Mt Shibi, Japan
eurycephalus	*Necrophorus* sp. *Thanatophilus* sp. *Hister* sp.	Erlangen, Germany
fukushimae	*Lucanus maculifemoratus*	Tadami, Japan
fissidentatus	Soil	Tatopani, Nepal
gallicus	Soil, humus	Switzerland
hoplostomus	Soil	Tokyo, Japan
iheringi	Damaged *Coffea* roots	São Paulo, Brazil

Table 4.1. *(Continued).*

Species epithet	Host or habitat	Type locality
inermis	Damaged *Allium* roots	Kiel, Germany
japonicus	Dead earthworm	Enoshima, Japan
lheritieri	Compost, soil, fungi	Vire, France
	Rotting plant material	
	Rhagium inquisitor	
	Geotrupes stercorosus	
linstowi	Not given	Naples, Italy
lucani	*Lucanus cervus*	Vailhauques, France
macrospiculum	Wood	Bex, Switzerland
marianneae	*Popillia japonica*	Geneva, NY, USA
maupasi	Soil	Norfolk, UK
	Melolontha spp.	
	Cetonia aurata	
maxplancki	*Lucanus maculifemoratus*	Fuzawa, Tadami, Japan
mayeri	*Hoplochelus marginalis*	Trois Bassins, La Réunion
	Adoretus sp.	
	Heteronychus licas	
	Hyposerica tibialis	
	H. vinsoni	
	Phyllophaga smithi	
micoletzkyi	Rotten wood	Schneeberg, Germany
microcercus	Rotten potatoes	Germany
migrans	*Leucotermes lucifugus*	France
obscuridens	Around roots	Congo
pacificus	Soil	Pasadena, CA, USA
	Various Scarabaeoidea	
	Elateridae	
	Cydnidae	
	Odontotermes formosanus	
	Riukiaria sp.	
	Pomace	
paramonovi	*Portulaca* root	Karakalpakstan, Uzbekistan

Table 4.1. *(Continued).*

Species epithet	Host or habitat	Type locality
pauli	*Lichnanthe vulpina*	Carver, MA, USA
pseudaerivorus	*Phyllophaga* spp.	Lincoln, NE, USA
quartusdecimus	*Exomala orientalis*	Amagasaki-shi, Japan
robustus	Soil	Hamma, Algeria
saccai	*Dianthus* roots	Brazil
trifomis	*Anoplotrupes stercorosus*	Tübingen, Germany
	Marronus borbonicus	
	Soil	
uniformis	*Leptinotarsa decemlineata*	Poland
	Melolontha melolontha	
	Amphimallon solstitiale	
	Geophilus sp.	
	Phyllophaga spp.	
	Rhizotrogus aestivus	
	Carabidae	
	Staphylinidae	
vidalae	*Diabrotica speciosa*	Salto, Argentina

1929), although the associations in these cases were not clearly separated from the surrounding environment in which the nematodes were discovered.

Recent efforts have uncovered a multitude of *Pristionchus* strains from insect hosts, but *Pristionchus* is also found in soil habitats. In the soil, strains have often been found associated with damaged roots or plant material, such as reported for *P. clavus*, *P. iheringi*, *P. inermis* and *P. microcercus*. Otherwise young and organically rich habitats such as compost have also served as sources of *Pristionchus*, including *P. gallicus* and *P. lheritieri*. The occurrence of *Pristionchus* in ephemeral habitats rich in microorganisms is not surprising, given the general prevalence of other Diplogastridae with such habitats (Bongers, 1999; Steel *et al.*, 2012; Kanzaki & Giblin-Davis, Chapter 3, this volume). It is likely that *Pristionchus* nematodes are able to find these habitats quickly by their phoresy with insects.

The widespread association of *Pristionchus* with beetles, as well as wood and soil, suggests a general model for the ecology of these nematodes. Namely, all insects shown to carry *Pristionchus* regularly are soil- or wood-inhabiting insects that live the large majority of their life in these habitats. Scarab beetles, for example, are soil-inhabiting insects with a short adult stage: the known host *M. melolontha* has a 4-year generation time but an adult flight period of only a few weeks. After this short interval for mating, females of the beetles (*e.g.*, *Melolontha*) lay their eggs in the soil and die immediately after oviposition (Balachowsky, 1962). Thus, the nematodes can be easily transmitted between cadaver, soil and living beetle stages. Stag beetles assume a similar life history to scarabs, instead laying their eggs in wood (Brechtel, 2002), thereby suggesting a similar cycle in which *Pristionchus* nematodes propagate and disperse. Consistent with an association with rotten wood, stag beetles and, in some cases, termites is the presence of cellulases in *Pristionchus* (Dieterich *et al.*, 2008). Cellulase genes are not only expressed but have undergone massive amplification and selection in the genus (Mayer *et al.*, 2011; Schuster & Sommer, 2012). However, the exact function of the enzymes is still unknown.

Other types of insects or habitats might also harbour *Pristionchus* nematodes, and the enormous success in collecting *Pristionchus* from scarab beetles may be due to the sampling bias toward those insects. Even centipedes and millipedes have been identified hosts for *Pristionchus* species (R. Rae, Liverpool John Moores University, pers. comm.; Kanzaki, unpubl. data). It is also apparent that at least some species are not host-specific (*e.g.*, *P. entomophagus*), even if certain strains or populations within a species may show preference for particular hosts (Morgan *et al.*, 2012). Indeed, phylogenetic studies of host associations have rejected congruence between lineages of *Pristionchus* nematodes and those of their carriers (Mayer *et al.*, 2009). Vertical transmission of nematodes with their insect hosts is known to occur in other Diplogastridae, namely in *Parasitodiplogaster* (Giblin-Davis *et al.*, 2004), *Teratodiplogaster* (Kanzaki *et al.*, 2009) and *Micoletzkya* (Susoy & Herrmann, 2014). By contrast, the terminal habitats of either phoretic or necromenic nematodes such as *Pristionchus* are relatively open. In such habitats, many potential hosts can meet, obviating the need for strict vertical transmission to find a new host. Considering the potentially broad host ranges for *Pristionchus*, collections of other hosts, such as other families of beetles or wood-associated insects, is likely to reveal even more new species.

Taxonomy

Pristionchus[4] Kreis, 1932
[*nomen protectum*][5]
= *Lycolaimus* Rahm, 1928: *L. iheringi* Rahm, 1928
= *Peronilaimus* Rahm, 1928: *P. saccai* Rahm, 1928
= *Paradiplogaster* Schuurmans Stekhoven & Teunissen, 1938: *P. aequidentatus* Schuurmans Stekhoven & Teunissen, 1938
= *Diplogasterium* Paramonov, 1952: *D. micoletzkyi* Hnatewytsch, 1929
= *Mesodiplogaster* Weingärtner, 1955 (Goodey, 1963): *Diplogaster lheritieri* Maupas, 1919
= *Paramonoviola* Blinova & Vosilite, 1976: *P. rhagii* Blinova & Vosilite, 1976
= *Isakis* Lespés, 1856 [*nomen oblitum*]: *I. migrans* Lespés, 1856
= *Chroniodiplogaster* Poinar, 1990: *Diplogaster aerivora* Cobb *in* Merrill & Ford, 1916; *C. formosiana* Poinar *et al.*, 2006

TYPE SPECIES

Pristionchus longicaudatus Kreis, 1932 (= a junior subjective synonym of *Diplogaster lheritieri* Maupas, 1919)[6]
= *Pristionchus lheritieri* (Maupas, 1919) Paramonov, 1952
= *Diplogaster longicauda apud* Bütschli, 1876, *nec* Claus, 1862
= *D. horticola* Fuchs, 1929
= *P. ottoi* Paramonov, 1952
= *Paramonoviola rhagii* Blinova & Vosilite, 1976
= *Mesodiplogaster pseudolheritieri* Geraert, 1984

[4] Etymology: 'shark-tooth' nematode, deriving from the Greek roots πρίστηζ ('sawfish, shark') + ὄνυξ ('claw', meant as 'tooth').
[5] Although the names *Lycolaimus* and *Peronilaimus* have taxonomic priority, the overwhelming usage of the name *Pristionchus* recommends fixation of the latter as valid under ICZN article 23.9.3.
[6] The type species of *Pristionchus* is, by monotypy/original designation, *P. longicaudatus* Kreis, 1932. This species was regarded as a junior subjective synonym of *Diplogaster lheritieri* Maupas, 1919 by Paramonov (1952) who also erroneously regarded the latter species as type. This is not the case as even though *longicaudatus* is now regarded as being synonymous with the older name, it remains as type of the genus.

OTHER SPECIES

Hermaphroditic species, whether tested or inferred by the rarity of males, are marked by an asterisk (*).

P. aequidentatus (Schuurmans Stekhoven & Teunissen, 1938) Andrássy, 1984
P. aerivorus (Cobb *in* Merrill & Ford, 1916) Chitwood, 1937
P. americanus Herrmann, Mayer & Sommer, 2006
P. arcanus Kanzaki, Ragsdale, Herrmann, Mayer & Sommer, 2012a
P. atlanticus Kanzaki, Ragsdale, Herrmann, Susoy & Sommer, 2013c
**P. biformis* (Hirschmann, 1951) Sudhaus & Fürst von Lieven, 2003
P. boliviae Kanzaki, Ragsdale, Herrmann, Susoy & Sommer, 2013c
**P. brachycephalus* (Steiner, 1943) Sudhaus & Fürst von Lieven, 2003
P. brevicauda (Kotlán, 1928) Paramonov, 1952
**P. breviflagellum* (Schuurmans Stekhoven, 1951) Sudhaus & Fürst von Lieven, 2003
P. bucculentus Kanzaki, Ragsdale, Herrmann, Röseler & Sommer, 2013a
P. bulgaricus Kanzaki, Ragsdale, Herrmann & Sommer, 2014a
**P. clausii* (Bütschli, 1873) Paramonov, 1952
P. clavus (von Linstow, 1901) Sudhaus & Fürst von Lieven, 2003
P. dentatus (Schneider, 1866) Sudhaus & Fürst von Lieven, 2003
P. elegans Kanzaki, Ragsdale, Herrmann & Sommer, 2012b
**P. entomophagus* (Steiner, 1929) Sudhaus & Fürst von Lieven, 2003
P. exspectatus Kanzaki, Ragsdale, Herrmann, Mayer & Sommer, 2012a
P. eurycephalus (Völk, 1950) Sudhaus & Fürst von Lieven, 2003
P. fukushimae Ragsdale, Kanzaki, Röseler, Herrmann & Sommer, 2013
P. fissidentatus Kanzaki, Ragsdale, Herrmann & Sommer, 2012b
**P. gallicus* (Steiner, 1914) Paramonov, 1952
= *Diplogaster minor* Maupas, 1900; *nec* Cobb, 1893
P. hoplostomus Ragsdale, Kanzaki, Röseler, Herrmann & Sommer, 2013
**P. iheringi chilensis* (Rahm, 1932) Sudhaus & Fürst von Lieven, 2003
**P. iheringi iheringi* (Rahm, 1928) Sudhaus & Fürst von Lieven, 2003
P. inermis (Bütschli, 1874) Paramonov, 1952
P. japonicus Kanzaki, Ragsdale, Herrmann, Mayer & Sommer, 2012a
**P. linstowi* (Potts, 1910) Paramonov, 1952
P. lucani Kanzaki, Ragsdale, Herrmann & Sommer, 2014a
P. macrospiculum (Altherr, 1938) Kanzaki, Ragsdale & Giblin-Davis, 2014c
P. marianneae Herrmann, Mayer & Sommer, 2006

*P. maupasi (Potts, 1910) Paramonov, 1952
P. maxplancki Kanzaki, Ragsdale, Herrmann, Röseler & Sommer, 2013b
*P. mayeri Kanzaki, Ragsdale, Herrmann, Susoy & Sommer, 2013c
*P. micoletzkyi (Hnatewytsch, 1929) Sudhaus & Fürst von Lieven, 2003
= *Diplogaster subterraneus* Hnatewytsch, 1929
*P. microcercus (Wollenweber, 1921) Paramonov, 1952
P. migrans (Lespés, 1856) Sudhaus & Fürst von Lieven, 2003
*P. obscuridens (Schuurmans Stekhoven, 1951) Sudhaus & Fürst von Lieven, 2003
*P. pacificus Sommer, Carta, Kim & Sternberg, 1996
= *Chroniodiplogaster formosiana* Poinar, Meikle & Mercadier, 2006 syn. nov.[7]
*P. paramonovi (Atakhanov, 1958) Sudhaus & Fürst von Lieven, 2003
P. pauli Herrmann, Mayer & Sommer, 2006
P. pseudaerivorus Herrmann, Mayer & Sommer, 2006
P. quartusdecimus Kanzaki, Ragsdale, Herrmann, Röseler & Sommer, 2013b
*P. robustus (Maupas, 1900) Paramonov, 1952
*P. saccai (Rahm, 1928) Sudhaus & Fürst von Lieven, 2003
*P. triformis Ragsdale, Kanzaki, Röseler, Herrmann & Sommer, 2013
P. uniformis Fedorko & Stanuszek, 1971
P. vidalae (Stock, 1993) Sudhaus & Fürst von Lieven, 2003

GENERAL MORPHOLOGY OF *PRISTIONCHUS*

Pristionchus has the typical morphology diagnostic of Diplogastridae, namely a well-developed anterior pharynx (corpus) and a glandular posterior pharynx (postcorpus) (Figs 4.2A; 4.3B). The genus is distinguished from other diplogastrid genera by its stomatal morphology: *i.e.*, *i*) six per- and interradial cheilostomatal plates (in some species secondarily divided) that each end in a rounded flap; *ii*) a relatively short, stout, barrel-like gymnostom; and *iii*) a relatively shallow stegostom with dorsal and right subventral teeth and with a left subventral ridge of denticles (Sudhaus & Fürst von Lieven, 2003; Figs 4.4, 4.5). Further distinguishing the genus are the presence of stomatal dimorphism (see Ragsdale,

[7] This synonymy is based on morphological and reproductive characters, but it is best justified by the exact match of a well-tested molecular diagnostic maker, a partial SSU rRNA sequence, between the type strain of *C. formosiana* and *P. pacificus* (Herrmann, unpubl. data).

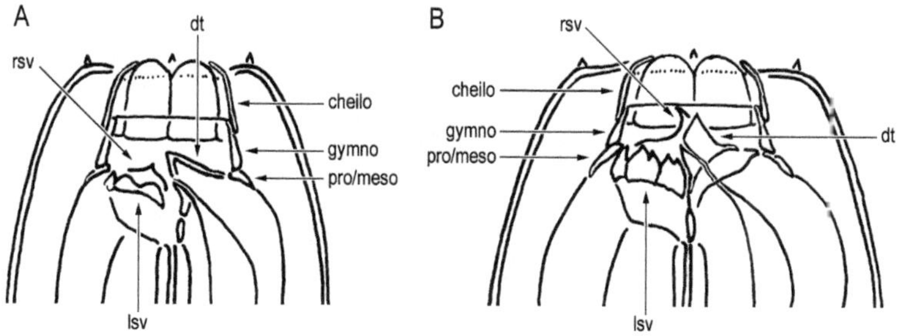

Fig. 4.4. *The stoma of* Pristionchus, *as drawn for* P. pacificus *as a representative of the genus. Stomatal structures are among the most diverse traits in* Pristionchus *and are diagnostic of species or species groups. A: Stenostomatous form in left lateral aspect; B: Eurystomatous form in left lateral aspect. Abbreviations: dt, dorsal tooth; lsv, left subventral denticles; rsv, right subventral ridge or tooth; cheilo, cheilostom; gymno, gymnostom; pro/meso, pro-/mesostegostom.*

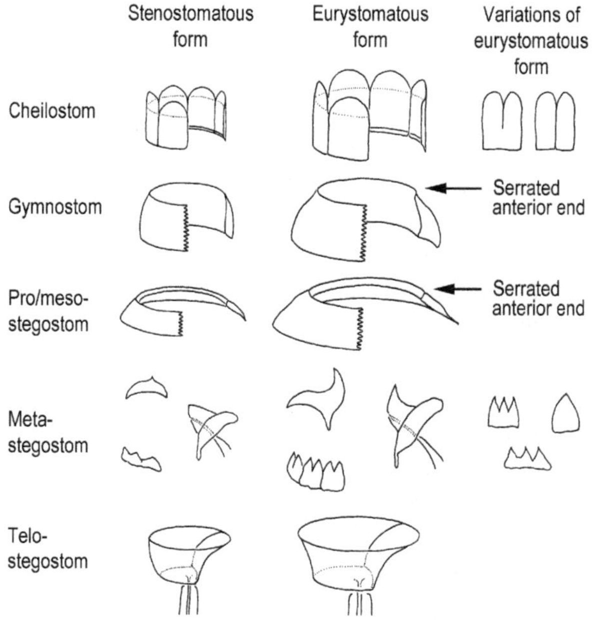

Fig. 4.5. *Schematic drawing of the stoma of* Pristionchus. *In addition to the basic composition of the stoma, variations to individual regions variously diagnose individual species or species groups (see Fig. 4.8).*

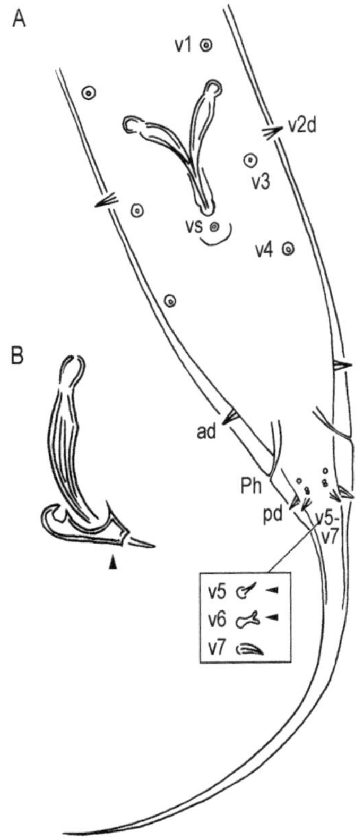

Fig. 4.6. *Informative male sexual morphology of* Pristionchus. *A: Tail, in ventral aspect; B: Spicule and gubernaculum, in right lateral aspect. The shape of the gubernaculum posterior to where it envelops the spicules (arrowhead) can be diagnostic of individual species. Genital papillae are labelled following terminology of Sudhaus & Fürst von Lieven (2003).*

Chapter 11, this volume; Fig. 4.4) and the lack of a telostegostomatal apodeme (Sudhaus & Fürst von Lieven, 2003), although the former trait is known to vary (Kanzaki *et al.*, 2012b, 2013a). The generic morphology of *Pristionchus* is summarised here (Figs 4.4-4.6).

Body shape and surface structure

The body is relatively stout (Figs 4.1A; 4.3A). The cuticle is marked by clear vertical striations, which each consist of two parallel lines of punctations (Fig. 4.2B), and transverse annulations. The deirids, which are small lateral pores appearing as two concentric circles at the body surface, are relatively clear compared with those of other Diplogastridae (Fig. 4.2B). Deirid position varies among individuals and even more so across species, but the pores are typically located from just posterior to the pharynx to the anterior part of the pharyngeal isthmus. Several (*ca* ten) minute secretory pores (Fig. 4.2B), each accompanied by a small gland cell, open on each lateral body surface and are either slightly dorsal or slightly ventral of the lateral midline, although they do not alternate regularly with respect to their dorsoventral orientation. The first of these sublateral pores (from anterior to posterior) is located anterior to the deirid and can be distinguished from the deirid by its smaller size. Additionally, the nematodes have two pairs of larger pores, which have been called 'postdeirids', one pair being located at mid-body, the other almost as posterior as the rectum. The function of the postdeirids is unknown. The excretory pore opens ventrally at the same level or slightly anterior to the deirids, namely at basal bulb level or slightly anterior.

Stomatal region

The lips of *Pristionchus* are not clearly offset from the rest of the anterior body wall cuticle (Fig. 4.1C). Each of the six lip sectors bears a short labial sensillum. Four cephalic papillae are present in males, as typical of Rhabditina, and are slightly posterior to the labial sensilla. The amphids, the main chemoreceptor organs, have oval openings that are located slightly posterior to the labial sensilla on both sides.

Stomatal structures are relatively variable among species. Additionally, most species have two distinct stomatal forms, a wide-mouthed (eurystomatous, Eu) and a narrow-mouthed (stenostomatous, St) form. The stoma consists of three parts, the cheilostom, gymnostom and stegostom, which are each associated with a particular type of underlying tissue (De Ley *et al.*, 1995; Baldwin *et al.*, 1997; Figs 4.4, 4.5). The cheilostom is separated by at least six adradial divisions, *i.e.*, into six per- and interradial cuticular plates, unless further divided. The anterior end of each plate is elongated to form a short, rounded flap that partially covers the stomatal opening. While most species have only six plates, in the *triformis* group of *Pristionchus* species (*P. fukushimae*, *P. hoplostomus*, *P. triformis*) each of these six primary plates may be partially or completely split into two smaller plates, such that the cheilostom can have 7-12 plates (Ragsdale *et al.*, 2013). The gymnostom is stout and barrel-shaped, distinguishing it from genera with narrow, tube-like ('rhabditiform') stomata. The anterior end of the gymnostom overlaps the posterior end of cheilostom medially (internally). The anterior edge is smooth in most species, but in the *triformis* group it has a serrated surface (Ragsdale *et al.*, 2013). The stegostom is separated into three elements, the pro- and mesostegostom, metastegostom, and telostegostom (Figs 4.4, 4.5). The pro-/mesostegostom is axially short and forms an indistinct ring that connects the gymnostom and metastegostom. The metastegostom bears a dorsal tooth, left subventral ridges or denticles and, in the Eu form, a right subventral tooth. The telostegostom is a sclerotised region connecting the metastegostom with the three radii of the pharynx.

Of the structures of the stoma, the teeth and denticles differ most between the St and Eu forms (see Ragsdale, Chapter 11, this volume), reflecting form-specific differences in feeding function (Serobyan *et al.*, 2014). In the St form, the dorsal tooth is flint-shaped or diamond-shaped in lateral view. By contrast, it is large, claw-like, and more heavily sclerotised in the Eu form. The right subventral tooth of the Eu form is large and claw-like (Fig. 4.1E), whereas the St form has a ridge with at most a small denticle or series of cusps (Fig. 4.1G). In both forms, the left subventral sector of the metastegostom bears a ridge weakly separated into three longitudinal parts (Figs 4.4, 4.5): two are approximately lateral, whereas the third is closer to ventral. In the St form this ridge has two or three cusps at its apex (Fig. 4.1F), whereas in the Eu form these ridges are host to large denticles (Fig. 4.1D), each of which may be further split into two or three tips, such that the ridge may bear as many as nine cusps. Additionally, some species have adventitious denticles in the left subventral sector, and in at least two species (*P. fukushimae*, *P. hoplostomus*) entire duplicate ridges can be present (Ragsdale *et al.*, 2013). Further details about stomatal structures diagnosing individual species or species groups are given below.

Digestive system

The corpus consists of a well-developed muscular procorpus and an even wider muscular metacorpus. The postcorpus consists of a non-muscular isthmus and a glandular basal bulb. The postcorpus is shorter than the corpus, although in fixed material the corpus may be shortened to nearly the length of the postcorpus. The nerve ring, which is conspicuous, encircles the isthmus. The pharyngo-intestinal junction (cardia) is well developed and easily observed by light microscopy (LM). The intestine comprises relatively large and flat cells, which store large quantities of lipid in well-fed animals. The rectum is conspicuous and separated from the intestine by a muscular constriction (sphincter). Three rectal glands, two subventral and one dorsal, empty into the intestinal-rectal junction. The anus is a dome-shaped slit, often with a slightly protuberant posterior lip and under LM appears oval in ventral view.

Female/hermaphrodite reproductive system

The morphology of the reproductive system of females and hermaphrodites is essentially identical. Sperm production in hermaphrodites is hologonic: the distal germ cell in each gonadal branch divides into sperm that mature as they move to the oviduct, which serves as a spermatheca.

The female reproductive system is didelphic (Rudel *et al.*, 2005; Rudel, Chapter 9, this volume). Although monodelphy has evolved multiple times in Diplogastridae, no monodelphic species has been reported in *Pristionchus*. The anterior and posterior gonads are located on the right and left of the intestine, respectively. Since they are identical to each other in structure and composition, only the anterior gonad is described here. From its distal tip, the gonad consists of the ovary (ovotestis in hermaphrodites), a short epithelial passage, oviduct, uterus, vagina and vulva (Fig. 4.2C). The ovary makes up the reflexed part of the gonad and connects to the proximal part by an antidromous reflexion, such that the proximal part of the reflexion turns away from the ventral body wall and the distal part turns back toward it (Figs 4.1B; 4.2C). In the flexure, small oogonia are arranged in multiple rows that give way to a single row of developed oocytes with large nuclei, such that the entire developing germ line is confined to the flexure. The stretch of young oogonia, including cells that would be in the pachytene stage, is thus shorter in *P. pacificus* compared with *C. elegans*. The proximal part of the reflexion forms an epithelial passage and is composed of offset, rounded cells, analogous in form to a crustaformeria, and connects the ovary and oviduct.

The oviduct is a simple tube, and the distal part near the epithelial passage functions as a spermatheca as there is no *receptaculum seminis* offset from the gonad. The oviduct gradually widens to a uterus in the proximal part of the gonad branch. The uterus is marked by flattened, often diamond-shaped, epithelial cells. The vagina is perpendicular to the body surface and is surrounded by relatively dark, distinct cells. The vulva is pore-like, and does not form a slit. Four vaginal glands and a circular sphincter muscle can be observed in either lateral or ventral view. In mature females, the uterus and oviduct contain many (sometimes more than 40) eggs of various developmental stages, and in such cases the eggs mask the detailed structure of the reproductive system.

Male reproductive system

Males of *Pristionchus* have a single reflexed testis (Fig. 4.1A). The testis is located on the right or ventral side of the intestine and is reflexed to the left. At the distal tip of the testis is a distinct cap cell. Spermatogonia are arranged in multiple rows in the flexure and distally in the main part of the testis, followed by well-developed spermatocytes arranged in two to three rows at the middle part of the testis, with mature sperm occupying the rest of the testis. The posterior part of the male gonad forms the *vas deferens* but the distinction between the testis and *vas deferens* is not clear in live nematodes. The posterior end of the *vas deferens* is fused with the posterior end of the rectum and forms a narrow cloaca. The cloaca opens through a dome-shaped slit.

The copulatory organ comprises a pair of symmetrical, unfused, protrusible spicules and a relatively stationary gubernaculum (Figs 4.2H; 4.3D, E; 4.6B). Each spicule is arcuate and separated into a rounded or ovoid manubrium, on which retractor and protractor muscles insert, and a shaft (calomus) and an arcuate blade (lamina) that form a complex in which the shaft is short and the blade is either slightly or not widened with respect to the shaft. The precise form of the manubrium varies among, and is sometimes diagnostic of, *Pristionchus* species. The gubernaculum has a tubular proximal (posterior) part that envelops the spicules and a rounded, expanded distal (anterior) part. Distal to the tubular part is a pair of laterally and ventrally directed processes, which when viewed in lateral aspect separate the distal gubernaculum into two serial arcs (Fig. 4.6B). The orientation and prominence of these processes, as well as the shape of the arcs and distal gubernaculum in general, can also be diagnostic of species.

Tail structure

The tail structure of females or hermaphrodites is simple but can vary within a species. The tail is typically conical and elongate (Fig. 4.2D, E) and in some species the tip is long and filiform (Fig. 4.3A). A pair of small, oval phasmid (chemoreceptor) openings are located usually one to two anal body diam. posterior to the anus.

The male tail has a single ventral and nine pairs of genital papillae, in addition to a pair of phasmids (Figs 4.2C; 4.3B; 4.6A). The arrangement of genital papillae and phasmids is divergent among species and is highly diagnostic of *Pristionchus* species. Following the nomenclature of Sudhaus & Fürst von Lieven (2003) to describe papillal arrangement, the ventral papillae from anterior to posterior are referred to as v1-v7, where either v2 or v3 is appended with 'd' to denote whether the more lateral ('dorsal') of these two are anterior (v2d) or posterior (v3d) to the other; the two anterior and posterior pairs of relatively dorsal papillae are referred to as 'ad' and 'pd', respectively. Diagnostic morphology is discussed below but some papilla characters are consistent for the genus. In most species, the posterior four pairs are located around the tail tip, three of the pairs (v5-v7) being ventral and the other (pd) being dorsolateral. The morphology of v5-v7 is characteristic for *Pristionchus*, and this morphology may diagnose the genus from most other Diplogastridae with the exception of *Parapristionchus*. Whereas the structure of v7 is similar to the other, more anterior papillae, v6 has a bifurcated tip and v5 is very small and borne in a socket-like depression. The lateral phasmid openings are usually between v5 and the next anterior pair of papillae (ad). Although stereotypic within species of *Pristionchus*, the number and arrangement of genital papillae can still vary somewhat among individuals, especially in hermaphroditic species (Kanzaki *et al.*, 2013c). Posterior to the papillae, the tail narrows abruptly into a tip, which ends in a spike (Fig. 4.2F) or filiform projection (Fig. 4.3F). The body-wall cuticle of the tail region is thick, and thus in ventral view the tail, although lacking true alae, is expanded to appear similar to a narrow leptoderan bursa (Fig. 4.2G).

Difficulties caused by older descriptions

We currently recognise 48 valid species of *Pristionchus*. However, most descriptions from the mid-20th century or earlier lack character information now known to be important for diagnostics, as has been determined over a series of phylogenetically supported studies (Kanzaki *et al.*, 2011, 2012a, c, 2013a, b, c, 2014a; Ragsdale *et al.*, 2013). Such characters were often neglected or not described in the detail necessary to distinguish even distantly related *Pristionchus* species. For example, informative stomatal structure and male papillae patterns are often simplified or missing in older descriptions. They might instead include only a few measurements and body ratios, many of which show as much intraspecific as interspecific variation in *Pristionchus*. Furthermore, no type material or live cultures are available for most species from older descriptions, making it impossible to confirm relevant characters. Compounded with the difficulties presented by incomplete morphological descriptions is the presence of cryptic biological species (Herrmann *et al.*, 2006b; Kanzaki *et al.*, 2012a, 2014a), any one of which might be the true bearer of the original name. Thus, even if some fixed or mounted materials were available, it may still be impossible to determine the species status of a newly isolated population based on morphology.

Because of the difficulties posed by strictly morphological and often inadequate descriptions, further hypotheses of species identity will rely mostly on host association and locality. However, host information is fraught with difficulties for accurately identifying species. Because the carrier range of a given species can be large and non-specific (Table 4.1), the type host or carrier cannot definitively diagnose an isolate. Locality information may be of greater use for identifying species, as the ranges of gonochoristic species are often predictable, at least at a continental scale (see below). However, range information will presumably be less useful for hermaphroditic species, as some are widely distributed due to their ease of dispersal (see below).

Despite the difficulties presented by incomplete descriptions, systematic collection efforts followed by biological, molecular, and detailed morphological examination can resolve the taxonomic system of *Pristionchus*. By increasingly saturating the collector's curve, a complete revision of the genus should soon be possible, in part by supporting the synonymies of valid names. For example, there are eight apparently hermaphroditic *Pristionchus* species described from Europe. However, intensive sampling efforts have repeatedly isolated only four biological species of hermaphrodites (*P. entomophagus*, *P. maupasi*, *P. pacificus* and *P. triformis*) from this continent. Likewise, worldwide collections have repeatedly discovered the same seven species of hermaphrodites (also including *P. boliviae*, *P. mayeri* and *P. fissidentatus*), in contrast to the 19 hermaphroditic species that are described. It is therefore likely that some names erected based on host associations or other inconsistent diagnostic characters belong to one of a smaller subset of tested biological species.

MOLECULAR SYSTEMATICS AND SPECIES CONCEPTS

Given the difficulties of conserved general morphology and a long taxonomic history, the establishment of a robust comparative system for *Pristionchus* has only been possible in combination with molecular phylogenetics. The independent character set provided by nuclear gene sequences has been essential to test hypotheses of species limits, morphological evolution and biogeography. In particular, a set of some 26 ribosomal protein-coding genes and a diagnostic fragment of the SSU rRNA gene (Table 4.2) have yielded high resolution and support to relationships in the genus (*e.g.*, Kanzaki *et al.*, 2014a; Fig. 4.7). The set of phylogenetic markers most often used in *Pristionchus* was first developed from transcriptomic studies of *Pristionchus* (Mayer *et al.*, 2007). These genes were highly and consistently expressed as RNA in all tested *Pristionchus* species, making them easy to recover and amplify from whole RNA extracts of pooled nematodes. A comparison of the fully sequenced genomes of *P. pacificus* and *C. elegans* shows that these ribosomal protein genes are highly conserved as orthologues, lessening

Table 4.2. *Nuclear protein-coding genes used for phylogenetic analysis of* Pristionchus. *Together with a diagnostic fragment of the SSU rRNA gene, this dataset of ribosomal protein genes has given high resolution and support to most relationships within the genus. The numbers of nucleotides given are from the aligned dataset of Kanzaki* et al. *(2014a). Amplification primers for protein-coding genes are given in Mayer* et al. *(2007).*

Gene	Aligned nucleotides	Gene	Aligned nucleotides
rpl-1	642	*rpl-34*	330
rpl-2	783	*rpl-35*	348
rpl-10	630	*rpl-38*	195
rpl-14	402	*rpl-39*	132
rpl-16	597	*rps-1*	759
rpl-23	408	*rps-8*	618
rpl-26	423	*rps-14*	435
rpl-27	396	*rps-20*	357
rpl-27a	423	*rps-21*	255
rpl-28	387	*rps-24*	396
rpl-30	336	*rps-25*	336
rpl-31	345	*rps-27*	255
rpl-32	381	*rpl-28*	201

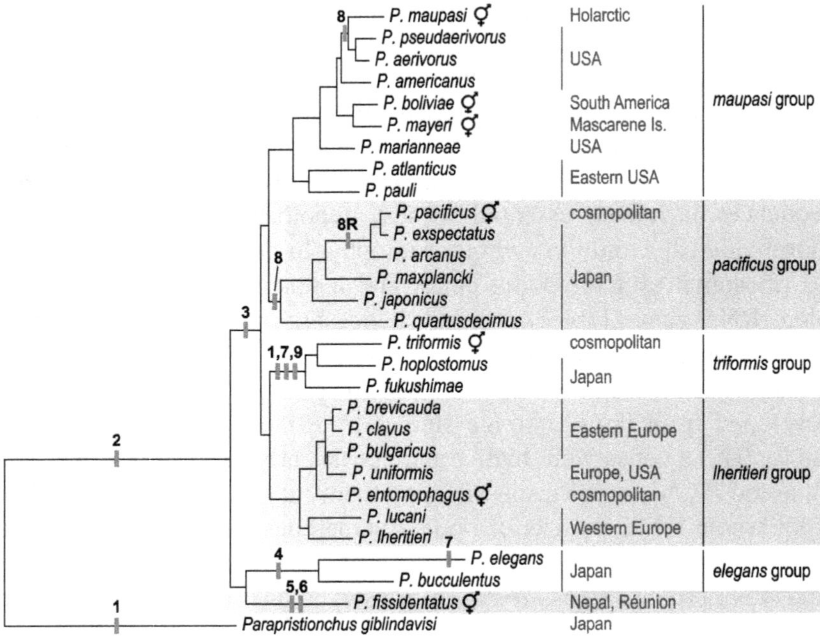

Fig. 4.7. *Phylogeny of* Pristionchus *species and its implications for morphological evolution and biogeography. Tree is modified from Kanzaki* et al. *(2014a) and was inferred from 27 ribosomal protein-coding genes and a diagnostic fragment of the SSU rRNA gene. Although the morphology of* Pristionchus *species is generally conserved, stomatal and male sexual traits are diverse enough to distinguish some clades within the genus: 1 = cheilostom with secondary divisions; 2 = St form with flint-shaped dorsal tooth; 3 = fourth pair of male papillae (v4) far posterior to cloaca; 4 = cheilostom vacuolated; 5 = Eu form with separate left subventral denticle; 6 = Eu form with separate right lateral ridge of denticles* (i.e., *in addition to tooth); 7 = anterior edge of gymnostom serrated; 8 = anterior lateral papillae are the third pair (= v3d); 8R = anterior lateral papillae are the second pair (= v2d); 9 = anterior edge of Eu stegostom serrated. Polarity of characters 1 and 2 were inferred using* Micoletzkya *spp. (claw-like St dorsal tooth, undivided cheilostom) as outgroup (Kanzaki* et al.*, 2014b; Susoy* et al.*, 2015). Known geographic ranges of taxa are given to the right of their names on the tree. Gonochoristic species of* Pristionchus *are hypothesised to have originated in East Asia, and gonochoristic lineages have since colonised Europe (*lheritieri *group) and North America (*maupasi *group,* P. uniformis*). Hermaphroditism has evolved at least six times independently in* Pristionchus, *and hermaphroditic species show cosmopolitan distributions or ranges disjunct with respect to those of closely related gonochoristic species.*

the risk that paralogy will mislead phylogenetic inference. Consistent with this prediction is the generally high congruence of several markers for the phylogeny of the genus. As a result, the phylogenetic framework for *Pristionchus* has emerged as highly robust and provides the necessary reliability for comparative biology.

The fragment of SSU that has been used to diagnose species in *Pristionchus* is approximately 470 bp in length and includes a barcode sequence developed for broad use across Nematoda (Floyd *et al.*, 2002). Because most of the recently isolated *Pristionchus* species can be kept in laboratory culture, mating experiments can test species boundaries: reproductive isolation of taxonomic units from each other indicates the independent evolutionary trajectories that separate unique species (Wiley, 1978; Adams, 1998). In some pairs of closely related species, interspecific crosses have produced viable F1 offspring, although, in all cases, self-sterility of hybrids has demonstrated reproductive isolation of putative species (Herrmann *et al.*, 2006b; Kanzaki *et al.*, 2012a). Based on crossing experiments, a working operational taxonomic unit has been established for *Pristionchus*. In the diagnostic fragment of the SSU rRNA gene, up to one nucleotide difference has usually correlated with membership to a single biological species (Kanzaki *et al.*, 2013c). One exception to this has been found in *P. aerivorus* and *P. maupasi*, which are identical in this genetic marker but differ in their mode of reproduction, as these species are gonochoristic and hermaphroditic, respectively (Herrmann *et al.*, 2006b). However, in most cases, differences in the diagnostic SSU rRNA sequence are more pronounced, in contrast to the relatively high sequence similarity found between species in other groups of nematodes such as *Caenorhabditis* (Kiontke *et al.*, 2011). Even among hybridising species, for example, *P. pacificus*, *P. exspectatus* and *P. arcanus*, these sequences can differ in as many as five nucleotides in the diagnostic fragment (Kanzaki *et al.*, 2012a).

Morphological characters for species identification

The backbone provided by molecular and reproductive characters tests which morphological characters are phylogenetically informative of groups above the species level. Because *Pristionchus* is largely conserved in its general morphology, morphological distinctions between some closely related biological species are nearly impossible. Nevertheless, a suite of diagnostic characters can at least distinguish phylogenetically supported 'groups' of species (Kanzaki *et al.*, 2013c; Ragsdale *et al.*, 2013), if not always the individual species within them. The most stable and informative suites of characters at the group level are stomatal structures and the position of the second to fourth pairs of male genital papillae (v2-v4). Within species groups, these and other sets of characters can often diagnose individual species or species complexes. The most informative characters are: *i*) the arrangement of the fifth to ninth pairs of genital papillae (ad, v5-v7, pd); *ii*) gubernaculum shape; and *iii*) tail shape. However, even these characters can vary within a population or overlap between species.

Based on relationships inferred from molecular data, the genus can be separated into six major clades: the *elegans* group; the *triformis* group; the *lheritieri* group; the *pacificus* group; the *maupasi* group; and *P. fissidentatus* (Kanzaki *et al.*, 2013c; Ragsdale *et al.*, 2013; Fig. 4.7). Of these clades, three (the *lheritieri*, *pacificus* and *maupasi* groups) are very similar to each other in most typological characters (Kanzaki *et al.*, 2012a, 2013b, c, 2014a; Ragsdale *et al.*, 2013). According to simple parsimony, and using *Parapristionchus giblindavisi* as outgroup, the common ancestor of *Pristionchus* is hypothesised to have had: *i*) a cheilostom with six undivided plates; *ii*) a non-vacuolated cheilostom; *iii*) a gymnostom with a smooth anterior margin; *iv*) a pro-/mesostegostom with a smooth anterior margin; *v*) a metastegostom with flint-like right dorsal tooth in the St form; *vi*) right subventral metastegostom with a single tooth (Eu form) or ridge (St form); *vii*) left subventral metastegostom with one ridge of left subventral denticles; *viii*) male papilla v2 lateral (= v2d) and v3 ventral; and *ix*) papillae v2-v4 close to the cloacal opening (Fig. 4.7). From these ancestral states, several characters have evolved to new states independently: an anteriorly serrated gymnostom is observed in both *P. elegans* (Kanzaki *et al.*, 2012b) and the *triformis* group (Ragsdale *et al.*, 2013); the anterior lateral papillae have become the third pair (= v3d) in several species of the *pacificus* and *maupasi* groups (Kanzaki *et al.*, 2012a, 2013b, c); a 12-plated cheilostom has evolved both in the *triformis* group and in *Parapristionchus giblindavisi* (Kanzaki *et al.*, 2012c; Ragsdale *et al.*, 2013).

In addition to group-diagnostic morphology, some characters can diagnose individual species or species pairs of *Pristionchus*. Characters or character combinations that distinguish clades or individual species within them are summarised here. Furthermore, some nominal species with no molecular vouchers might also be identified by morphology, specifically male tail characters, so potentially diagnostic characters are given for those species as well. Variations in stomatal and male papillae morphology are illustrated in Figure 4.8 and in Figures 4.9, 4.10, respectively.

elegans group: vacuolated cheilostom (Figs 4.3C; 4.8I); tail long (>8 anal body diam. in female; Fig. 4.3A); v2 papillae lateral (= v2d); v4 close to cloacal opening

- *P. bucculentus*: thick cheilostomatal walls; anteriorly smooth gymnostom; a series of three conical left subventral denticles; right subventral tooth present (species presumed to have only a Eu form); lateral processes of gubernaculum far (1/2 gubernaculum length) from distal opening of proximal tube; v1 anterior, such that v3 is closer to v4 than v1
- *P. elegans*: membranous cheilostomatal walls; anteriorly serrated gymnostom; right subventral tooth absent (species presumed to have only a St form); lateral processes of gubernaculum close (<1/2 gubernaculum length) to distal opening of proximal tube; v1, v3, and v4 close together and equidistant

lheritieri group: v4 far from cloacal opening

- *P. brevicauda*: ad anterior, such that Ph is mid-way between ad and v5
- *P. bulgaricus*: right subventral tooth of Eu form sometimes with two cusps; ad anterior, such that Ph is mid-way between ad and v5
- *P. clavus*: ad posterior, such that Ph is closer to ad than to v5
- *P. entomophagus*: hermaphroditic; males rare and in many strains unknown
- *P. lheritieri*: pd overlaps v5-v7; male tail spike shorter than distance from cloacal opening to root of spike
- *P. lucani*: pd overlaps v5-v7; male tail spike shorter than distance from cloacal opening to root of spike
- *P. uniformis*: dorsal tooth of St form large, extending into gymnostom

maupasi group: v4 far from cloacal opening

- *P. aerivorus*, *P. pseudaerivorus*: v2 clearly anterior to v3d; ad far (2/3 cloacal body diam.) from v4; pd posterior to v7
- *P. americanus*, *P. marianneae*: v2d clearly anterior to v3; v4 1/2 cloacal body diam. from v3; pd posterior to v7
- *P. atlanticus*: v2d clearly anterior to v3; v4 very close (<1/4 cloacal body diam.) to v3d; pd posterior to v7

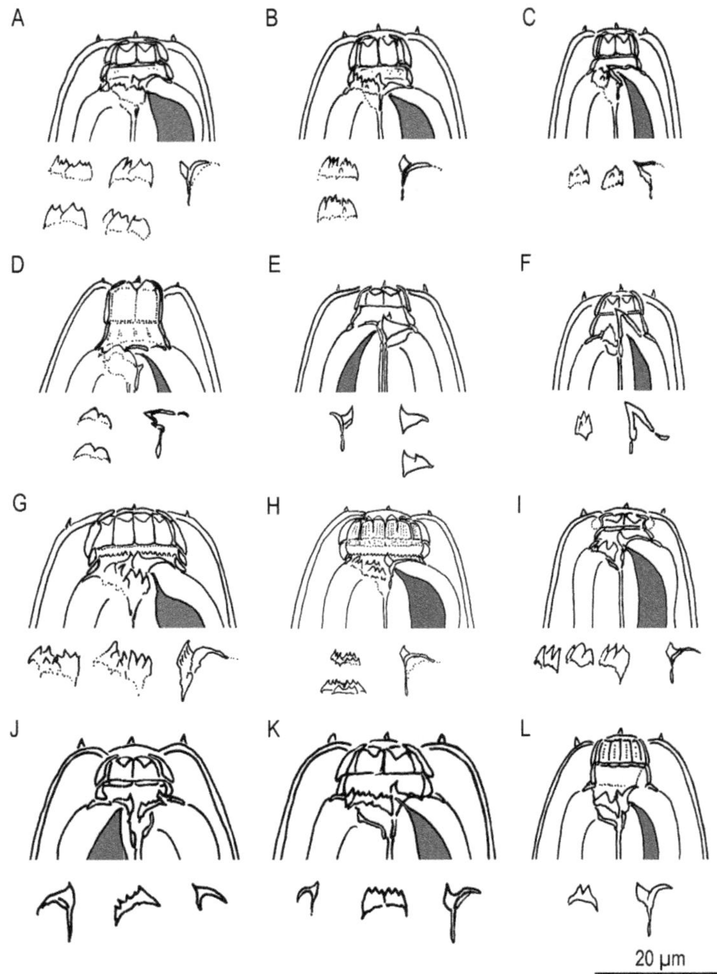

Fig. 4.8. *Comparative stomatal morphology of* Pristionchus *species. Examples of morphology distinguishing* Pristionchus *species or species groups are presented. Stomata are depicted in left lateral aspect unless otherwise indicated. Illustrations are modified from Kanzaki* et al. *(2012a, b, c, 2013a, b, c, 2014a) and Ragsdale* et al. *(2013). A:* P. exspectatus, *eurystomatous (Eu) form. Four variations of left subventral ridge, which can be variable within* Pristionchus *species, are shown below; B:* P. japonicus, *Eu form. The left subventral ridge often has an adventitious denticle; C:* P. maxplancki, *stenostomatous (St) form. The left subventral ridge of the St form has one or more adventitious denticles; D:* P. boliviae, *St form. The stoma and dorsal tooth of the St form are typically large in many species of the* maupasi *group of* Pristionchus*; E:* P. bulgaricus,

P. boliviae: hermaphroditic; v2d clearly anterior to v3; pd overlaps v5-v7
P. maupasi: hermaphroditic; v2 clearly anterior to v3d; ad close to (1/3 cloacal body diam.) from v4; pd overlaps v5-v7
P. mayeri: hermaphroditic; v2d and v3 at about same level; pd overlaps v5-v7
P. pauli: v2d clearly anterior to v3; v4 1/2 cloacal body diam. from v3; pd posterior to v7

pacificus group: v4 far from cloacal opening
P. arcanus, *P. exspectatus*: v2 and v3 at same level; v3 and v4 relatively close (1/2 cloacal body diam. apart); pd overlaps v5-v7
P. japonicus: Eu form sometimes with adventitious denticles; tail spike sometimes with constriction; proximal tube of gubernaculum opens distally at a narrow (30°) angle; v3 and v4 relatively far (a cloacal body diam.) apart; pd posterior to v7
P. maxplancki: left subventral ridge of St form with adventitious denticles; proximal tube of gubernaculum opens distally at a wide (60°) angle; v2 far anterior to (1/3 cloacal body diam. from) v3d; pd posterior to v7
P. pacificus: hermaphroditic; v2 and v3 at same level; pd overlaps v5-v7

Eu form, right lateral aspect. The right subventral tooth is often host to a second cusp; F: P. uniformis, *St form. The St dorsal tooth is uniquely large, extending anteriad of the gymnostom; G:* P. triformis, *12-plated Eu ('megastomatous') form. Cheilostomatal divisions, where present, are complete; H:* P. hoplostomus, *Eu form. As also in* P. fukushimae, *secondary cheilostomatal divisions are incomplete. This species often has secondary left subventral ridge of denticles; I:* P. bucculentus, *Eu (the only known) form. As in* P. elegans, *the cheilostom is vacuolated. A series of three conical denticles distinguish the left subventral sector of this species; J:* P. fissidentatus, *Eu form, right lateral aspect. In addition to a right subventral tooth (below, left), the stoma has a right subventral ridge of denticles (below, centre); K:* P. fissidentatus, *Eu form. Just left of the ventral midline is a separate denticle (below, left); L:* Parapristionchus giblindavisi, *St form. The cheilostom, as also in the Eu form, is made up of 12 complete plates. The St dorsal tooth is claw-like, unlike the flint-shaped St dorsal tooth of* Pristionchus *species.*

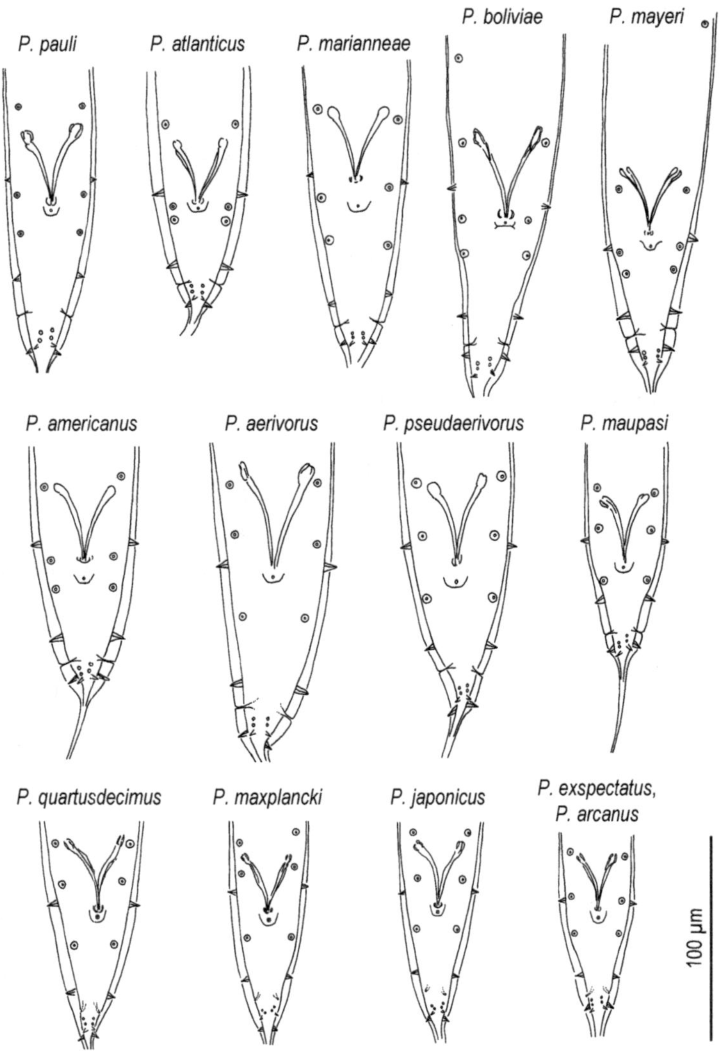

Fig. 4.9. *Male genital papillae in the* maupasi *and* pacificus *groups of* Pristionchus *species. Arrangements of papillae can be informative of groups or individual species of* Pristionchus. *Tails are shown in ventral aspect. Illustrations are modified from Kanzaki* et al. *(2013b, c).*

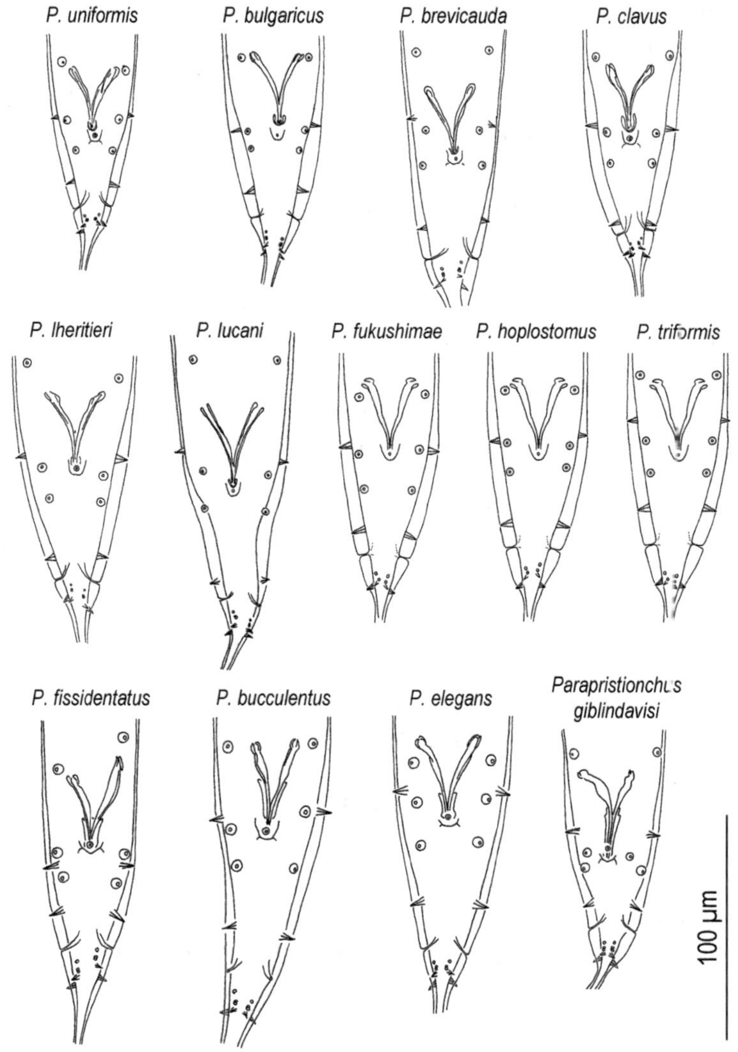

Fig. 4.10. *Male genital papillae in the* lheritieri, triformis *and* elegans *groups of* Pristionchus *species,* P. fissidentatus *and* Parapristionchus giblindavisi. *Tails are shown in ventral aspect. Illustrations are modified from Kanzaki* et al. *(2012b, c, 2013a, 2014a) and Ragsdale* et al. *(2013).*

P. quartusdecimus: proximal tube of gubernaculum opens distally at a narrow (30°) angle; v2 far anterior to (1/3 cloacal body diam. from) v3d; v3 and v4 relatively far (a cloacal body diam.) apart; pd posterior to v7

triformis group: cheilostom with six to 12 complete or incomplete plates; gymnostom and stegostom with serrated anterior margins; v2 papillae lateral (= v2d)

P. fukushimae: cheilostomatal plates incompletely divided; adventitious left subventral plate, where present, with fewer denticles; v3 and v4 far (1/2 cloacal body diam.) apart

P. hoplostomus: cheilostomatal plates incompletely divided; often with adventitious left subventral plate with many denticles; v3 and v4 close together (1/3 cloacal body diam. apart)

P. triformis: hermaphroditic; cheilostomatal plates are always completely divided; adventitious left subventral plate absent

P. fissidentatus is distinguished by additional denticles in the Eu form, including a separate left subventral denticle and a right subventral ridge in addition to the right subventral tooth. Furthermore, the right subventral ridge of the St form has multiple cusps, a feature unique to this species.

Other nominal species circumscribed by male tail characters

P. eurycephalus: papillae arranged as ⟨v1, v2, (v3d, C), 4, ad, Ph, (v5, v6, v7), pd⟩; v2 far anterior to (1/3 cloacal body diam. from) v3d; v4 close to (1/3 cloacal body diam. from) v3; ad anterior, 2/3 cloacal body diam. anterior to v5; pd posterior to v7

P. linstowi: papillae arranged as ⟨v1, (v2, v3d), C, v4, ad, (v5, v6, v7), pd⟩; ad close to (1/2 cloacal body diam. from) v4; v2 clearly anterior to v3d; similar to *P. maupasi*

P. macrospiculum: papillae arranged as ⟨v1, v2d, v3, C, v4, ad, (v5, v6, v7, pd)⟩; v2d slightly anterior to v3; v4 close to (1/3 cloacal body diam. from) v3; pd overlaps v5-v7

P. robustus: papillae arranged as ⟨v1, (v2d, v3), C, v4, ad, (v5, v6, v7, pd)⟩; extra anterior papillae 'v0'; v2d and v3 at about same level; v4 close to (1/3 cloacal body diam. from) v3; ad mid-way between v4 and v5; pd overlaps v5-v7; very similar to *P. mayeri*

P. vidalae: papillae arranged as ⟨v1, v2d, v3, C, v4, ad, Ph, (v5, v6, v7, pd)⟩; v2d far anterior (1/3 cloacal body diam.) from v3; ad far (2/3 cloacal body diam.) from v4; pd overlaps v5-v7

PRISTIONCHUS TAXONOMY IN A DIGITAL AGE

Given recent refinements of the taxonomic system of *Pristionchus*, we are now making efforts to distribute information on the genus by developing a database for *Pristionchus* taxonomy using the 'Scratchpads' interface (scratchpads.eu), a developed open-access repository for taxonomic information (Herrmann *et al.*, unpubl.). When functional, the *Pristionchus* Scratchpad will provide a regularly updated status of the taxonomy of the genus and will include available literature (downloads or links), image vouchers, molecular markers, illustrations of diagnostic morphology and current distribution maps. The Scratchpad will also allow users to incorporate new information under the management of an editor. This interface should make *Pristionchus* taxonomy transparent and globally accessible, while also honouring the practice of proper description that provides the biological information useful for future research. It is our hope that such an interface will serve as a model for both enhancing the service of nematode taxonomy and keeping up with discoveries of nematode diversity.

Biogeography

The biogeography of terrestrial nematode taxa often eludes inference. Through the global movement of soil systems, human activity has dispersed nematodes across otherwise formidable barriers, causing distributions to appear disjunct. Moreover, many terrestrial habitats and insect hosts are still largely undersampled (Porazinska *et al.*, 2010), so distribution maps for particular taxa are generally far from complete. Known distributions of many nematodes therefore obscure biogeographic patterns. Nevertheless, it is possible for free-living nematodes to assume a vicariant distribution, or geographic ranges congruent with plate tectonics (Ferris *et al.*, 1976). What impedes biogeographic inferences for most nematode taxa is missing range information or a lack of phylogenetic infrastructure.

Like most free-living nematode taxa, *Pristionchus* presumably still suffers from incomplete range information. However, the increasingly comprehensive sampling of *Pristionchus* species, as well as their inclusion in a well-supported phylogeny, has helped to provide a clear biogeographic history (Herrmann *et al.*, 2006b; Mayer *et al.*, 2007; Kanzaki *et al.*, 2013b, c, 2014a) (Fig. 4.7). The remaining exceptions to the inferred patterns are androdioecious species. Because selfing species require the dispersal of a single hermaphrodite of any life-stage to establish a new population, they are thought to disperse more easily than obligate out-crossers over geographic obstacles (Herrmann *et al.*, 2010; Morgan *et al.*, 2012). In contrast, due to the more stringent conditions for gonochorists to colonise new localities, the ranges of gonochoristic species are predicted to more closely reflect vicariance. Consistent with this expectation, all molecularly characterised gonochorists of *Pristionchus* belong to clades, each corresponding to a geographic region (Fig. 4.7). Since there are no inferred cases of gonochoristic species evolving from a hermaphroditic ancestor (Mayer *et al.*, 2007, 2009; Kanzaki *et al.*, 2012a, 2013c; Ragsdale *et al.*, 2013), gonochorists are thought to have shaped these ancient geographic patterns.

All major clades of *Pristionchus*, with the exception of androdioecious species, fall into one of three geographic groups: Eastern Palaearctic (East Asia), Western Palaearctic (Europe), and Nearctic (North America). The ancestral distribution of *Pristionchus* was likely in East Asia, as is supported by the Japanese localities of the outgroup *Parapristionchus giblindavisi* and the *elegans* group of *Pristionchus* species (Kanzaki *et al.*, 2012b, c, 2013a). Two additional clades of *Pristionchus* have since radiated in East Asia, namely the *pacificus* and *triformis* groups (Kanzaki *et al.*, 2012a, 2013b; Ragsdale *et al.*, 2013). From this putative ancestral range, the ancestors of two clades colonised other regions. The *lheritieri* group is found in Europe (Kanzaki *et al.*, 2014a), while the *maupasi* group is restricted to North America (Herrmann *et al.*, 2006b). Beyond this basic pattern, biogeographical structure is sometimes also apparent within continents. For example, gonochorists of the *lheritieri* group fall into two clades that generally correspond to eastern and western Europe (Kanzaki *et al.*, 2014a; Fig. 4.7). An exception to this distribution was the wider range discovered for *P. uniformis*, which was systematically collected from the scarab beetles and the Colorado potato beetle from both Europe and North America (D'Anna & Sommer, 2011). However, a detailed investigation of that exception proved the rule: population genetics analysis of *P. uniformis* confirmed its initial range to be in Europe and the range expansion of *P. uniformis* could be attributed to the movement of its beetle hosts.

Gonochoristic *Pristionchus* species are currently best known from the Holarctic. However, further sampling in this region, for example in central Asia, is likely to reveal more about the movement of *Pristionchus* from its putative ancestral range. Moreover, gonochorists of *Pristionchus* are still almost unknown from the southern hemisphere. Missing range information may be attributed in part to sampling bias. For example, preliminary evidence suggests that gonochoristic *Pristionchus* species do exist in South America (F. Brown, University of São Paulo, pers. comm.). Persistent collecting efforts here as well as in Africa, Australia, and continental Southeast Asia will be needed to test global distribution patterns of *Pristionchus*. Given the pace of discoveries over the past decade, it is likely that the next will see many more.

Acknowledgements

We thank Alessandro Minelli (International Commission on Zoological Nomenclature) for his suggestions on nomenclatural procedure.

References

ADAMS, B.J. (1998). Species concepts and the evolutionary paradigm in modern nematology. *Journal of Nematology* 30, 1-21.

ALTHERR, E. (1938). La faune des mines de Bex, avec étude special des nématodes. *Revue Suisse de Zoologie* 45, 567-720.

ANDRÁSSY, I. (1984). *Klasse Nematoda*. Stuttgart, Germany, Gustav Fischer Verlag.

ATAKHANOV, S.A. (1958). Sem novykh vidov nematode. *Rabot po gel'mintologii 80-let. Akademia K.I. Skrabina (Moskva)*, pp. 42-49.

BALACHOWSKY, A.S. (1962). *Entomologie Appliquée à l'Agriculture, Traité, Vol. 1, Coléoptères*. Paris, France, Masson.

BALDWIN, J.G., GIBLIN-DAVIS, R.M., EDDLEMAN, C.D., WILLIAMS, D.S., VIDA, J.T. & THOMAS, W.K. (1997). The buccal capsule of *Aduncospiculum halicti* (Nemata: Diplogasterina): an ultrastructural and molecular phylogenetic study. *Canadian Journal of Zoology* 75, 407-423.

BLAXTER, M.L. (2004). The promise of a DNA taxonomy. *Philosophical Transactions of the Royal Society B: Biological Sciences* 359, 669-679.

BLINOVA, S.L. & VOSILITE, B.S. (1976). *Paramonoviola rhagii* gen. et sp. n. (Rhabditida, Diplogasteridae) – parazit usacha *Rhagium inquisitor*. *Zoologichesky Zhurnal* 55, 131-133.

BONGERS, T. (1999). The Maturity Index, the evolution of nematode life history traits, adaptive radiation and cp-scaling. *Plant and Soil* 212, 13-22.

BOVIEN, P. (1937). Some types of association between nematodes and insects. *Videnskabelige Meddelelser fra Dansk Naturhistorik Forening* 101, 1-114.

BRECHTEL, F. (2002). Biologie und Ökologie der Pracht- und Hirschkäfer. In: Brechtel, F. & Kostenbader, H. (Eds). *Die Pracht- und Hirschkäfer Baden-Württembergs*. Stuttgart, Germany, Verlag Eugen Ulmer, pp. 92-153.

BÜTSCHLI, O. (1873). Beiträge zur Kenntnis der freilebenden Nematoden. *Nova Acta der Kaiserlich Leopoldinisch-Carolinischen Deutschen Akademie der Naturforscher* 36, 1-144.

BÜTSCHLI, O. (1874). Zur Kenntniss der freilebenden Nematoden insbesondere der des Kieler Hafens. *Abhandlungen der Senckenbergischen naturforschenden Gesellschaft* 9, 1-56.

BÜTSCHLI, O. (1876). Untersuchungen über freilebende Nematoden und die Gattung *Chaetonotus*. *Zeitschrift für wissenschaftliche Zoologie* 26, 363-413.

CHITWOOD, B.G. (1937). Cephalic structure and stoma. In: Chitwood, B.G. & Chitwood, M.B. (Eds). *An introduction to nematology*. Baltimore, MA, USA, Monumental Printing Co., p. 55.

CINKORNPUMIN, J.K., WISIDAGAMA, D.R., RAPOPORT, V., GO, J.L., DIETERICH, C., WANG, X., SOMMER, R.J. & HONG, R.L. (2014). A host beetle pheromone regulates development and behavior in the nematode *Pristionchus pacificus*. *eLife* 3, e03229. DOI:10.7554/eLife.03229.

CLAUS, C. (1862). Über einige im Humus lebende Anguillulinen. *Zeitschrift für wissenschaftliche Zoologie* 12, 354-359.

COBB, N.A. (1893). Nematodes, mostly Australian and Fijian. *Department of Agriculture, New South Wales, Miscellaneous Publications (Sydney)* 13, 1-59.

D'ANNA, I. & SOMMER, R.J. (2011). *Pristionchus uniformis* – should I stay or should I go? Recent host range expansion in an European nematode. *Ecology and Evolution* 1, 468-478.

DARWIN, C. (1859). *On the origin of species by means of natural selection, or the preservation of favoured races in the struggle for life*. London, UK, John Murray.

DE LEY, P. & BLAXTER, M.L. (2002). Systematic position and phylogeny. In: Lee, D.L. (Ed.). *The biology of nematodes*. London, UK, Taylor & Francis, pp. 1-30.

DE LEY, P., VAN DE VELDE, M.C., MOUNPORT, D., BAUJARD, P. & COOMANS, A. (1995). Ultrastructure of the stoma in Cephalobidae, Panagrolaimidae, and Rhabditidae, with a proposal for a revised stoma terminology in Rhabditida (Nematoda). *Nematologica* 41, 153-182.

DIETERICH, C., CLIFTON, S.W., SCHUSTER, L., CHINWALLA, A., DELEHAUNTY, K., DINKELACKER, I., FULTON, L., FULTON, R., GODFREY, J., MINX, P. *ET AL*. (2008). The *Pristionchus pacificus* genome provides a unique perspective on nematode lifestyle and parasitism. *Nature Genetics* 40, 1193-1198.

EIZINGER, A. & SOMMER, R.J. (1997). The homeotic gene *lin-39* and the evolution of nematode epidermal cell fates. *Science* 278, 452-455.

FEDORKO, A. & STANUSZEK, S. (1971). *Pristionchus uniformis* sp. n. (Nematoda, Rhabditida, Diplogasteridae), a facultative parasite of *Leptinotarsa decemlineata* Say and *Melolontha melolontha* L. in Poland. Morphology and biology. *Acta Parastiologica* 19, 95-112.

FÉLIX, M.-A., BRAENDLE, C. & CUTTER, A.D. (2014). A streamlined system for species diagnosis in *Caenorhabditis* (Nematoda: Rhabditidae) with name designations for 15 distinct biological species. *PLoS ONE* 9, e94723.

FERRIS, V.R., GOSECO, C.G. & FERRIS, J.M. (1976). Biogeography of free-living nematodes from the perspective of plate tectonics. *Science* 193, 508-510.

FLOYD, R., ABEBE, E., PAPERT, A. & BLAXTER, M. (2002). Molecular barcodes for soil nematode identification. *Molecular Ecology* 11, 839-850.

FUCHS, G. (1929). *Diplogaster horticola* n. sp. *Zoologische Anzeiger* 81, 229-237.

GERAERT, E. (1984). Morphology and morphometry of *Mesodiplogaster pseudolheritieri* n. sp. (Nematoda: Rhabditida). *Nematologica* 29, 284-297.

GIBLIN-DAVIS, R.M., DAVIES, K.A., TAYLOR, G.S. & THOMAS, W.K. (2004). Entomophilic nematode models for studying biodiversity and cospeciation. In: Chen, Z.X., Chen, S.Y. & Dickson, D.W. (Eds). *Nematology: advances and perspectives*, Vol. 1. Wallingford, UK, CAB International Publishing, pp. 493-540.

GOODEY, T. (1963). *Soil and freshwater nematodes*, 2nd edition (revised by Goodey, J.B.). London, UK, Methuen & Company.

HENNIG, W. (1966). *Phylogenetic systematics*. Urbana, IL, USA, University of Illinois Press.

HERRMANN, M., MAYER, E.W. & SOMMER, R.J. (2006a). Nematodes of the genus *Pristionchus* are closely associated with scarab beetles and the Colorado potato beetle in western Europe. *Zoology* 109, 96-108.

HERRMANN, M., MAYER, E.W. & SOMMER, R.J. (2006b). Sex, bugs and Haldane's rule: the nematode genus *Pristionchus* in the United States. *Frontiers in Zoology* 3, 14.

HERRMANN, M., MAYER, W., HONG, R., KIENLE, S., MINASAKI, R. & SOMMER, R.J. (2007). The nematode *Pristionchus pacificus* (Nematoda: Diplogastridae) is associated with the Oriental beetle *Exomala orientalis* (Coleoptera: Scarabaeidae) in Japan. *Zoological Science* 24, 883-889.

HERRMANN, M., KIENLE, S., ROCHAT, J., MAYER, W. & SOMMER, R.J. (2010). Haplotype diversity of the nematode *Pristionchus pacificus* on Réunion in the Indian Ocean suggests multiple independent invasions. *Biological Journal of the Linnean Society* 100, 170-179.

HIRSCHMANN, H. (1951). Über das Vorkommen zweier Mundhöhlentypen bei *Diplogaster lheritieri* Maupas und *Diplogaster biformis* n. sp. und die Entstehung dieser hermaphroditischen Art aus *Diplogaster lheritieri*. *Zoologische Jahrbücher, Abteilung für Systematik* 80, 132-170.

HNATEWYTSCH, B. (1929). Die Fauna der Erzgruben von Schneeberg im Erzgebirge. *Zoologische Jahrbücher, Abteilung für Systematik* 56, 173-261.

HONG, R.L. & SOMMER, R.J. (2006). Chemoattraction in *Pristionchus* nematodes and implication for insect host recognition. *Current Biology* 16, 2359-2365.

HONG, R.L., WITTE, H. & SOMMER, R.J. (2008a). Natural variation in *P. pacificus* insect pheromone attraction involves the protein kinase EGL-4. *Proceedings of the National Academy of Sciences of the United States of America* 105, 7779-7784.

HONG, R.L., SVATOS, A., HERRMANN, M. & SOMMER, R.J. (2008b). The species-specific recognition of beetle cues by *Pristionchus maupasi*. *Evolution & Development* 10, 273-279.

KANZAKI, N., GIBLIN-DAVIS, R.M., DAVIES, K., YE, W., CENTER, B.J. & THOMAS, W.K. (2009). *Teratodiplogaster fignewmani* gen. nov., sp. nov. (Nematoda: Diplogastridae) from the syconia of *Ficus racemosa* in Australia. *Zoological Science* 26, 569-578.

KANZAKI, N., TAKI, H., MASUYA, H., OKABE, K., TANAKA, R. & ABE, F. (2011). Diversity of stag beetle-associated nematodes in Japan. *Environmental Entomology* 40, 281-288.

KANZAKI, N., RAGSDALE, E.J., HERRMANN, M., MAYER, W.E. & SOMMER, R.J. (2012a). Description of three *Pristionchus* species (Nematoda: Diplogastridae) from Japan that form a cryptic species complex with the model organism *P. pacificus*. *Zoological Science* 29, 403-417.

KANZAKI, N., RAGSDALE, E., HERRMANN, M. & SOMMER, R.J. (2012b). Two new species of *Pristionchus* (Rhabditida: Diplogastridae) *P. fissidentatus* n. sp. from Nepal and La Réunion Island and *P. elegans* n. sp. from Japan. *Journal of Nematology* 44, 80-90.

KANZAKI, N., RAGSDALE, E., HERRMANN, M., MAYER, W.E., TANAKA, R. & SOMMER, R.J. (2012c). *Parapristionchus giblindavisi* n. gen., n. sp. (Rhabditida: Diplogastridae) isolated from stag beetles (Coleoptera: Lucanidae) in Japan. *Nematology* 14, 933-947.

KANZAKI, N., RAGSDALE, E., HERRMANN, M., RÖSELER, W. & SOMMER, R.J. (2013a). *Pristionchus bucculentus* n. sp. (Rhabditida: Diplogastridae) isolated from a shining mushroom beetle (Coleoptera: Scaphidiidae) in Hokkaido, Japan. *Journal of Nematology* 45, 78-86.

KANZAKI, N., RAGSDALE, E., HERRMANN, M., RÖSELER, W. & SOMMER, R.J. (2013b). Two new species of *Pristionchus* (Nematoda: Diplogastridae) support the biogeographic importance of Japan for the evolution of the genus *Pristionchus* and the model system *P. pacificus*. *Zoological Science* 30, 680-692.

KANZAKI, N., RAGSDALE, E., HERRMANN, M., SUSOY, V. & SOMMER, R.J. (2013c). Two androdioecious and one dioecious new species of *Pristionchus* (Nematoda: Diplogastridae): new reference points for the evolution of reproductive mode. *Journal of Nematology* 45, 172-194.

KANZAKI, N., RAGSDALE, E., HERRMANN, M. & SOMMER, R.J (2014a). Two new and two recharacterized species resulting from a radiation of *Pristionchus* (Nematoda: Diplogastridae) in Europe. *Journal of Nematology* 46, 60-74.

KANZAKI, N., TAKI, H., MASUYA, H. & OKABE, K. (2014b). Isolation of *Pristionchus bucculentus* from the large mushroom beetle, *Encaustes praenobilis*. *Bulletin of the Forestry and Forest Products Research Institute* 13, 27-28.

KANZAKI, N., RAGSDALE, E.J. & GIBLIN-DAVIS, R.M. (2014c). Revision of *Koerneria* Meyl, 1960 and resurrection of two other genera of Diplogastridae (Nematoda). *ZooKeys* 442, 17-30.

KIONTKE, K.C., FÉLIX, M.-A., AILION, M., ROCKMAN, M.V., BRAENDLE, C., PÉNIGAULT, J.-B. & FITCH, D.H.A. (2011). A phylogeny and molecular barcodes for *Caenorhabditis*, with numerous new species from rotting fruits. *BMC Evolutionary Biology* 11, 339.

KOTLÁN, A. (1928). *Diplogaster brevicauda* sp. n., a possible nematode parasite of the larvae of *Pyrausta nubialis*. In: *Scientific Reports, International Corn Borer Investigations*. Chicago, IL, USA, International Live Stock Exposition, pp. 179-183.

KREIS, H.A. (1932). Beiträge zur Kenntnis pflanzenparasitischer Nematoden. *Zeitschrift für Parasitenkunde* 5, 184-194.

LESPÉS, C. (1856). Notes sur un nématoïde parasite des termites. *Annales des Sciences Naturelles, Zoologie (Paris) (Sér. 4)* 5, 335-336.

MAUPAS, É. (1900). Modes et formes de reproduction des nématodes. *Archive Zoologique Expérimentelle et Générale* 8, 463-624.

MAUPAS, É. (1919). Essais d'hybridation chez des nématodes. *Bulletin Biologique de la France et de la Belgique* 52, 466-498.

MAYER, M.G. & SOMMER, R.J. (2011). Natural variation in *Pristionchus pacificus* dauer formation reveals cross-preference rather than self-preference of nematode dauer pheromones. *Proceedings of the Royal Society B: Biological Sciences* 278, 2784-2790.

MAYER, W., HERRMANN, M. & SOMMER, R.J. (2007). Phylogeny of the nematode genus *Pristionchus* and implications for biodiversity, biogeography and the evolution of hermaphroditism. *BMC Evolutionary Biology* 7, 104.

MAYER, W., HERRMANN, M. & SOMMER, R.J. (2009). Molecular phylogeny of beetle associated diplogastrid nematodes suggests host switching rather than nematode – beetle coevolution. *BMC Evolutionary Biology* 9, 212.

MAYER, W.E., SCHUSTER, L., BARTELMES, G., DIETERICH, C. & SOMMER, R.J. (2011). Horizontal gene transfer of microbial cellulases into nematode genomes is associated with functional assimilation, high gene turnover and positive selection. *BMC Evolutionary Biology* 11, 13.

MERRILL, J.H. & FORD, A.L. (1916). Life history and habits of two new nematodes parasitic on insects. *Journal of Agricultural Research* 6, 115-127.

MORGAN, K., MCGAUGHRAN, A., WITTE, H., BARTELMES, G., VILLATE, L., HERRMANN, M., ROCHAT, J. & SOMMER, R.J. (2012). Multi-locus analysis of *Pristionchus pacificus* on La Réunion Island reveals an evolutionary history shaped by multiple introductions, constrained dispersal events, and rare out-crossing. *Molecular Ecology* 21, 250-266.

PARAMONOV, A.A. (1952). Opyt ekologicheskoi klassifikatsii fitonematod. *Trudy Gelmintologicheskoi Laboratorii, Akademia Nauk SSSR (Moskva)* 6, 338-369.

POINAR JR, G.O. (1990). Redescription of *Chroniodiplogaster aerivora* (Cobb) gen. n., comb. n. (Rhabditida: Diplogasteridae) from termites. *Journal of the Helminthological Society of Washington* 57, 26-30.

POINAR JR, G.O. (2011). *The evolutionary history of nematodes, as revealed in stone, amber and mummies. Nematology Monographs and Perspectives 9* (Series editors David J. Hunt & Roland N. Perry). Leiden, The Netherlands, Brill.

POINAR JR, G.O., MEIKLE, W. & MERCADIER, G. (2006). *Chroniodiplogaster formosiana* sp. n. (Rhabditida: Diplogastridae) from Chinese populations of *Odontotermes formosanus* Shiraki (Isoptera: Termitidae). *Journal of Nematology* 38, 181-186.

PORAZINSKA, D.L., GIBLIN-DAVIS, R.M., ESQUIVEL, A., POWERS, T.O., SUNG, W. & THOMAS, W.K. (2010). Ecometagenomics confirms high tropical rainforest nematode diversity. *Molecular Ecology* 19, 5521-5530.

POTTS, F.A. (1910). Notes on the free-living nematodes. *Quarterly Journal of Microscopical Science* 55, 433-485.

POULIN, R. (2007). Are there general laws in parasite ecology? *Parasitology* 134, 763-776.

RAE, R., RIEBESELL, M., DINKELACKER, I., WANG, Q., HERRMANN, M., WELLER, A.M., DIETERICH, C. & SOMMER, R.J. (2008). Isolation of naturally associated bacteria of necromenic *Pristionchus* nematodes and fitness consequences. *Journal of Experimental Biology* 211, 1927-1936.

RAGSDALE, E., KANZAKI, N., ROESELER, W., HERRMANN, M. & SOMMER, R.J. (2013). Three new species of *Pristionchus* (Nematoda: Diplogastridae) show morphological divergence through evolutionary intermediates of a novel feeding-structure polymorphism. *Zoological Journal of the Linnean Society* 168, 671-698.

RAHM, G. (1928). Alguns nematodes parasitas e semiparasitas das plantas culturães do Brasil. *Archivos do Instituto Biológico de Defesa Agricola e Animal (São Paulo)* 1, 239-251.

RAHM, G. (1932). Freilebende Nematoden, Rotatorien und Tardigraden aus Südamerika (besonders aus Chile). *Zoologischer Anzeiger* 98, 94-128.

RUDEL, D., RIEBESELL, M. & SOMMER, R.J. (2005). Gonadogenesis in *Pristionchus pacificus* and organ evolution: development, adult morphology and cell/cell interactions in the hermaphrodite gonad. *Developmental Biology* 277, 200-221.

SCHNEIDER, A. (1866). *Monographie der Nematoden*. Berlin, Germany, Reimer.

SCHUSTER, L.N. & SOMMER, R.J. (2012). Expressional and functional variation of horizontally acquired cellulases in the nematode *Pristionchus pacificus*. *Gene* 506, 274-282.

SCHUURMANS STEKHOVEN, J.H. (1951). Nématodes saprozoaires et libres du Congo Belge. *Mémoires de l'Institut Royal des Sciences Naturelles de Belgique, Série 2* 39, 3-79.

SCHUURMANS STEKHOVEN, J.H. & TEUNISSEN, R.J.H. (1938). Nématodes libres terrestres. Exploration du Parc National Albert. *Mission G. F. de Witte (1933-1935)* 22, 1-229.

SEROBYAN, V., RAGSDALE, E.J. & SOMMER, R.J. (2014). Adaptive value of a predatory mouth-form in a dimorphic nematode. *Proceedings of the Royal Society B: Biological Sciences* 281, 20141334.

SOMMER, R.J., WITTE, H. & SCHLAK, I. (2001). Towards a microevolutionary approach in evolutionary developmental biology: biogeography of the nematode genus *Pristionchus* (Diplogastridae). *Zoology* 103, 91-98.

SRINIVASAN, J., SINZ, W., LANZ, C., BRAND, A., NANDAKUMAR, R., RADDATZ, G., WITTE, H., KELLER, H., KIPPING, I., PIRES-DASILVA, A. *ET AL.* (2002). A BAC-based genetic linkage map of the nematode *Pristionchus pacificus*. *Genetics* 162, 129-134.

SRINIVASAN, J., SINZ, W., JESSE, T., WIGGERS-PEREBOLTE, L., JANSEN, K., BUNTJER, J., VAN DER MEULEN, M. & SOMMER, R.J. (2003). An integrated physical and genetic map of the nematode *Pristionchus pacificus*. *Molecular Genetics and Genomics* 269, 715-722.

STEEL, H., VANDECASTEELE, B., WILLIKENS, K., SABBE, K., MOENS, T. & BERT, W. (2012). Nematode communities and macronutrients in composts and compost-amended soils as affected by feedstock composition. *Applied Soil Ecology* 61, 100-112.

STEINER, G. (1914). Freilebende Nematoden aus der Schweiz. 2. Teil. *Archiv für Hydrobiologie* 9, 420-438.

STEINER, G. (1929). *Diplogaster entomophaga* n. sp., a new *Diplogaster* (Diplogasteridae, Nematodes) found on a *Pamphilius stellatus* (Christ) (Tentheredinidae, Hymenoptera). *Zoologischer Anzeiger* 80, 143-145.

STEINER, G. (1943). New nematodes associated with a disease of the papaya in Chile. *Boletín Departamento de Sanidad Vegetal (Santiago)* 3, 95-116.

STOCK, S.P. (1993). *Micoletzkya vidalae* n. sp. (Nematoda: Diplogasteridae), a facultative parasite of *Diabrotica speciosa* (Coleoptera: Chrysomelidae) from Argentina. *Research and Reviews in Parasitology* 53, 109-112.

SUDHAUS, W. (2010). Preadaptive plateau in Rhabditida (Nematoda) allowed the repeated evolution of zooparasites, with an outlook on evolution of life cycles within Spiroascaridae. *Palaeodiversity* 3(Supplement), 117-130.

SUDHAUS, W. & FÜRST VON LIEVEN, A. (2003). A phylogenetic classification and catalogue of the Diplogastridae (Secernentea, Nematoda). *Journal of Nematode Morphology and Systematics* 6, 43-90.

SUDHAUS, W. & SCHULTE, F. (1989). *Rhabditis* (*Rhabditis*) *necromena* sp. n. (Nematoda: Rhabditidae) from South Australian Diplopoda with notes on its sibling *R. myriophila* Poinar, 1986 and *R. caulleryi* Maupas, 1919. *Nematologica* 35, 15-24.

SUSOY, V. & HERRMANN, M. (2014). Preferential host switching and codivergence shaped radiation of bark beetle symbionts, nematodes of *Micoletzkya* (Nematoda: Diplogastridae). *Journal of Evolutionary Biology* 27, 889-898.

SUSOY, V., RAGSDALE, E.J., KANZAKI, N. & SOMMER, R.J. (2015). Rapid diversification associated with a macroevolutionary pulse of developmental plasticity. *eLife*, in press.

VÖLK, J. (1950). Die Nematoden der Regenwürmer und aasbesuchenden Käfer. *Zoologische Jahrbücher, Abteilung für Systematik* 79, 1-70.

VON LINSTOW, O. (1901). Beobachtungen an Helminthen des Senckenbergischen naturhistorischen Museums, des Braslauer zoologischen Instituts und anderen. *Archiv für mikroskopische anatomie* 58, 182-198.

WELLER, A., MAYER, W., RAE, R. & SOMMER, R.J. (2010). Quantitative assessment of the nematode fauna present on *Geotrupes* dung beetles reveals species-rich communities with a heterogenous distribution. *Journal of Parasitology* 96, 525-531.

WILEY, E.O. (1978). The evolutionary species concept reconsidered. *Systematic Zoology* 27, 17-26.

WOLLENWEBER, H.W. (1921). Beiträge zur Älchenfauna der Kartoffel. *Mitteilungen aus der Biologischen Reichsanstalt für Land- und Forstwirtschaft, Berlin-Dahlem* 21, 258-266.

Nematology Monographs & Perspectives, 2015, Vol. 11, 121-140

Chapter 5

The laboratory model: genetics, genetic mapping and transgenics

Laura AURILIO and Jagan SRINIVASAN

Department of Biology and Biotechnology, 60 Gateway Park, Life Sciences and Bioengineering Center, Worcester Polytechnic Institute, Worcester, MA 01545, USA

laurilio@wpi.edu

jsrinivasan@wpi.edu

Introduction

Nothing in biology makes sense except in the light of evolution.

Theodosius Dobzhansky (1973).

The evolution of complex life forms involves marked changes in development, morphology and the evolution of new characters (Gerhart & Kirschner, 1997; Raff, 2000; Wilkins, 2002). The success of developmental genetics in the late 1960s led to the emergence of model systems like the fly, *Drosophila melanogaster*, the nematode, *Caenorhabditis elegans*, the house mouse, *Mus musculus*, the fish, *Danio rerio*, and the frog, *Xenopus laevis*, as well as the weed, *Arabidopsis thaliana* (Wolpert, 2002). However, these organisms are all so distantly related that evolutionary insight resulting from comparisons among these organisms is limited to those patterns and processes conserved over large evolutionary distances. In the last decade, comparative studies of closely related organisms have become increasingly popular to understand developmental diversity. These studies focus on the element of variation and the source of molecular variation that give rise to phenotypic change (Raff, 2000; Stern, 2000; Simpson, 2002). Hence, comparing closely related species spanning a smaller evolutionary scale underlies the foundation of evolutionary developmental biology (evo-devo) (Raff, 2000; Stern, 2000; Simpson, 2002). The aim of this chapter is to describe the status of *Pris-*

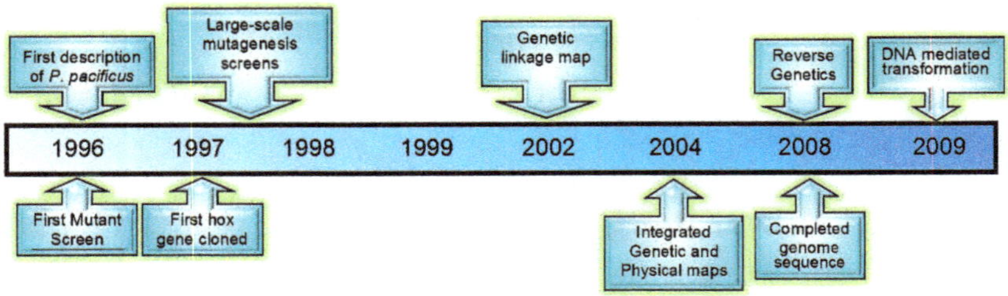

Fig. 5.1. *Timeline of technology development in* Pristionchus pacificus. *Since the first description in 1996, the organism has become a chosen model for comparative developmental, behavioural and ecological studies.*

tionchus pacificus as a laboratory model organism, providing up-to-date information of the genetic and genomic resources used to understand development, behaviour and ecology of *P. pacificus*.

Pristionchus pacificus: beginnings of a laboratory model system

Pristionchus pacificus was chosen in the mid-1990s as a new 'satellite' species to understand the evolution of developmental diversity (Sommer & Sternberg, 1996; Sommer *et al.*, 1996; Eizinger & Sommer, 1997; Sommer, 1997, 2000). During the initial period of research on *P. pacificus*, a strong foundation was laid in the form of technology development to put *P. pacificus* on the map of comparative biology (Sommer, 2000, 2001, 2012) (Fig. 5.1). Experiments in genetics and developmental biology have largely been performed using the *P. pacificus* 'wild type' isolate PS312 from California (Sommer *et al.*, 1996). This strain was isolated from soil in downtown Pasadena, CA, USA, in 1988 and has been cultured ever since. It can be requested from the *Caenorhabditis* Genetic Center (CGC) as well as the *Pristionchus* Stock Center in Tübingen, Germany (www.pristionchus.org).

General description and genetics

Pristionchus pacificus is a free-living nematode that belongs to the Diplogastridae (Sommer *et al.*, 1996; Fürst von Lieven, 2005) and shows a large evolutionary divergence from *C. elegans*. Physically, it is approximately 1 mm long and has a distinct body morphology when compared to *C. elegans*. The latter species comprises two sexes, a self-fertilising hermaphrodite, which forms 99.9% of most populations, and naturally occurring males, allowing for cross-fertilisation. Similarly, hermaphrodites of *P. pacificus* (like those of *C. elegans*) are only self-fertilising and not cross-fertilising, representing an important difference to other known hermaphrodites, such as in annelids and molluscs. Spontaneous generation of males has been investigated in several *P. pacificus* wild isolates and is mostly in the range of 0.1% of populations (Click *et al.*, 2009).

Pristionchus pacificus hermaphrodites are modified females and have two X chromosomes (XX genotype). They are phenotypically larger than males and produce sperm for a short period during their post-embryonic development. Males are smaller in size and have only one X chromosome (XO genotype). Whilst males occur in low frequency, they can be easily obtained and maintained under laboratory conditions and are used in classical genetic experiments (Sommer & Sternberg, 1996; Rae *et al.*, 2008). Progeny from self-fertilisation is genotypically >99% XX, whereas progeny from outcrossing with males follows Mendelian inheritance with 50% hermaphrodites and 50% males. The typical life cycle of *P. pacificus* at 20°C lasts approximately 4 days. This life cycle includes four juvenile stages, J1-J4, and an adult stage; however, the first moult occurs in the egg (Fürst von Lieven, 2005) (Fig. 2.4). Adult hermaphrodites lay an average of 190 eggs during their lifetime. The 4-day life cycle and the large number of progeny allow culture densities similar to microbes, an important pre-requisite for any type of large-scale study of a laboratory model organism (Fig. 5.2). Developmentally speaking, *P. pacificus* is very similar to *C. elegans*; however, on the cellular and molecular levels there are clear divergences in post-embryonic development (Pires-daSilva & Sommer, 2003). *Pristionchus pacificus*, like *C. elegans*, has six chromosomes as seen in squashed preparations (Sommer *et al.*, 1996). Molecular comparisons revealed that the six chromosomes are largely homologous to those of *C. elegans*, although the exact gene order is not conserved (Lee *et al.*, 2003). Thus, these chromosomes show macrosynteny but not microsynteny. The ability to self- and cross-fertilise makes *P. pacificus* ideal for genetic analysis using conventional genetic crosses (Sommer & Sternberg, 1996).

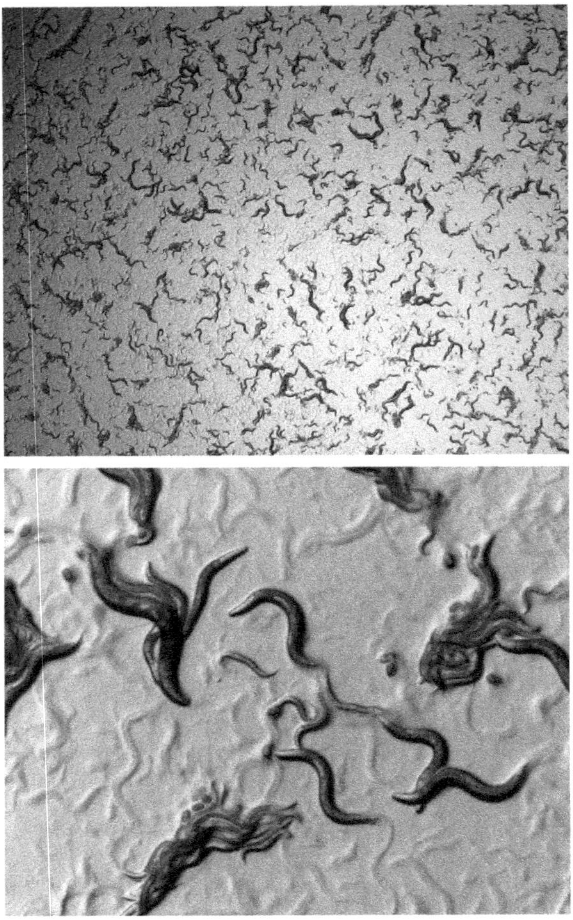

Fig. 5.2. *A standard plate of* Pristionchus pacificus *exhibiting different life stages.*

Forward genetic screens in *P. pacificus*

The success of a genetic model system relies on the ability to create a large and diverse array of mutants defective in diverse developmental and behavioural processes. Such an approach is essential for understanding how biological processes function. In *P. pacificus*, standard forward genetic screens have been performed using different mutagens such as Ethyl methanesulfonate (EMS) and Trimethylpsoralen/UV (TMP-UV). The former mutagen yields preferentially point mutations, whereas the latter yields mainly small deletions. For a detailed description of performing mutagenesis in *P. pacificus*, we refer to the WORMBOOK method chapter by Pires-daSilva (2013). Screens performed using these mutagens have yielded a wide variety of mutants with defects in different biological processes, such as the egg-laying system (Jungblut & Sommer, 1998, 2001), muscle formation (Photos *et al.*, 2006), sex determination (Pires-daSilva & Sommer, 2004), dauer formation (Ogawa *et al.*, 2011; Ragsdale *et al.*, 2013), behaviour (Hong & Sommer, 2006), olfaction (Hong *et al.*, 2008) and gonad formation (Rudel *et al.*, 2008).

An additional group of mutants obtained in these early studies were morphological mutants. Morphological mutants are often used as markers in crossing experiments in order to distinguish self- from cross-progeny and to quantify the success of a mutagenesis screen by the number of times a certain phenotype recurs. In *P. pacificus*, two types of morphological mutants have been identified and are commonly used. They have characteristic phenotypes, for example, a 'shorter and fatter' body size than wild type are termed as 'dumpy' (Dpy) (Kenning *et al.*, 2004), whilst others have a characteristic uncoordinated motion and are termed 'uncoordinated movement' (Unc). Characterisation of more than 40 morphological mutations was carried out and these mutations fell into 12 Dpy genes and one Roller gene that represent morphological markers for all six *P. pacificus* chromosomes (Kenning *et al.*, 2004).

In order to prevent confusion in nomenclature with *C. elegans* mutants of related phenotypes but distinct molecular identity, *P. pacificus* mutants with a Dpy phenotype have been called *pdl* for *Pristionchus dumpy-like* (also see Appendix to this chapter) (Sommer, 2012; Sommer & McGaughran, 2013). As for all other mutants, renaming occurs only after the identification of the molecular lesions: if a mutant phenotype is caused by a mutation in a gene that is 1 : 1 orthologous to a corresponding gene in *C. elegans*, the *P. pacificus* mutant is renamed accordingly. The only exception to this rule is *P. pacificus unc-1*, which was named before these rules were adopted and which represents the Twitchin gene *Cel-unc-22* (Sommer *et al.*, 1996). Finally, to prevent confusion, *P. pacificus* genes are named using the '*Ppa-*' prefix (Sommer *et al.*, 1996; Pires-daSilva, 2013).

Initial success in cloning *P. pacificus* mutants was largely achieved using a 'candidate gene approach' based on the knowledge of *C. elegans* mutant phenotypes. For example, some mutations in known conserved genes, such as the homeodomain genes *lin-39*, *mab-5* and *vab-7*, were identified based on certain similarities to the corresponding *C. elegans* mutants (Eizinger & Sommer, 1997; Jungblut & Sommer, 1998, 2001). However, this approach was largely unsuccessful in cloning other mutants suggesting that the candidate gene approach falls off with phylogenetic distance and that many genes and their functions evolve at a faster rate (Eizinger *et al.*, 1999). Therefore, to alleviate this bottleneck a 'genomics' or 'map-based approach' was embarked upon.

Positional cloning approaches and integrated maps

In 2002, as a first step towards positional cloning, a genetic linkage map of *P. pacificus* was generated (Fig. 5.3) to allow mapping of mutants onto the various chromosomes (Srinivasan *et al.*, 2002). To achieve this goal, characterisation of natural isolates of *P. pacificus* was essential. Initial experiments identified a highly polymorphic strain PS1843 from Port Angeles (Washington State, USA) that differs substantially from the "wild type" strain PS312 (Schlak *et al.*, 1997). In addition, early experiments identified two more isolates and it was found that males from all these strains can mate productively with hermaphrodites of any of the other strains, although all strains are highly polymorphic (Srinivasan *et al.*, 2001). To date, hundreds of *P. pacificus* strains have been isolated and they show a much larger genetic differentiation in comparison to what is known from *C. elegans* wild isolates, indicating that both species have gone through different types of historical processes (Rodelsperger *et al.*, 2014).

For genetic mapping, different methodologies were applied at different times. Initially, amplified restriction fragment length polymorphism (AFLP) analysis of *P. pacificus* PS312 and PS1843 was carried out and the high rate of polymorphisms indicated that these strains were suitable for the construction of a genetic linkage map (Srinivasan *et al.*, 2001). For this, two BAC libraries were constructed with the PS312 strain and end sequenced. The sequence information from the BAC ends was used to find polymorphisms between the PS312 and the PS1843 strains using the single-stranded conformational polymorphism technique (SSCP) (Orita *et al.*, 1989). These molecular markers were assigned to six linkage groups of *P. pacificus* and contained a total of more than 600 loci (Fig. 5.3) (Srinivasan *et al.*, 2002).

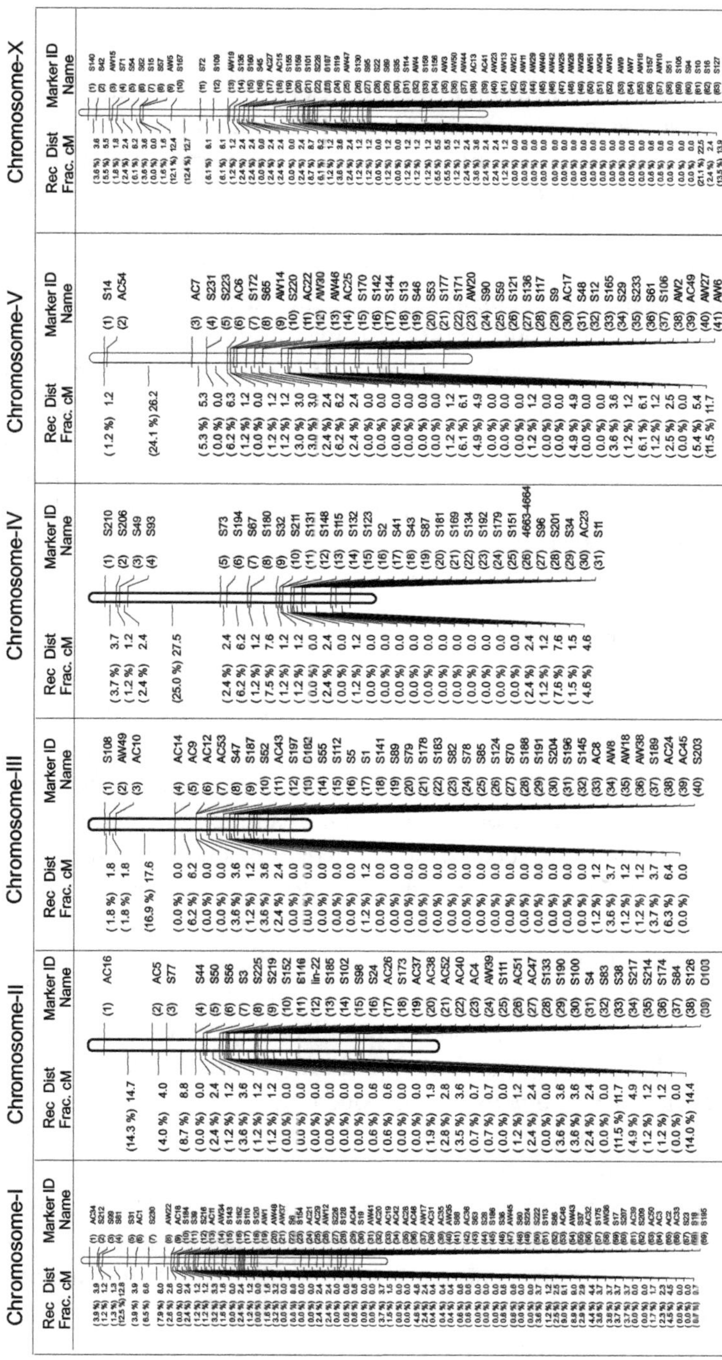

Fig. 5.3. *The genetic linkage map of* Pristionchus pacificus *generated using single stranded conformation polymorphism (SSCP markers) from BAC end sequences (Srinivasan* et al.*, 2002).*

While the genetic linkage map of *P. pacificus* confirmed the macrosynteny to *C. elegans* chromosomes, a surprising finding was the occurrence of multiple cross-overs in a single meiosis. Correspondingly, the *P. pacificus* chromosomes have a larger size when compared with *C. elegans*. Specifically, *P. pacificus* chromosomes vary between approximately 100 and 200 cM, whereas in *C. elegans* all chromosomes have a nearly fixed size of 50 cM due to the phenomenon of interference (Brenner, 1974). Together, the current genetic size of *P. pacificus* is above 1000 cM.

A physical map of *P. pacificus* was generated using AFLP-based fingerprints of the whole BAC library (Srinivasan *et al.*, 2003). Specifically, overlapping fingerprints of different clones were used to generate a physical map of *P. pacificus*. Since markers on the genetic map are derived from BAC end sequences, they were anchored to the physical map. The opposite is true wherein physical map clones can be anchored to the genetic map, resulting in an integrated genetic and physical map (Srinivasan *et al.*, 2003).

Post-genome era *P. pacificus*

All of the resources and technology described above underwent a major transformation with the publication of the draft genome of *P. pacificus* in 2008 (Fig. 5.1) (Dieterich *et al.*, 2008). The *P. pacificus* genome project resulted from a National Institute of Health (NIH) initiative in 2004, was executed at the NIH sequencing centre in St Louis, MO, USA, and provided a detailed draft of the genome of this nematode. The *P. pacificus* genome is approximately 170 Mb in size, thus also physically nearly double the size of the *C. elegans* genome. Gene models and predictions are still in flux with many estimates being in the range of 24 000 to 26 000 genes (Rodelsperger *et al.*, 2014). As for any other model organism, the draft genome marked a major landmark event for *P. pacificus* biology, opening new avenues in terms of resources and tools to increase understanding of the biology and ecology of *P. pacificus*. Although the genome sequence and its implications is a subject of discussion of another chapter (Diederich & Rödelsperger, Chapter 6, this volume), the genome sequence, in combination with the integrated genetic and physical maps, puts cloning of genetic mutations in *P. pacificus* in a better perspective. Genetic and genomic resources of *P. pacificus* and related *Pristionchus* species are available online at www.pristionchus.org. Gene models are also found at WORMBASE (www.wormbase.org) and other databases. Also, the availability of the genome sequence led to the development of novel technologies, as discussed below.

Reverse genetics

The availability of the genome sequence in *P. pacificus* allowed targeted gene-based screens to be undertaken. The first reverse genetic method of isolating mutants, 'deletion library screening', relied on the sequence information of the gene of interest combined with a mutagenesis of worms with TMP/UV. A library of mutated worms is then screened with primers spanning the gene of interest using polymerase chain reactions (Pires-daSilva, 2013). This approach has been successfully used in *P. pacificus* to identify different ligands and receptors of the Wnt signalling pathway (Tian *et al.*, 2008).

Transient methods of knocking down gene expression in *P. pacificus* were of limited success and have been applied only in a few studies. First, morpholino oligonucleotides were used that can block translation of specific mRNAs (Pires-daSilva & Sommer, 2004; Zheng *et al.*, 2005). Second, RNA interference did not, for reasons still unknown, result in clear phenotypes marking a stark difference to what is known in *C. elegans*. One exception was a study on the sex determination gene *tra-1*, the RNAi phenotpye of which is strong and robust to various experimental variations (Pires-daSilva & Sommer, 2004; Zheng *et al.*, 2005). To alleviate this surprising discrepancy compared with *C. elegans*, Cinkornpumin & Hong (2011) published a report showing the efficacy of RNA interference in *P. pacificus*. They used the dominant marker *Pristionchus* roller-like gene, *Ppa-prl-1*, as a case study and designed a double-stranded RNA (dsRNA) to target this gene. After injection of the dsRNA at different concentrations into the gonad of these mutant worms, the resulting progeny had more than 80% non-roller phenotypes, suggesting that the dsRNA was able to knockdown the function of this gene and thereby reverted the neomorphic phenotype.

More recently, studies have involved reverse genetic knockout approaches and genome editing using zinc-finger nucleases and the transcription activator-like effector nuclease (TALEN) technology (Lo *et al.*, 2013). The TALEN nucleases are engineered site-specific nucleases that induce DNA double-strand breaks at specifically designed genomic locations (Chapman *et al.*, 2012). In *P. pacificus*, the TALEN technology has first been used to generate *Ppa-unc-119* mutants that can be used as co-injection markers (Lo *et al.*, 2013). Second, Lo and co-workers mutagenised the *Ppa-unc-119* locus using TALEN and single-stranded oligonucleotides, which were designed to insert a *Hin*dIII restriction site. With this insertion strategy, an HA-tagged version of the UNC-119 protein has been generated, indicating the ability to perform genome editing in *P. pacificus* (Lo *et al.*, 2013).

The most recent novel method of gene disruption that has currently gained popularity is the CRISPR/Cas9 system. CRISPR stands for Clustered Regularly Interspaced Short Palindromic Repeats and Cas9 is an RNA-guided endonuclease system that targets specific genes using guide RNAs (Chiu *et al.*, 2013; Cho *et al.*, 2013; Friedland *et al.*, 2013). This method, essentially following the approach of Cho and co-workers, has been successfully applied to knock out the conserved Hox gene *mab-5* and the morphological marker *dpy-1* in *P. pacificus* (Witte *et al.*, 2015), indicating that this methodology can be applied with the same efficacy as in *C. elegans*.

Transgenics and gene function

DNA-mediated transformation is another essential tool in the study of gene function. Studying the expression of a gene of interest in its native cell or tissue, or ectopically expressing it in other tissues, can provide important functional insights. In *C. elegans*, transgenes can be obtained by microinjection of plasmid DNA into the gonad resulting in the injected worms' progeny carrying extrachromosomal arrays, which are concatamers of injected DNA molecules (Mello *et al.*, 1991). Unfortunately, initial attempts to establish the transformation technology using standard *C. elegans* protocols in *P. pacificus* did not result in the production of transgenic animals. To overcome this bottleneck, Schlager and co-workers (Schlager *et al.*, 2009) were able to generate dominant roller transgenics using complex arrays, which are propagated in *P. pacificus*. Complex arrays are generated by injecting linearised plasmids of a gene of interest at low concentrations mixed with higher-concentration of digested genomic DNA. The resulting extrachromosomal arrays are not silenced in F1 animals as they are less repetitive (Schlager *et al.*, 2009). One of the interesting findings was that transgenic lines are obtained only when a complex array was formed with genomic DNA digested with the same restriction enzyme as the plasmid DNA.

In addition, these studies also generated fluorescent transcriptional reporters for individual genes using *P. pacificus*-specific regulatory elements (Schlager *et al.*, 2009). Using these extrachromosomal arrays, characterisation of gene expression and function can be carried out in a spatial and temporal manner. Also, transgenes can be used to rescue gene mutations by providing a wild type copy of the gene in a mutant strain, thereby confirming the identification of the molecular lesion causing the mutant phenotype (Wang & Sommer, 2011; Ragsdale *et al.*, 2013). Finally, gene overexpression studies can be carried out to test the effects of too much gene function or to test whether the same gene has similar effects in other naturally occurring strains (Ragsdale *et al.*, 2013). In-depth protocols for generating transgenics and the different types of transformation markers in *P. pacificus* are described in the following references (Schlager *et al.*, 2009; Cinkornpumin & Hong, 2011; Wang & Sommer, 2011; Kienle & Sommer, 2013).

In-situ hybridisation and immunohistochemistry

In-situ hybridisation allows for analysis of gene expression in a spatiotemporal manner during juvenile development. Probes for *in-situ* in *P. pacificus* use labelled RNA, as they are more specific and give less background than DNA-labelled probes. Using these methods in *P. pacificus*, studies have demonstrated the functions of genes in signalling cascades (Tian *et al.*, 2008; Schlager *et al.*, 2009; Wang & Sommer, 2011). Although immunohistochemistry of *P. pacificus* has not been popular, one study has shown the efficacy of this approach in understanding protein localisation (Tian *et al.*, 2008).

Whole genome sequencing and mRNA quantification using next-generation sequencing (NGS) technologies

The advent of modern re-sequencing technologies has revolutionised the understanding of gene and genome function in *P. pacificus* (Kienle & Sommer, 2013; Ragsdale *et al.*, 2013; Rodelsperger *et al.*, 2014). Modern day gene cloning approaches use whole genome sequencing

(WGS) to identify molecular lesions in genes in *C. elegans* and other organisms (Doitsidou *et al.*, 2010; Hobert, 2010; Sarin *et al.*, 2010). Deep sequencing technology platforms, such as Illumina's Genome Analyzer or ABI's SOLiD, allow a faster and high throughput sequencing of the entire model system's genome, resulting in the detection of mutagen-induced sequence alterations compared to a non-mutagenised reference genome. Such a technology can alleviate most of the issues faced by traditional gene cloning approaches, such as mapping onto chromosomes and then fine mapping the gene of interest onto a chromosome. WGS is cost-effective, reliable and fast compared to traditional gene mapping and is gaining popularity in various model systems. In *P. pacificus*, the Illumina Genome Sequencer Technology has been very effective to clone genes affecting developmental plasticity and defecation processes obtained *via* traditional genetic screens (Rae *et al.*, 2012; Ragsdale *et al.*, 2013).

In addition to cloning genes, NGS technology can be utilised to quantify transcript levels. Ragsdale and co-workers employed this method to compare transcriptomes of wild isolates of *P. pacificus* and mutant lines and were also able to detect differences in transcript levels between males and hermaphrodites (Ragsdale *et al.*, 2013). Hence, next-generation technology has proven essential in the understanding of the molecular processes governing developmental processes in *P. pacificus*.

Natural variation and recombinant inbred line studies

Another application of this methodology is to understand microevolutionary variation that governs population level processes during genome evolution. Currently there are over 600 strains of *P. pacificus* (Morgan *et al.*, 2012; McGaughran *et al.*, 2013) and such a database forms a perfect platform for understanding the evolutionary forces that cause this diversity. In a new study, 104 natural isolates of *P. pacificus* together with the outgroup *P. exspectatus* were sequenced using next-generation sequencing technology (Rodelsperger *et al.*, 2014). Analysis of whole genome sequence data from this dataset revealed extensive population structure diversity in this species, suggesting that background selection plays a major role in generating genetic diversity (Rodelsperger *et al.*, 2014; see Rödelsperger & Dieterich, Chapter 6, this volume).

Relying on the available re-sequencing data, the genetic basis of phenotypic differences between natural isolates of *P. pacificus* can be studied by applying the well-established recombinant inbred line (RIL) approach. For this, worms from the isolates that differ in phenotype are crossed and F1 hybrids are clonally propagated by single worm bottlenecks for ten generations, resulting in RILs of unique genetic composition. Usually, genetic and phenotypic analysis of around 100-500 RILs can result in the identification of causative genes and quantitative trait loci (QTL). This approach has been successfully applied in *P. pacificus* (Zauner *et al.*, 2007; Hong *et al.*, 2008; Mayer & Sommer, pers. comm.). Most recently, this approach was also used to identify cryptic variation between strains (Kienle & Sommer, 2013). All of the applications of modern technology described above still rely on the foundation of traditional mapping tools to a certain extent. However, it is evident that in the years to come, data generated using this technology will surpass the necessity of traditional mapping. Thus, the development of the technological tools and its applications makes the era ripe for a greater understanding of the biology and ecology of *P. pacificus*. As will be seen in the following chapters of this book, this organism stands at the cusp of integrating diverse areas of research, such as evolutionary biology, population genetics and ecology, to aid our understanding of nematode evolution.

Acknowledgements

The authors wish to acknowledge Waltraud Roeseler, Hanh Witte and other members of the genetic mapping team. We also thank the anonymous reviewer for their constructive suggestions that helped develop the content of this book chapter. JS is supported by startup funds from the Worcester Polytechnic Institute.

References

BRENNER, S. (1974). The genetics of *Caenorhabditis elegans*. *Genetics* 77, 71-94.

CHAPMAN, J.R., TAYLOR, M.R. & BOULTON, S.J. (2012). Playing the end game: DNA double-strand break repair pathway choice. *Molecular Cell* 47, 497-510.

CHIU, H., SCHWARTZ, H.T., ANTOSHECHKIN, I. & STERNBERG, P.W. (2013). Transgene-free genome editing in *Caenorhabditis elegans* using CRISPR-Cas. *Genetics* 195, 1167-1171.

CHO, S.W., LEE, J., CARROLL, D., KIM, J.S. & LEE, J. (2013). Heritable gene knockout in *Caenorhabditis elegans* by direct injection of Cas9-sgRNA ribonucleoproteins. *Genetics* 195, 1177-1180.

CINKORNPUMIN, J.K. & HONG, R.L. (2011). RNAi mediated gene knockdown and transgenesis by microinjection in the necromenic nematode *Pristionchus pacificus*. *Journal of Visualized Experiments* 56, e3270.

DIETERICH, C., ROESELER, W. & SRINIVASAN, J. (2006). *Pristionchus pacificus* genomics: from genetics to genome sequence. In: The *C. elegans* Research Community (Ed.). *WormBook*. DOI:/10.1895/wormbook.1.107.1.

DIETERICH, C., ROESELER, W., SOBETZKO, P. & SOMMER, R.J. (2007). Pristionchus.org: a genome-centric database of the nematode satellite species *Pristionchus pacificus*. *Nucleic Acids Research* 35, D498-D502.

DIETERICH, C., CLIFTON, S.W., SCHUSTER, L.N., CHINWALLA, A., DELEHAUNTY, K., DINKELACKER, I., FULTON, L., FULTON, R., GODFREY, J., MINX, P. *ET AL.* (2008). The *Pristionchus pacificus* genome provides a unique perspective on nematode lifestyle and parasitism. *Nature Genetics* 40, 1193-1198.

DOBZHANSKY, T. (1973). Nothing in biology makes sense except in the light of evolution. *American Biology Teacher* 35, 125-129.

DOITSIDOU, M., POOLE, R.J., SARIN, S., BIGELOW, H. & HOBERT, O. (2010). *C. elegans* mutant identification with a one-step whole-genome-sequencing and SNP mapping strategy. *PLoS ONE* 5, e15435.

EIZINGER, A. & SOMMER, R.J. (1997). The homeotic gene *lin-39* and the evolution of nematode epidermal cell fates. *Science* 278, 452-455.

EIZINGER, A., JUNGBLUT, B. & SOMMER, R.J. (1999). Evolutionary change in the functional specificity of genes. *Trends in Genetics* 15, 197-202.

FRIEDLAND, A.E., TZUR, Y.B., ESVELT, K.M., COLAIACOVO, M.P., CHURCH, G.M. & CALARCO, J.A. (2013). Heritable genome editing in *C. elegans* via a CRISPR-Cas9 system. *Nature Methods* 10, 741-743.

FÜRST VON LIEVEN, F. (2005). The embryonic molt in diplogastrids (Nematoda) – homology of developmental stages and heterochrony as a prerequisite for morphological diversity. *Zoologischer Anzeiger* 244, 79-91.

GERHART, J. & KIRSCHNER, M. (1997). *Cells, embryos, and evolution: toward a cellular and developmental understanding of phenotypic variation and evolutionary adaptability*. Malden, MA, USA, Blackwell Science.

HOBERT, O. (2010). The impact of whole genome sequencing on model system genetics: get ready for the ride. *Genetics* 184, 317-319.

HONG, R.L. & SOMMER, R.J. (2006). Chemoattraction in *Pristionchus* nematodes and implications for insect recognition. *Current Biology* 16, 2359-2365.

HONG, R.L., WITTE, H. & SOMMER, R.J. (2008). Natural variation in *Pristionchus pacificus* insect pheromone attraction involves the protein kinase EGL-4. *Proceedings of the National Academy of Sciences of the United States of America* 105, 7779-7784.

JUNGBLUT, B. & SOMMER, R.J. (1998). The *Pristionchus pacificus mab-5* gene is involved in the regulation of ventral epidermal cell fates. *Current Biology* 8, 775-778.

JUNGBLUT, B. & SOMMER, R.J. (2001). The nematode even-skipped homolog *vab-7* regulates gonad and vulva position in *Pristionchus pacificus*. *Development* 128, 253-261.

KENNING, C., KIPPING, I. & SOMMER, R.J. (2004). Isolation of mutations with dumpy-like phenotypes and of collagen genes in the nematode *Pristionchus pacificus*. *Genesis* 40, 176-183.

KIENLE, S. & SOMMER, R.J. (2013). Cryptic variation in vulva development by cis-regulatory evolution of a HAIRY-binding site. *Nature Communications* 4, 1714.

LO, T.W., PICKLE, C.S., LIN, S., RALSTON, E.J., GURLING, M., SCHARTNER, C.M., BIAN, Q., DOUDNA, J.A. & MEYER, B.J. (2013). Precise and heritable genome editing in evolutionarily diverse nematodes using TALENs and CRISPR/Cas9 to engineer insertions and deletions. *Genetics* 195, 331-348.

MCGAUGHRAN, A., MORGAN, K. & SOMMER, R.J. (2013). Unraveling the evolutionary history of the nematode *Pristionchus pacificus*: from lineage diversification to island colonization. *Ecology and Evolution* 3, 667-675.

MELLO, C.C., KRAMER, J.M., STINCHCOMB, D. & AMBROS, V. (1991). Efficient gene transfer in *C. elegans*: extrachromosomal maintenance and integration of transforming sequences. *EMBO Journal* 10, 3959-3970.

MORGAN, K., MCGAUGHRAN, A., VILLATE, L., HERRMANN, M., WITTE, H., BAERELMES, G., ROCHAT, J. & SOMMER, R.J. (2012). Multi locus analysis of *Pristionchus pacificus* on La Réunion Island reveals an evolutionary history shaped by multiple introductions, constrained dispersal events and rare out-crossing. *Molecular Ecology* 21, 250-266.

OGAWA, A., BENTO, G., BARTELMES, G., DIETERICH, C. & SOMMER, R.J. (2011). *Pristionchus pacificus daf-16* is essential for dauer formation but dispensable for mouth form dimorphism. *Development* 138, 1281-1284.

ORITA, M., SUZUKI, Y., SEKIYA, T. & HAYASHI, K. (1989). Rapid and sensitive detection of point mutations and DNA polymorphisms using the polymerase chain reaction. *Genomics* 5, 874-879.

PHOTOS, A., GUTIERREZ, A. & SOMMER, R.J. (2006). *sem-4/spalt* and *egl-17/* FGF have a conserved role in sex myoblast specification and migration in *P. pacificus* and *C. elegans*. *Developmental Biology* 293, 142-153.

PIRES-DASILVA, A. (2013). *Pristionchus pacificus* protocols. The *C. elegans* Research Community (Ed.). *WormBook*. DOI:10.1895/wormbook.1.114.2.

PIRES-DASILVA, A. & SOMMER, R.J. (2003). The evolution of signalling pathways in animal development. *Nature Reviews Genetics* 4, 39-49.

PIRES-DASILVA, A. & SOMMER, R.J. (2004). Conservation of the global sex determination gene *tra-1* in distantly related nematodes. *Genes & Development* 18, 1198-1208.

RAE, R., SCHLAGER, B. & SOMMER, R.J. (2008). *Pristionchus pacificus*: a genetic model system for the study of evolutionary developmental biology and the evolution of complex life-history traits. *Cold Spring Harbor Protocols* 2008. DOI:10.1101/pdb.emo102.

RAE, R., WITTE, H., RODELSPERGER, C. & SOMMER, R.J. (2012). The importance of being regular: *Caenorhabditis elegans* and *Pristionchus pacificus* defecation mutants are hypersusceptible to bacterial pathogens. *International Journal for Parasitology* 42, 747-753.

RAFF, R.A. (2000). Evo-devo: the evolution of a new discipline. *Nature Reviews Genetics* 1, 74-79.

RAGSDALE, E.J., MULLER, M.R., RODELSPERGER, C. & SOMMER, R.J. (2013). A developmental switch coupled to the evolution of plasticity acts through a sulfatase. *Cell* 155, 922-933.

RODELSPERGER, C., NEHER, R.A., WELLER, A.M., EBERHARDT, G., WITTE, H., MAYER, W.C., DIETERICH, C. & SOMMER, R.J. (2014). Characterization of genetic diversity in the nematode *Pristionchus pacificus* from population-scale resequencing data. *Genetics* 196, 1153-1165.

RUDEL, D., TIAN, H. & SOMMER, R.J. (2008). Wnt signaling in *Pristionchus pacificus* gonadal arm extension and the evolution of organ shape. *Proceedings of the National Academy of Sciences of the United States of America* 105, 10826-10831.

SARIN, S., BERTRAND, V., BIGELOW, H., BOYANOV, A., DOITSIDOU, M., POOLE, R.J., NARULA, S. & HOBERT, O. (2010). Analysis of multiple ethyl methanesulfonate-mutagenized *Caenorhabditis elegans* strains by whole-genome sequencing. *Genetics* 185, 417-430.

SCHLAGER, B., WANG, X., BRAACH, G. & SOMMER, R.J. (2009). Molecular cloning of a dominant roller mutant and establishment of DNA-mediated transformation in the nematode *Pristionchus pacificus*. *Genesis* 47, 300-304.

SCHLAK, I., EIZINGER, A. & SOMMER, R.J. (1997). High rate of restriction fragment length polymorphisms between two populations of the nematode *Pristionchus pacificus* (Diplogastridae). *Journal of Zoological Systematics and Evolutionary Research* 35, 137-142.

SIMPSON, P. (2002). Evolution of development in closely related species of flies and worms. *Nature Reviews Genetics* 3, 907.

SOMMER, R.J. (1997). Evolution and development – the nematode vulva as a case study. *Bioessays* 19, 225-231.

SOMMER, R.J. (2000). Comparative genetics: a third model nematode species. *Current Biology* 10, R879-R881.

SOMMER, R.J. (2001). As good as they get: cells in nematode vulva development and evolution. *Current Opinion in Cell Biology* 13, 715-720.

SOMMER, R.J. (2006). *Pristionchus pacificus*. In: The *C. elegans* Research Community (Ed.). *WormBook*. DOI:10.1895/wormbook.1.102.1.

SOMMER, R.J. (2012). Evolution of regulatory networks: nematode vulva induction as an example of developmental systems drift. *Advances in Experimental Medicine and Biology* 751, 79-91.

SOMMER, R.J. & MCGAUGHRAN, A. (2013). The nematode *Pristionchus pacificus* as a model system for integrative studies in evolutionary biology. *Molecular Ecology* 22, 2380-2393.

SOMMER, R.J. & STERNBERG, P.W. (1996). Apoptosis and change of competence limit the size of the vulva equivalence group in *Pristionchus pacificus*: a genetic analysis. *Current Biology* 6, 52-59.

SOMMER, R.J., CARTA L.K., KIM S.-Y. & STERNBERG, P.W. (1996). Morphological, genetic and molecular description of *Pristionchus pacificus* sp. n. (Nematoda: Neodiplogastridae). *Fundamental and Applied Nematology* 19, 511-521.

SRINIVASAN, J., PIRES-DASILVA, A., GUTIERREZ, A., ZHENG, M., JUNGBLUT, B., WITTE, H., SCHLAK, I. & SOMMER, R.J. (2001). Microevolutionary analysis of the nematode genus *Pristionchus* suggests a recent evolution of redundant developmental mechanisms during vulva formation. *Evolution & Development* 3, 229-240.

SRINIVASAN, J., SINZ, W., LANZ, C., BRAND, A., NANDAKUMAR, R., RADDATZ, G., WITTE, H., KELLER, H., KIPPING, I., PIRES-DASILVA, A. *ET AL.* (2002). A bacterial artificial chromosome-based genetic linkage map of the nematode *Pristionchus pacificus*. *Genetics* 162, 129-134.

SRINIVASAN, J., SINZ, W., JESSE, T., WIGGERS-PEREBOLTE, L., JANSEN, K., BUNTJER, J., VAN DER MEULEN, M. & SOMMER, R.J. (2003). An integrated physical and genetic map of the nematode *Pristionchus pacificus*. *Molecular Genetics and Genomics* 269, 715-722.

STERN, D.L. (2000). Evolutionary developmental biology and the problem of variation. *Evolution* 54, 1079-1091.

TIAN, H., SCHLAGER, B., XIAO, H. & SOMMER, R.J. (2008). Wnt signaling induces vulva development in the nematode *Pristionchus pacificus*. *Current Biology* 18, 142-146.

WANG, X. & SOMMER, R.J. (2011). Antagonism of LIN-17/Frizzled and LIN-18/Ryk in nematode vulva induction reveals evolutionary alterations in core developmental pathways. *PLoS Biology* 9, e1001110.

WILKINS, A.S. (2002). *The evolution of developmental pathways*. Sunderland, MA, USA, Sinauer Associates.

WITTE, H., MORENO, E., RÖDELSPERGER, C., KIM, J., KIM, J.-S., STREIT, A. & SOMMER, R.J. (2015). Gene inactivation using the

CRISPR/Cas9 system in the nematode *Pristionchus pacificus*. *Development, Genes and Evolution* 225, 55-62.

WOLPERT, L. (2002). *Principles of development*. Oxford, UK, Oxford University Press.

ZAUNER, H., MAYER, W.E., HERRMANN, M., WELLER, A., ERWIG, M. & SOMMER, R.J. (2007). Distinct patterns of genetic variation in *Pristionchus pacificus* and *Caenorhabditis elegans*, two partially selfing nematodes with cosmopolitan distribution. *Molecular Ecology* 16, 1267-1280.

ZHENG, M., MESSERSCHMIDT, D., JUNGBLUT, B. & SOMMER, R.J. (2005). Conservation and diversification of Wnt signaling function during the evolution of nematode vulva development. *Nature Genetics* 37, 300-304.

Appendix

Genomic resources for cloning genes of *P. pacificus*

Most of the resources mentioned in this chapter are available on the web portal www.pristionchus.org (Dieterich *et al.*, 2007).

Gene nomenclature in *P. pacificus*

Gene names in *P. pacificus* are classified by a species prefix '*Ppa*' and follow the rules that have been established for *C. elegans*. This system ensures a unique classification of genes isolated in *P. pacificus*. Nomenclature of genes cloned based on molecular similarity to *C. elegans* genes is carried by adding the prefix to the known *C. elegans* gene (*e.g.*, *Ppa-mab-5*). Genes identified from classical screens are named using a standard three-letter code; for instance *Ppa-prl* refers to a *Pristionchus r*oller-*l*ike gene (Sommer, 2006).

Genetic maintenance of *P. pacificus*

Pristionchus pacificus maintenance and culture methods, together with mutagenesis of *P. pacificus* using either EMS or Psoralen, creation of deletion libraries for generating gene knockouts and reverse genetic approaches, such as RNA interference and morpholinos, have been described in detail at http://wormbook.org/chapters/www_ppageneticprotocols.2/ppageneticprotocols.html.

Mapping of mutations obtained from screens onto the six linkage groups in *P. pacificus*, is primarily done using molecular markers on a meiotic mapping panel based on batch segregant analysis (Srinivasan *et al.*, 2002) or one can use morphological markers, such as *Pristionchus* dumpies or *pdl* for *Pristionchus dumpy-like* markers, to map mutants onto a particular chromosome (Kenning *et al.*, 2004). Details of mapping mutations on to chromosomes can be found at http://diamond.tuebingen.mpg.de/mediawiki/index.php/Mapping and http://diamond.tuebingen.mpg.de/mediawiki/index.php/Mapping_with_SSLP_markers.

Libraries for genomic cloning

Three different library-types have been created to gain a better overall understanding of the genome of *P. pacificus*. An expressed sequenced tag (EST) library contains over 14 000 sequences grouped into approximately 2600 clusters. This library was generated from the *P. pacificus* isolate from California (PS312) and is stage-specific (Srinivasan *et al.*, 2002). Two different bacterial artificial chromosome (BAC) libraries exist. One library was created through genomic DNA partial digestions with *Hin*dIII while the other was created using *Eco*RI. The average size of the BAC clones in the *Hin*dIII library, 126 kb, is larger than the average of the *Eco*RI library, 70 kb. For each of these libraries, the ends of both the 5′ and 3′ sequences are sequenced. These ends are approximately 700 bp each. There are over 31 000 BAC end sequences now know from both libraries (Dieterich *et al.*, 2006). The third type of library that exists for *P. pacificus* is a fosmid library. This fosmid library is of the California isolate and contains 50 000 clones. Half of those clones have been end sequenced on both ends. These fosmid ends were on average 500 bp long. Slightly over 48 500 fosmid end sequences are now available (Dieterich *et al.*, 2008). Clones from these libraries can be requested from the Sommer Laboratory at the Max Planck Institute for Developmental Biology in Tuebingen, Germany (http://www.eb.tuebingen.mpg.de/research/departments/evolutionary-biology.html).

Sequence information of libraries and genome sequence

All the sequence information including the integrated genetic and physical maps of *P. pacificus* can be accessed online at the web portal www.pristionchus.org (Dieterich *et al.*, 2007). The genome of wild type *P. pacificus* California strain (PS312) has been performed using whole genome shotgun technique with 9× coverage along with a 1× coverage of the reference strain from Washington (PS1843) (Dieterich *et al.*, 2008). All of these sequence data can be accessed at www.pristionchus.org. Gene annotations from *P. pacificus* are currently available at Wormbase (www.wormbase.org), a centralised portal for all nematode genomes.

Nematology Monographs & Perspectives, 2015, Vol. 11, 141-165

Chapter 6

Comparative and functional genomics

Christian RÖDELSPERGER [1] and Christoph DIETERICH [2]

[1] *Department for Evolutionary Biology, Max-Planck Institute for Developmental Biology Tübingen, 72076 Tübingen, Germany*
christian.roedelsperger@tuebingen.mpg.de
[2] *Max-Planck Institute for the Biology of Ageing, Joseph-Stelzmann Straße 9b, D-50931 Köln/Cologne, Germany*
christoph.dieterich@age.mpg.de

Introduction

> *There's millions and millions of unsolved problems. Biology is so digital, and incredibly complicated, but incredibly useful. The trouble with biology is that, if you have to work as a biologist, it's boring. Your experiments take you three years and then, one night, the electricity goes off and all the things die! You start over. In computers we can create our own worlds.*
>
> D. E. Knuth, interviewed by Computer Literacy Bookshops, 1993.

The genome is defined as the entirety of an organism's hereditary information. With the exception of epigenetic effects such as heritable modifications at histone tails, which define the so-called epigenome, the hereditary information is encoded in the DNA sequence of unicellular and multicellular organisms. The genome of the nematode *Caenorhabditis elegans* was the first sequenced genome of a multicellular organism (The *C. elegans* Sequencing Consortium, 1998). During the last decade, it was followed by more than a dozen nematode genomes including *Pristionchus pacificus* (Stein *et al.*, 2003; Dieterich *et al.*, 2008; Opperman *et al.*, 2008; Kikuchi *et al.*, 2011; Mitreva *et al.*, 2011; Desjardins *et al.*, 2013). Since the publication of the first *P. pacificus* draft assembly, its genome sequence has proven to be an invaluable resource for a number of studies of molecular and genome evolution and population genetics, and also for developmental biology. In particular, the availability of the reference genome sequence in combination with high-throughput sequencing had a tremendous impact on the identification of associations between genotype and phenotype based on forward genetic screens, *i.e.*, the identification of mutations by whole genome sequencing for subsequent cloning of candidate genes (Ogawa *et al.*, 2011; Rae *et al.*, 2012; Ragsdale *et al.*, 2013). While it is impossible to decode the complete biological information of every single nucleotide out of the hundreds of millions by linking genotypes to the phenotypic level, it is possible to analyse the complete genome by associating other types of experimental data at the genomic level or simply by comparison with the sequences from other organisms. Such global genome analyses are widely known as the research fields of comparative and functional genomics. 'Comparative genomics' studies the evolution of gene families, transposons and other genomic features across species borders and may generate hypotheses about the role of certain gene classes in the diversification of distinct lineages and their adaptations to specific environments. By contrast, the field of 'functional genomics' employs genomic, transcriptomic and proteomic tools to improve the annotation of coding and non-coding functional elements within a single genome, such as protein-coding genes, small RNAs, long non-coding RNAs and regulatory elements. Moreover, functional genomic approaches may help to gather further insights into temporal and spatial profiles of gene activities. This is done by analysing samples obtained from different developmental stages and tissues or by analysing samples from worms that have undergone different experimental treatments. The combination of comparative and functional genomic approaches facilitates addressing questions such as whether differences in lifestyles of *C. elegans* and *P. pacificus* are reflected by divergence in genes that are associated with specific developmental stages or with responses to environmental stimuli (Sinha *et al.*, 2012a, b).

With *C. elegans* being considered as a free-living nematode and *P. pacificus* living in necromenic association with scarab beetles, recent investigations have revealed largely divergent expression patterns in their development as well as in response to environmental stimuli (Sinha *et al.*, 2012a, b). The comparison of *P. pacificus* to *C. elegans* and other nematodes has revealed a number of divergent patterns, but these comparisons reflect rather large evolutionary distances spanning hundreds of million years. Given the difficulties in estimating divergence times between nematodes because of the lack of informative fossil records and the uncertainty in generation times that can vary from a matter of days to up to a year (Mayer & Sommer, 2011), the most

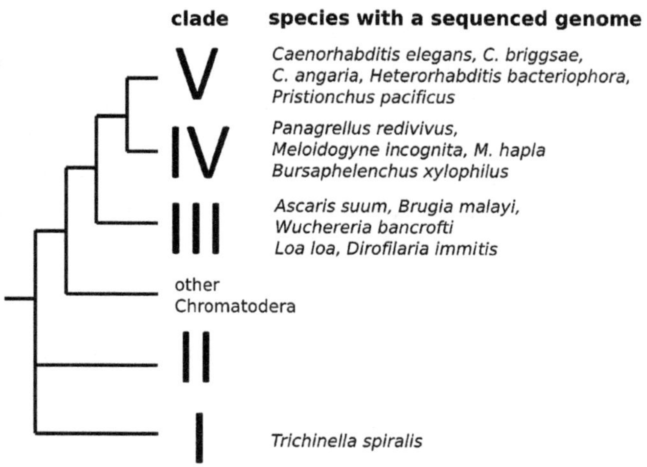

Fig. 6.1. *Schematic phylogeny of nematode clades according to Blaxter* et al. *(1998). Nematode species with a sequenced genome are indicated on the right.*

recent estimations suggest that *P. pacificus* and *C. elegans* diverged from their last common ancestor between 280 and 430 MYA (Dieterich *et al.*, 2008). However, these dates are based on previous estimates of the divergence times between *C. elegans* and the two nematodes *C. briggsae* and *Brugia malayi* (Fig. 6.1), which were computed under the assumption that the split between nematodes and arthropods occurred 800-1000 MYA. By contrast, more recent studies have suggested that the common ancestor of all bilaterian animals lived 643-733 MYA (Peterson *et al.*, 2008). Despite these substantial uncertainties in divergence times, it is obvious that genome-wide comparisons at evolutionary distances such as represented by comparisons of the currently available nematode genomes (Fig. 6.1) need to be complemented by further studies at much smaller timescales for gaining a more detailed view of nematode genome evolution. In summary, the availability of additional nematode genome sequences in combination with supplementary genome, transcriptome and proteome analyses form powerful means to extend our knowledge and ideas about the evolution and biology of *P. pacificus*.

Genome sequence

The genome of *P. pacificus* was sequenced by the Genome Sequencing Center (GSC), Washington University, St Louis, MO, USA, and the first draft assembly of the *P. pacificus* genome was published in 2008 (Dieterich *et al.*, 2008). The initial genome-sequencing project involved the high coverage whole genome shotgun sequencing of an isolate from California (PS312) to assemble a first reference draft genome. For the *de novo* assembly of the reference genome sequence, multiple libraries of variable insert sizes were prepared in order to resolve highly repetitive regions. Highly repetitive regions typically cause a high degree of fragmentation within the resulting *de novo* assembly, as they cannot be assembled unambiguously. All genomic libraries were sequenced mostly from both ends using the Sanger technology. The first *de novo* assembly of the reference sequence was later improved using Roche 454 platform sequencing data (Borchert *et al.*, 2010). This assembly spans 153 megabases (Mb) of assembled sequences and 173 Mb including gaps that were estimated from the average insert size of the plasmid, fosmid and bacterial artificial chromosome (BAC) libraries. However, recent re-sequencing projects have indicated that assembly gap sizes are mostly overestimated and that the genome assembly is >99% complete. Overall, the GC content of the assembly is 42% as compared to 38% in *C. elegans*. In protein-coding sequences, average GC content is even around 50%. The genome assembly is split into around 18 000 fragments that are usually referred to as supercontigs or scaffolds. The distribution of supercontig sizes is commonly used as a measure for the contiguity of a genome assembly. The N50 value of the *P. pacificus* assembly, which represents the minimal supercontig size, in the set of largest supercontigs that together represent 50% of the total assembly, is 1.2 Mb. Thus, the *P. pacificus* assembly exhibits a higher level of contiguity than all currently published nematode genomes that are solely based on high-throughput sequencing technologies (Rödelsperger *et al.*, 2013).

In addition to the sequencing of the *P. pacificus* reference strain PS312, one *P. pacificus* strain from Washington (PS1843) was sequenced at much lower coverage in order to detect polymorphic markers that could be used for genetic mapping. The PS1843 strain from Washington was already known to be polymorphic to the California strain and was used for the generation of the genetic linkage map (Srinivasan *et al.*, 2002). In addition to the sequencing of a second *P. pacificus* strain, two other species of the genus *Pristionchus*, *P. maupasi* and *P. entomophagus*, were sequenced at low coverage for comparative genomic analyses.

Protein-coding genes and operons

The *P. pacificus* genome assembly was annotated using multiple gene prediction algorithms that were trained on available expressed sequence tag (EST) data (Dieterich, 2008). The best-performing prediction algorithm, SNAP, which achieved 95% specificity and 77% sensitivity at nucleotide level, was later trained on larger transcriptome data sets to further improve the gene annotation (Borchert *et al.*, 2010; Sinha *et al.*, 2012a). The number of predicted protein-coding genes ranged between 24 000 and 30 000. While the majority of gene predictions is supported by transcriptome evidence, analysis of the proteome and phosphoproteome by mass spectrometry could detect evidence for translation of roughly 4000 genes of which around 2500 could even be detected as being phosphorylated (Borchert *et al.*, 2010, 2012). In total, all protein-coding exons cover between 18 and 21% of the assembled genome sequence.

As with *C. elegans*, a fraction of the *P. pacificus* genes is organised in polycistronic transcriptional units that are post-transcriptionally trans-spliced to a short RNA fragment (spliced leader) resulting in multiple mRNAs. These transcriptional units have been called operons (Blumenthal & Gleason, 2003); however, nematode operons are neither evolutionary nor functionally related to bacterial operons. The main difference at the molecular level is that in nematodes the individual coding units are split into separate mRNAs (by trans-splicing and poly-adenylation). By contrast, in bacteria the different coding units of bacterial operons are translated from the polycistronic mRNA. Furthermore, operon genes in nematodes are, in general, not functionally related and more frequently contain internal promoters. The process of trans-splicing and the associated presence of operons have been found in a number of diverse nematode species, including *P. pacificus* (Lee & Sommer, 2003). A more recent genome-wide analysis of trans-splicing in *P. pacificus* provided experimental validation for 2219 operons by RNA-seq data (Sinha *et al.*, 2014). While only 128 of the 1288 operons of *C. elegans* are conserved in *P. pacificus*, the same study revealed an enrichment of germline-expressed genes and dauer exit genes in both species. More generally, however, it should be noted that it is still an open question why operons evolved in nematodes in the first place. One recent study, supported by analysis of several gene expression data sets, proposed that operons arose as an evolutionary innovation in order to optimise the usage of limited transcriptional resources during the recovery from growth-arrested stages (Zaslaver *et al.*, 2011).

Repetitive elements and transposons

So far, all nematodes, including *P. pacificus*, exhibit a strongly reduced genome size and repeat content when compared to mammalian genomes (Rödelsperger *et al.*, 2013). Whether or not nematodes are more efficient than mammals in suppressing the activity of transposable elements is currently not known. In comparison to the human genome, where transposons are one of the major sources of repetitive sequences that make up almost half of the genome, different repeat detection methods identify only up to 17% of the *P. pacificus* genome assembly as repetitive or being of transposon origin (Dieterich *et al.*, 2008). Transposons are generally subdivided into two major classes (Wicker *et al.*, 2007): *i*) DNA transposons, for which transposition events resemble a 'cut and paste' mechanism; and *ii*) *RNA* transposons, for which transposition events function *via* an RNA intermediate, resulting in a 'copy and paste'-like mechanism. Both major classes have been identified in the *P. pacificus* genome, but so far no evidence for transposon activity in current *P. pacificus* populations has been reported. Interestingly, RNA transposons of the Rte-1 family have been identified as being horizontally transferred between insects and *P. pacificus* (Rödelsperger & Sommer, 2011). Given that *P. pacificus* Rte-1 family members show the highest similarities to sequences from multiple insect species, this suggests that the direction of the horizontal transfer was from insects to the nematode. This is plausible in the context of the close association between *P. pacificus* nematodes and scarab beetles. In addition, horizontal gene transfer of Rte-1-like transposons has been reported previously between plants and fish and between arthropods and reptiles (Zupunski *et al.*, 2001).

Role and evolution of miRNA families

Post-transcriptional regulation has emerged as a key factor in controlling eukaryotic gene expression by affecting virtually every aspect of RNA metabolism. Non-coding RNAs were characterised as potent regulators of RNA availability and abundance decades ago (Dieterich &

Stadler, 2013). In particular, miRNAs were described as an entirely new class of regulators for the first time in *C. elegans* (Lee *et al.*, 1993). miRNA precursors originate from stem-loop containing primary transcripts (pri-miRNAs) and are processed into short double-stranded hairpin RNAs (pre-miRNAs) by Drosha, an RNAse III enzyme. The pre-miRNA is then cleaved by Dicer, another RNAse III enzyme, into an RNA duplex. Either the 5′ or 3′ arm of this duplex is loaded into an effector Protein-RNA complex. This complex binds primarily to the 3′ untranslated region (3′UTR) of partially complementary target mRNAs, causing translational repression and/or mRNA destabilisation (Krol *et al.*, 2010). Whilst miRNAs are usually only partially complementary to their mRNA targets, they exhibit a 6-8 nucleotide core motif, the seed region, that is mostly exactly complementary to the target mRNA.

Generally, miRNAs are paradoxical regulators. On the one hand, they are extremely conserved throughout the animal kingdom and control key steps in development; on the other hand, many miRNAs seem to be dispensable as they have little effect on development. From studies in *C. elegans*, it is known that miRNAs influence nematode lifespan (de Lencastre *et al.*, 2010) as well as J1 and dauer diapause decisions (Zhang *et al.*, 2011), which are important traits in nematode evolution. At the time of writing, no miRNA null allele has been described in *P. pacificus*.

In a recent study, Ahmed *et al.* (2013) carried out a comparative approach to advance the understanding of the role of miRNAs in nematode evolution. In this study, the small RNA complement of three nematodes species with different life styles was characterised: the free-living *C. elegans*, the necromenic *P. pacificus*, and the true parasite, *Strongyloides ratti*. It was hypothesised that the sequence of key miRNA regulators of homologous developmental transitions would be conserved across these species and that some candidates may even show 'conserved' expression patterns in a comparison of dauer/infective larval stages to mixed non-dauer stages. Ahmed *et al.* (2013) tripled the known miRNA gene set for *P. pacificus* to 362 miRNAs, and for the first time described the miRNA gene set in a *Strongyloides* parasite (106 genes). Surprisingly, only a limited set of 24 conserved miRNA families across these three species could be identified. By integrating expression data into phylogenetic analysis, conserved post-transcriptional regulators with similar expression signatures in dauer *vs* non-dauer fates could be detected. In a more detailed analysis of *mir-34* and *mir-71*, which are both important regulators of stress response and aging in *C. elegans* (Boulias & Horvitz, 2012; Liu *et al.*, 2012), it could be shown, on the one hand, that the *mir-71* family is a well-conserved post-transcriptional regulator with coherent expression across all three species and, on the other hand, that the *mir-34* family may represent a case of convergent gene evolution in *P. pacificus*. Herein, unrelated miRNA precursors with identical or almost identical (off by one substitution) seed sequences show similar expression patterns in the dauer fate as the reference family.

Evidently, an understanding of the role of post-transcriptional networks in evolution, ecology and development is just at the beginning. So far, most datasets and functional studies originate from *C. elegans* work but similar studies need to be extended to satellite species (*e.g.*, *P. pacificus*) as comparative approaches could inform about key characteristics of post-transcriptional network evolution in nematodes.

Evolution of gene families

Comparing *C. elegans* and *P. pacificus*, only around 7000 genes can be identified as direct 1 : 1 orthologues (Sinha *et al.*, 2012a). For all other genes, a direct 1 : 1 correspondence cannot be identified easily due to losses and duplication events since the divergence from their last common ancestor. Thus, in order to extend cross-species comparisons beyond the level of orthologous genes for which a 1 : 1 correspondence exists, usually the sizes of gene families are compared (Fig. 6.2). Such comparisons have been widely used to generate hypotheses about the importance of certain gene families in the adaptation to distinct ecological niches. However, it has to be noted that, even when the numbers of genes within a family is more or less the same, nevertheless, numerous gene loss and duplication events may have occurred in both lineages but have finally resulted in similar overall numbers, yet such cases of gene losses and duplications can only be detected by detailed phylogenetic analysis. One such example is the detailed comparison of detoxification enzymes that reveal pervasive gene losses and duplications and the near absence of true 1 : 1 orthologues between *P. pacificus* and *C. elegans* (Markov *et al.*, 2015).

When comparing the size of gene families, which are usually defined by the presence of a certain protein domain between *C. elegans* and *P. pacificus*, one interesting observation has been that the most

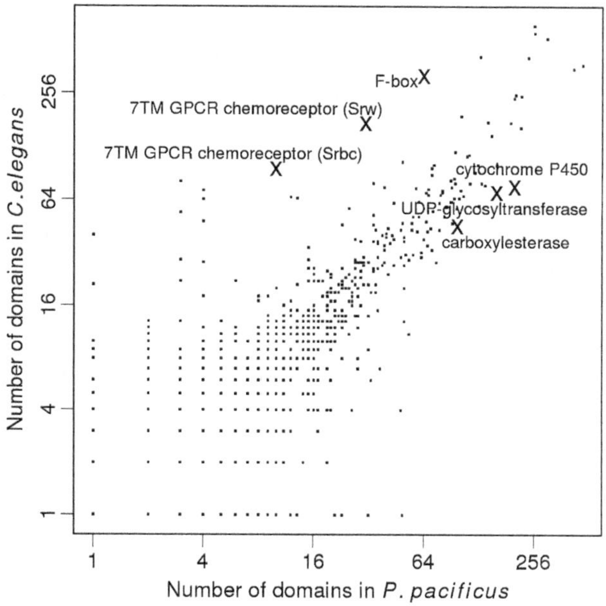

Fig. 6.2. *Numbers of predicted protein domains in* Pristionchus pacificus *and* Caenorhabditis elegans. *Most protein domains have similar counts in both species, suggesting limited gene duplications and losses. The most extreme outliers (only one occurrence in one species but several occurrences in the other) may be a result of misannotation and have to be regarded with caution. However, a number of protein domain counts show highly robust changes. GPCR chemoreceptors and F-box proteins are strongly depleted in* P. pacificus *relative to* C. elegans. *By contrast,* P. pacificus *shows an enrichment of three protein domains (cytochrome P450, UDP-glycosyltransferases and carboxylesterases) that are associated with detoxification of xenobiotics (Dieterich* et al., *2008).*

pronounced gene family expansions found in the genome of *P. pacificus* concern three families (Fig. 6.2) that are known to have a key role in the detoxification of xenobiotics (Dieterich *et al.*, 2008). This could potentially suggest that these gene families have been expanded in *Pristionchus* nematodes during adaptation to the diverse ecosystems that are associated with beetle environment. Consistently, *P. pacificus* shows strongly increased survival on most pathogens as compared to *C. elegans* (Sinha *et al.*, 2012b). In contrast to the expansion in xenobiotic metabolism-related gene families as a potential result of constant exposure to a diverse range of pathogens, other quantitative changes in gene families have less obvious explanations. For example,

P. pacificus shows a strong expansion of ribosomal proteins with respect to *C. elegans*, a finding that is difficult to interpret (Dieterich *et al.*, 2008). One potential explanation might be that the higher number of genes in *P. pacificus* necessitates larger quantities of the corresponding translational machineries, which could be achieved by gene duplication. More recently, it has been reported that *P. pacificus* and also other nematodes have lost substantial parts of the purine synthesis pathway (Desjardins *et al.*, 2013). This lack of ability to synthesise purine nucleotides *de novo* may be compensated by the ability to generate all required nucleotides by means of interconversion from pyrimidines or externally acquired purines. It is currently unknown why these changes in very elementary processes like protein translation or the synthesis of nucleotides have occurred during nematode evolution and, hopefully, future studies will bring more insights into these questions.

Orphan genes

In contrast to 1 : 1 orthologues and genes that are members of known protein families, there exists a class of genes that completely impedes comparative analysis based on traditional homology detection methods. Approximately one third of all predicted *P. pacificus* genes have no recognisable homologues in other nematode genomes (Borchert *et al.*, 2010). In some cases, homologues may exist outside the nematode phylum, which may be an indication of horizontal gene transfer (see below). In other nematode sequencing projects, these genes, which are often referred to as orphan or pioneer genes, constitute up to 48% of all predicted protein-coding sequences (Fig. 6.3). Although the fraction of orphan genes depends on the phylogenetic sampling of the analysed genomes, and may also partially be explained by artifactual gene predictions, analysis of ESTs from more than two dozen nematode species shows a robust signal of expressed sequences with no similarity to other nematode genomes and without any signal of saturation with increasing number of sampled nematode transcriptomes (Fig. 6.4). This observation suggests that the sequencing of nematode genomes and transcriptomes is far from being representative for the whole phylum. While traditional homology detection methods based on protein or nucleotide similarities fail to reveal any relationship with known protein families across species, intra-species comparisons have shown that

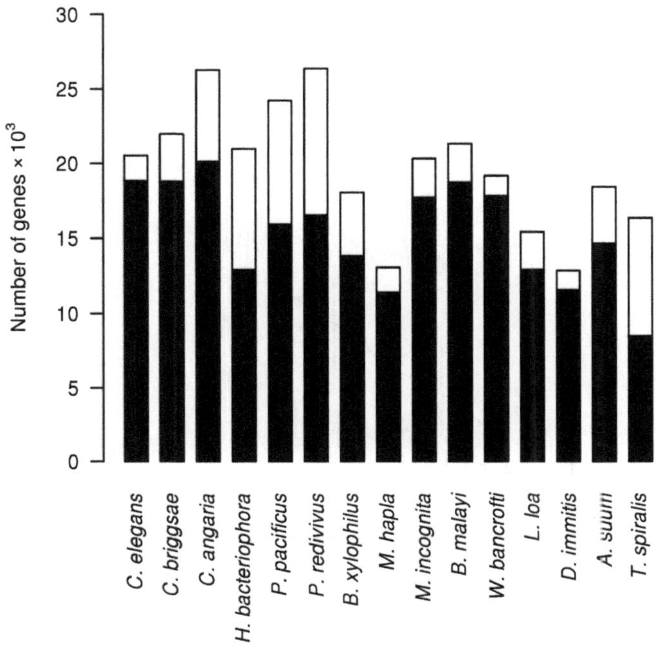

Fig. 6.3. *Fraction of orphan genes in 15 published nematode genomes. Based on all pairwise homology searches using the software BLASTP, all genes of a given species were tested to see whether homologues could be found in at least one other nematode genome (black) or not (white). As the estimated fraction of orphan genes largely depends on the phylogenetic resolution of the sampling, the fraction of orphan genes is markedly reduced within the* Caenorhabditis *genus. Nevertheless, species-specific, or at least genus-specific, genes constitute substantial fractions of nematode genomes. Genera abbreviations (left to right):* C = Caenorhabditis, H = Heterorhabditis, P = Pristionchus, P = Panagrellus, B = Bursaphelenchus, M = Meloidogyne, B = Brugia, W = Wuchereria, L = Loa, D = Dirofilaria, A = Ascaris, T = Trichinella.

orphan genes may be part of larger gene families of which individual members do have homologues in other nematode species (Dieterich *et al.*, 2008; Borchert *et al.*, 2010). This observation suggests that a fraction of orphan genes belongs to rapidly evolving gene families of which individual members might have diverged so much that no homologous sequences can be identified across species. It has been proposed that duplications allow for the development of novel protein functions in one of the two copies while the original function is still retained by the other

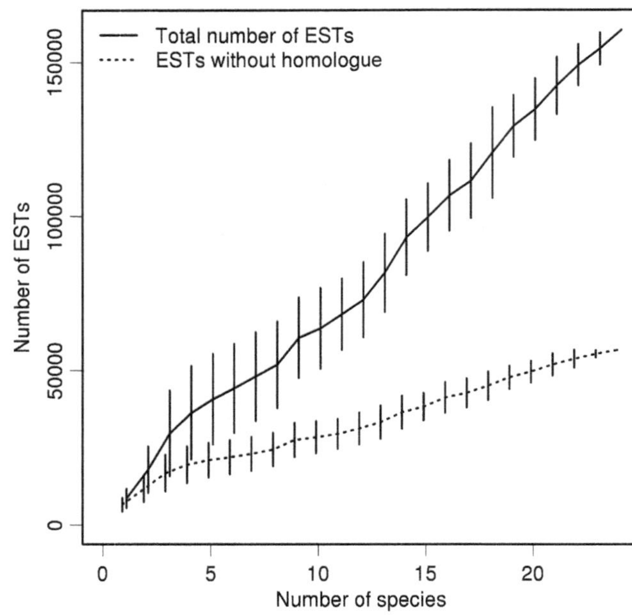

Fig. 6.4. *Analysis of ESTs from 24 nematode species shows no evidence for saturation of the number of orphan ESTs with the number of analysed species. This suggests that the transcriptome sequencing of the nematode phylum is far from constituting a representative sample.*

copy (Force *et al.*, 1999; Katju & Lynch, 2006). Such a scenario of rapid evolution subsequent to duplication events is consistent with the observed lack of sequence similarity of individual members of larger gene families. An alternative hypothesis for the generation of orphan genes has been proposed as *de novo* formation of genes from non-coding sequences. Such a process has only been reported in a small number of species without any examples in nematodes (Heinen *et al.*, 2009; Knowles & McLysaght, 2009; Li *et al.*, 2010), yet future phylogenomic analysis at much higher resolution has the potential to reveal whether *de novo* formation of novel genes did occur in the genus *Pristionchus* as well as in other nematodes.

Horizontal gene transfer of cellulase genes

Comparison of orphan genes with sequences from organisms outside the nematode phylum has revealed a number of homology relationships that are inconsistent with general metazoan phylogeny. Such cases can only be explained by horizontal gene transfer (HGT), which denotes the transfer of genes across organisms by a mechanism other than genetic inheritance from parent to offspring. The only alternative scenarios would be multiple gene losses in other nematode genomes or convergent evolution, which are both considered to be more unlikely than HGT. While HGT is known to be frequent in bacteria, it was considered to be rare in sexually reproducing eukaryotes (Andersson, 2005). Thus, the finding of numerous independent HGT events in nematodes involving sequences from multiple hosts has been one of the most striking findings of nematode comparative genomic analyses.

The finding of genes encoding cell wall degrading enzymes in the genome of *P. pacificus* was one of the most surprising outcomes of the initial sequencing of the *P. pacificus* genome (Dieterich *et al.*, 2008). The finding was totally unexpected because it represented the first finding of cellulase genes in a non-plant-parasitic nematode. In plant-parasitic nematodes, cellulase enzymes are used to penetrate plant cell walls during infection, whereas they have no obvious functions in *P. pacificus* (Schuster & Sommer, 2012). In addition, the cellulase genes identified in *P. pacificus* have most likely been acquired by an HGT event independent of their counterparts in plant-parasitic nematodes. It has been found that, even within plant parasites, cellulase genes have been acquired by independent HGT events. While cellulases found in the plant-parasitic *Meloidogyne* species belong to the glycoside hydrolase family 5 (GHF5) and presumably derive from an intron-less ancestral gene acquired from bacterial donors (Kyndt *et al.*, 2008), the pinewood nematode *Bursaphelenchus xylophilus* has independently acquired cellulases of a different family (GHF45) from fungi (Fig. 6.5) (Kikuchi *et al.*, 2004). Despite the fact that the enzymes found in *P. pacificus* also belong to the GHF5 family, the corresponding sequences are still substantially diverged from the GHF5 of *Meloidogyne* and are most similar to cellulase genes found in the slime mould *Dictyostelium discoideum* (Fig. 6.5) (Dieterich *et al.*, 2008; Mayer *et al.*, 2011). Thus, at least three different donor species have been reported for the cellulase genes that were identified in nematodes, which make the example of cellulases one of the most prominent cases of HGT in eukaryotes. However, they neither form the only known example of HGT in nematodes (Dunning Hotopp *et al.*, 2007), nor the only examples in *P. pacificus* (Dieterich *et al.*, 2008; Rödelsperger & Sommer, 2011). More detailed phylogenetic analysis employing diplo-

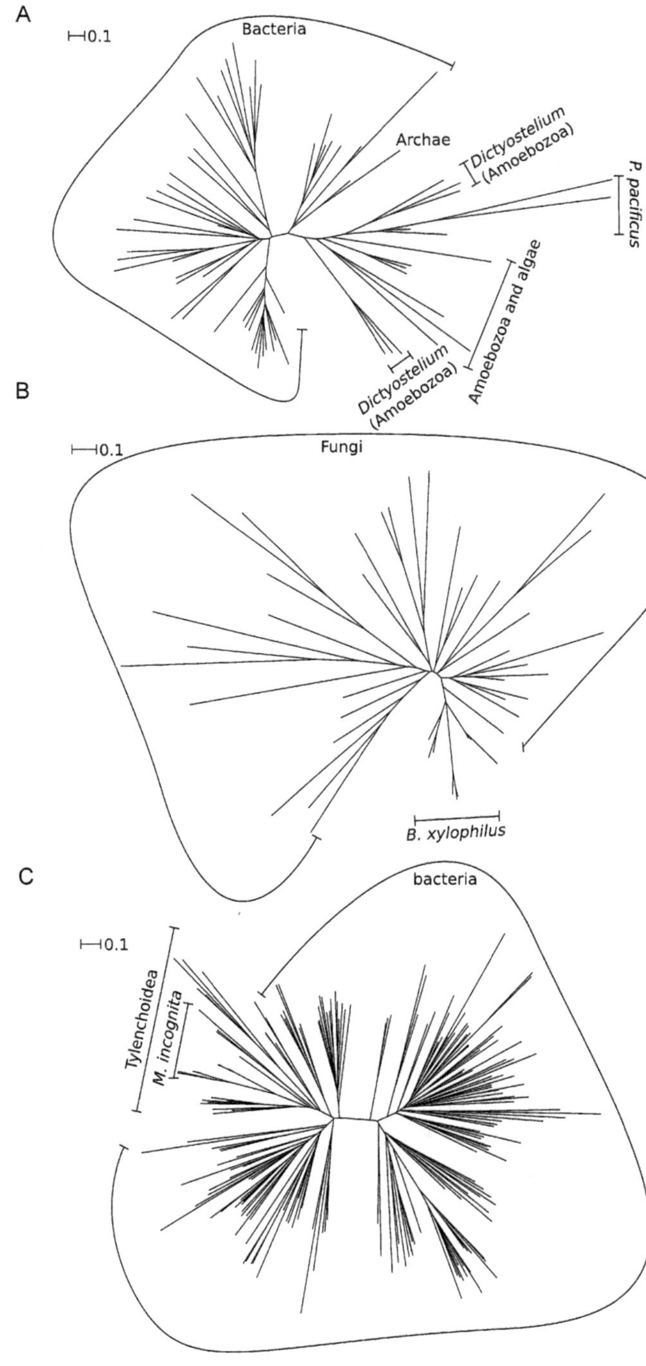
A
0.1
Bacteria
Archae
Dictyostelium (Amoebozoa)
P. pacificus
Amoebozoa and algae
Dictyostelium (Amoebozoa)
B
0.1
Fungi
B. xylophilus
C
bacteria
0.1
Tylenchoidea
M. incognita

gastrid transcriptome sequencing further identified cellulase genes in six other *Pristionchus* species and detected signals for site-specific positive selection in one of the cellulases (Mayer *et al.*, 2011). One unsolved question within studies of HGT is the identity of the vector for the transfer. While current sequence databases comprise sufficient data to narrow down the most likely origin of the horizontally acquired genes, it remains unclear whether the transfer occurred directly between donor and acceptor species, or whether intermediate vectors such as viruses and transposons might be involved (Tanaka *et al.*, 2003; Rödelsperger & Sommer, 2011).

In total, seven cellulase genes have been identified in the *P. pacificus* genome. Cellulase activity has been identified by Congo red-polysaccharide interaction assays in the supernatant of *P. pacificus* mixed stage cultures (Dieterich *et al.*, 2008; Mayer *et al.*, 2011). In the transcriptomes identified by Mayer *et al.* (2011), evidence for cellulase expression correlated strictly with experimentally determined cellulase activity. Within *P. pacificus*, spatial analysis of gene expression by *in situ* hybridisation showed expression of one of the cellulases in the posterior pharynx and anterior part of the intestine (Dieterich *et al.*, 2008). More detailed molecular and functional characterisation of the *P. pacificus* cellulases indicated that expression of all cellulases was overall low and feeding different carbohydrate sources could not induce the associated cellulase activity (Schuster & Sommer, 2012). However, expression profiling through all developmental stages showed that two cellulases

Fig. 6.5. *Nematodes have acquired cellulase genes by independent Horizontal Gene Transfers (HGTs). The three panels show phylogenetic trees that indicate the different origin of nematode cellulases. Cellulases from different nematodes are so diverged that cellulases from the other nematode species would branch as outgroups in the individual trees. A: Cellulases that are found in* Pristionchus pacificus *show highest similarity to sequences from* Dictyostelium discoideum *(Dieterich* et al., *2008); B: Cellulases from fungi exhibit the highest similarity to cellulases from* Bursaphelenchus xylophilus *(Kikuchi* et al., *2004); C: The plant-parasitic nematode* Meloidogyne incognita *and also other nematodes of the suborder Tylenchoidea have cellulase genes that show highest similarity to sequences from bacteria (Kyndt* et al., *2008). Phylogenetic trees represent neighbour-joining trees of representative nematode cellulase proteins and their best BLAST hits within the NCBI non-redundant database. Branch lengths represent the number of substitutions under the JTT model.*

were developmentally regulated with an onset of expression after the first day of development. In addition, those two cellulases not only contained the cellulase domain but also a carbohydrate-binding module. Moreover, peptides of those two enzymes could be detected by mass spectrometry in enzymatically-active worm secretions (Schuster & Sommer, 2012). Taken together, these results strongly support that *Pristionchus* genomes contain functional cellulases that have been conserved over multiple speciation events. Integration into the host biology and evolutionary permanence have been suggested as being the most important criteria for successful HGT (Blaxter, 2007). The *Pristionchus* cellulases are thus a good example, fulfilling both these criteria. However, the precise role of those genes in the ecology of *P. pacificus* remains unresolved.

Comparative functional genomics of the dauer stage

As comparative genomics focuses on genome-wide comparisons across species and functional genomics integrates other sources of high-throughput experimental data, there is no contradiction in combining the two disciplines to gain more insights into the evolution of developmental processes and gene regulatory networks. One recent study that combined comparative and functional genomics was by Sinha *et al.* (2012a), who compared the expression profiles of *C. elegans* and *P. pacificus* dauer larvae with the expression profiles of dauer exit worms that have resumed their development after having exited from the growth arrested dauer stage. Intra-species comparison of expression profiles identified candidate gene sets of 900 and 5000 differentially regulated genes in *C. elegans* and *P. pacificus*, respectively. Comparing the identified candidate gene sets for dauer-specific gene regulation showed that only 184 1 : 1 orthologous gene pairs could be identified as differentially expressed in both species. Surprisingly, around one third of those genes showed discordant trends across the two species, such as up-regulation in dauers of *C. elegans* and down-regulation in dauers of *P. pacificus* or *vice versa*. Although the overlap between differentially expressed genes in both species was still found to be significantly greater than expected, probably reflecting the evolutionary relationship between the two species, the limited conservation of expression profiles was unexpected given that parts of the regulation of dauer development are well conserved between *C. elegans* and *P. pacificus* (Ogawa *et al.*, 2009).

Evolutionary comparisons at shorter time-scales

The comparative studies described above represent comparisons at rather large evolutionary distances. A deeper phylogenetic sampling of the *Pristionchus* genus would lay the basis for future work to study the formation of orphan genes or to identify evolutionary constraints that act upon the horizontally-transferred cellulase genes. Recently, a solid phylogenetic as well as population genomic framework for *P. pacificus* was established by population-scale re-sequencing of globally sampled natural isolates and the assembly of *P. exspectatus* (Rödelsperger *et al.*, 2014), the gonochoristic sister species of *P. pacificus* (Kanzaki *et al.*, 2012). This genome-wide catalogue of natural variation was used to characterise even more fundamental processes of genome evolution, such as the mutational spectrum and the effect of mutations on fitness.

Base substitution mutations are a major source of novelty and represent the genetic basis for natural selection to act upon. However, until recently, the understanding of the mutational processes was rather limited. It was not known how often base substitutions occur, whether the genome harbours any mutational hotspots, and whether certain nucleotides are more likely to mutate than others. Most importantly, it was still unclear whether mutational processes are conserved across species borders. In order to study the processes of *de novo* mutation experimentally, mutation accumulation (MA) lines were proposed (Mukai *et al.*, 1964). In the case of *P. pacificus*, such an experiment was carried out by starting multiple independent MA lines with a founder individual and allowing only a single offspring to continue its line (Molnar *et al.*, 2011, 2012). Such an experimental setting minimises the effect of natural selection and facilitates the analysis of the frequency of spontaneous mutations that were accumulated throughout the duration of the experiment (in the case of *P. pacificus* 142 generations). Previous studies focused on mitochondrial and microsatellite regions to estimate frequencies of *de novo* mutations and identified in the order of 10^{-7} mutations per nucleotide per generation for mitochondria and 10^{-4} mutations per locus per generation for microsatellites (Molnar *et al.*, 2011, 2012). These findings suggested different mutational mechanisms for microsatellites such as errors during recombination, and polymerase slippage during DNA replication (Schlötterer & Tautz, 1992) or repair (Strand *et al.*, 1993). In a recent study, Weller and co-workers (Weller *et al.*, 2014) re-sequenced genomic DNA of 22 MA lines and calculated a genome-wide mutation rate in the order of 10^{-9} per nucleotide per generation, which is exactly the same order of magnitude as in nematodes of the genus *Caenorhabditis* (Denver *et al.*, 2012). The distribution of mutations across the chromosomes suggested the absence of large-scale mutational hotspots. In addition, the ratio between non-synonymous and silent substitutions was compatible with a neutral model of evolution (Weller *et al.*, 2014).

In agreement with previous analyses of MA lines in other organisms (Lynch *et al.*, 2008; Keightley *et al.*, 2009; Denver *et al.*, 2012), a strong tendency for spontaneous mutations to increase the A/T content of the genome was found. This A/T bias was also observed to a lesser extent in natural populations of *P. pacificus*, but seemed to vanish with increasing age of a single nucleotide variant (SNV), as measured by the derived allele frequency (Fig. 6.6). The decay of the A/T bias was not stronger in protein-coding exons than in other parts of the genome, suggesting that the loss of A/T driving variants is not due to selection on protein-coding genes. Therefore, a genome-wide process such as GC-biased gene conversion was proposed as the more likely explanation for the loss of A/T bias over evolutionary time (Weller *et al.*, 2014).

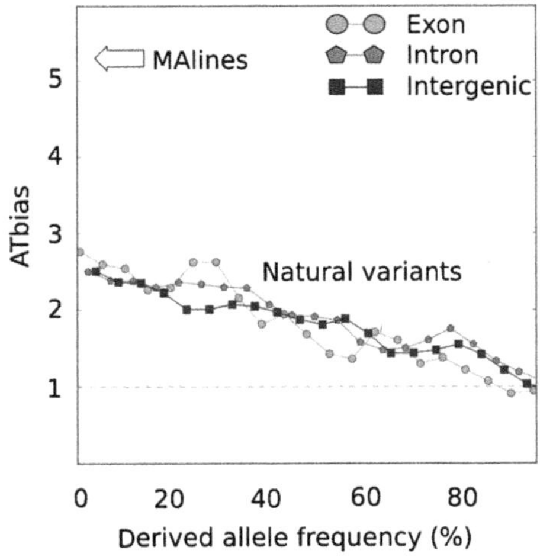

Fig. 6.6. *Loss of AT bias in natural populations of* Pristionchus pacificus. *AT bias of naturally occurring variants is shown as a function of derived allele frequency, which serves as an indicator of the age of an individual allele. The AT bias in natural populations is much weaker when compared with MA lines, and is completely lost in derived alleles that reach fixation.*

While the loss of A/T driving variants seemed not to be due to selection on protein-coding genes, other types of mutations revealed strong evidence for purifying selection in natural populations of *P. pacificus*. The overall ratio between non-synonymous and silent substitutions δ_{ns}/δ_{si} was 0.32 in natural populations as compared to $\delta_{ns}/\delta_{ns} = 1$ for MA lines (Rödelsperger *et al.*, 2014; Weller *et al.*, 2014). This indicates that around two-thirds of all non-synonymous substitutions are eliminated from *P. pacificus* populations. Measuring δ_{ns}/δ_{si} as a function of neutral distance between individual strains (δ_{si}) showed that half of all non-synonymous substitutions were eliminated so early that they are not found in a typical population sample of *P. pacificus* (104 strains). With increasing separation between strains, another 20% of non-synonymous substitutions are gradually lost (Fig. 6.7). These findings, in addition to

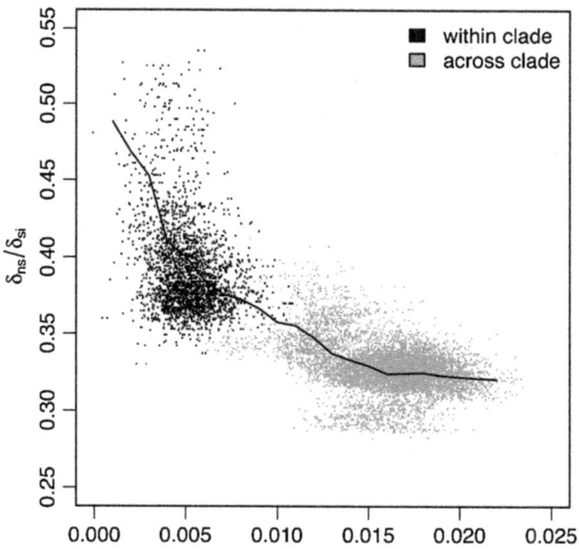

Fig. 6.7. *Purifying selection in the genome of* Pristionchus pacificus. *Individual dots represent ratios between non-synonymous and silent substitutions (δ_{ns}/δ_{si}) of all pairwise comparisons of 104* P. pacificus *strains. The* x*-axis represents different timescales as measured in presumably neutral distance δ_{si}. The line represents a smoothed average δ_{ns}/δ_{si} ratio of all measurements at a given δ_{si}. δ_{ns}/δ_{si} ratios decrease with distance (δ_{si}) from 0.5 to 0.3, suggesting that 50% of non-synonymous mutations have been selected against at very short time periods and were not observed as variable. In addition, 20% of non-synonymous variations are pruned with increasing distance.*

the observation of megabase-sized haplotype blocks with high linkage disequilibrium and an excess of derived alleles at high frequencies, support a scenario of background selection as a major factor in shaping genetic diversity in *P. pacificus* (Rödelsperger *et al.*, 2014).

Conclusions

> *I can't be as confident about computer science as I can about biology. Biology easily has 500 years of exciting problems to work on, it's at that level.*
>
> D. E. Knuth, interviewed by Computer Literacy Bookshops, 1993.

Despite its great value as a resource, the genome of *P. pacificus* still harbours millions of unsolved questions. The methods of comparative and functional genomics are both well suited to give a general characterisation of a genome's content and are able to generate hypotheses about the relevance of individual genes, gene families, and pathways related to the adaptation to novel ecological niches. However, focusing on the analysis of complete genomes comes with the drawback that the results are of rather general nature. To show definitely that the identified candidate gene sets are truly important for the biology of *P. pacificus* will be up to the experimentalists.

Acknowledgements

The authors thank Dr Gabriel Markov for carefully reading this manuscript.

References

AHMED, R., CHANG, Z., YOUNIS, A.E., LANGNICK, C., LI, N., CHEN, W., BRATTIG, N. & DIETERICH, C. (2013). Conserved miRNAs are candidate post-transcriptional regulators of developmental arrest in free-living and parasitic nematodes. *Genome Biology and Evolution* 5, 1246-1260.

ANDERSSON, J.O. (2005). Lateral gene transfer in eukaryotes. *Cellular and Molecular Life Sciences* 62, 1182-1197.

BLAXTER, M. (2007). Symbiont genes in host genomes: fragments with a future? *Cell Host & Microbe* 2, 211-213.

BLAXTER, M.L., DE LEY, P., GAREY, J.R., LIU, L.X., SCHELDEMAN, P., VIERSTRAETE, A., VANFLETEREN, J.R., MACKEY, L.Y., DORRIS, M., FRISSE, L.M. *ET AL.* (1998). A molecular evolutionary framework for the phylum Nematoda. *Nature* 392, 71-75.

BLUMENTHAL, T. & GLEASON, K.S. (2003). *Caenorhabditis elegans* operons: form and function. *Nature Reviews Genetics* 4, 112-120.

BORCHERT, N., DIETERICH, C., KRUG, K., SCHÜTZ, W., JUNG, S., NORDHEIM, A., SOMMER, R.J. & MACEK, B. (2010). Proteogenomics of *Pristionchus pacificus* reveals distinct proteome structure of nematode models. *Genome Research* 20, 837-846.

BORCHERT, N., KRUG, K., GNAD, F., SOMMER, R.J. & MACEK, B. (2012). Phosphoproteome of *Pristionchus pacificus* provides insight into signaling networks in nematode models. *Molecular Cell Proteomics* 11, 1631-1639.

BOULIAS, K. & HORVITZ, H.R. (2012). The *C. elegans* microRNA *mir-71* acts in neurons to promote germline-mediated longevity through regulation of DAF-16/FOXO. *Cell Metabolism* 15, 439-450.

DE LENCASTRE, A., PINCUS, Z., ZHOU, K., KATO, M., LEE, S.S. & SLACK, F.J. (2010). MicroRNAs both promote and antagonize longevity in *C. elegans*. *Current Biology* 20, 2159-2168.

DENVER, D.R., WILHELM, L.J., HOWE, D.K., GAFNER, K., DOLAN, P.C. & BAER, C.F. (2012). Variation in base-substitution mutation in experimental and natural lineages of *Caenorhabditis* nematodes. *Genome Biology and Evolution* 4, 513-522.

DESJARDINS, C.A., CERQUEIRA, G.C., GOLDBERG, J.M., DUNNING HOTOPP, J.C., HAAS, B.J., ZUCKER, J., RIBEIRO, J.M., SAIF, S., LEVIN, J.Z., FAN, L. *ET AL.* (2013). Genomics of *Loa loa*, a *Wolbachia*-free filarial parasite of humans. *Nature Genetics* 45, 495-500.

DIETERICH, C. & STADLER, P.F. (2013). Computational biology of RNA interactions. *Wiley Interdisciplinary Reviews: RNA* 4, 107-120.

DIETERICH, C., CLIFTON, S.W., SCHUSTER, L., CHINWALLA, A., DELEHAUNTY, K., DINKELACKER, I., FULTON, L., FULTON, R., GODFREY, J., MINX, P. *ET AL.* (2008). The *Pristionchus pacificus* genome provides a unique perspective on nematode lifestyle and parasitism. *Nature Genetics* 40, 1193-1198.

DUNNING HOTOPP, J.C., CLARK, M.E., OLIVEIRA, D.C., FOSTER, J.M., FISCHER, P., MUNOZ TORRES, M.C., GIEBEL, J.D., KUMAR, N., ISHMAEL, N., WANG, S. *ET AL.* (2007). Widespread lateral gene transfer from intracellular bacteria to multicellular eukaryotes. *Science* 317, 1753-1756.

FORCE, A., LYNCH, M., PICKETT, F.B., AMORES, A., YAN, Y.-L. & POSTLETHWAIT, J. (1999). Preservation of duplicate genes by complementary, degenerative mutations. *Genetics* 151, 1531-1545.

HEINEN, T.J., STAUBACH, F., HAMING, D. & TAUTZ, D. (2009). Emergence of a new gene from an intergenic region. *Current Biology* 19, 1527-1531.

KANZAKI, N., RAGSDALE, E., HERRMANN, M., MAYER, W.E. & SOMMER, R.J. (2012). Description of three *Pristionchus* species (Nematoda: Diplogastridae) from Japan that form a cryptic species complex with the model organism *P. pacificus*. *Zoolological Science* 29, 403-417.

KATJU, V. & LYNCH, M. (2006). On the formation of novel genes by duplication in the *Caenorhabditis elegans* genome. *Molecular Biology and Evolution* 23, 1056-1067.

KEIGHTLEY, P.D., TRIVEDI, U., THOMSON, M., OLIVER, F., KUMAR, S. & BLAXTER, M.L. (2009). Analysis of the genome sequences of three *Drosophila melanogaster* spontaneous mutation accumulation lines. *Genome Research* 19, 1195-1201.

KIKUCHI, T., JONES, J.T., AIKAWA, T., KOSAKA, H. & OGURA, N. (2004). A family of glycosyl hydrolase family 45 cellulases from the pine wood nematode *Bursaphelenchus xylophilus*. *FEBS Letters* 572, 201-205.

KIKUCHI, T., COTTON, J.A., DALZELL, J.J., HASEGAWA, K., KANZAKI, N., MCVEIGH, P., TAKANASHI, T., TSAI, I.J., ASSEFA, S.A., COCK, P.J. *ET AL.* (2011). Genomic insights into the origin of parasitism in the emerging plant pathogen *Bursaphelenchus xylophilus*. *PLoS Pathogens* 7, e1002219.

KNOWLES, D.G. & MCLYSAGHT, A. (2009). Recent *de novo* origin of human protein-coding genes. *Genome Research* 19, 1752-1759.

KROL, J., LOEDIGE, I. & FILIPOWICZ, W. (2010). The widespread regulation of microRNA biogenesis, function and decay. *Nature Reviews Genetics* 11, 597-610.

KYNDT, T., HAEGEMAN, A. & GHEYSEN, G. (2008). Evolution of GHF5 endoglucanase gene structure in plant-parasitic nematodes: no evidence for an early domain shuffling event. *BMC Evolutionary Biology* 8, 305.

LEE, K.-Z. & SOMMER, R.J. (2003). Operon structure and trans-splicing in the nematode *Pristionchus pacificus*. *Molecular Biology and Evolution* 20, 2097-2103.

LEE, R.C., FEINBAUM, R.L. & AMBROS, V. (1993). The *C. elegans* heterochronic gene *lin-4* encodes small RNAs with antisense complementarity to *lin-14*. *Cell* 75, 843-854.

LI, D., DONG, Y., JIANG, Y., JIANG, H., CAI, J. & WANG, W. (2010). A *de novo* originated gene depresses budding yeast mating pathway and is repressed by the protein encoded by its antisense strand. *Cell Research* 20, 408-420.

LIU, N., LANDREH, M., CAO, K., ABE, M., HENDRIKS, G.J., KENNERDELL, J.R., ZHU, Y., WANG, L.S. & BONINI, N.M. (2012). The microRNA miR-34 modulates ageing and neurodegeneration in *Drosophila*. *Nature* 482, 519-523.

LYNCH, M., SUNG, W., MORRIS, K., COFFEY, N., LANDRY, C.R., DOPMAN, E.B., DICKINSON, W.J., OKAMOTO, K., KULKARNI, S., HARTL, D.L. *ET AL.* (2008). A genome-wide view of the spectrum of spontaneous mutations in yeast. *Proceedings of the National Academy of Sciences of the United States of America* 105, 9272-9277.

MARKOV, G. & SOMMER, R.J. (2015). The same or not the same: lineage-specific gene expansions and homology relationship in multigene families in nematodes. *Journal of Molecular Evolution* 80, 18-36.

MAYER, M.G. & SOMMER, R.J. (2011). Natural variation in *Pristionchus pacificus* dauer formation reveals cross-preference rather than self-preference of nematode dauer pheromones. *Proceedings of the Royal Society B: Biological Sciences* 278, 2784-2790.

MAYER, W.E., SCHUSTER, L., BARTELMES, G., DIETERICH, C. & SOMMER, R.J. (2011). Horizontal gene transfer of microbial cellulases into nematode genomes is associated with functional assimilation, high gene turnover and positive selection. *BMC Evolutionary Biology* 11, 13.

MITREVA, M., JASMER, D.P., ZARLENGA, D.S., WANG, Z., ABUBUCKER, S., MARTIN, J., TAYLOR, C.M., YIN, Y., FULTON, L., MINX, P. *ET AL.* (2011). The draft genome of the parasitic nematode *Trichinella spiralis*. *Nature Genetics* 43, 228-235.

MOLNAR, R., BARTELMES, G., DINKELACKER, I., WITTE, H. & SOMMER, R.J. (2011). Mutation rates and intra-specific divergence of the mitochondrial genome of *Pristionchus pacificus*. *Molecular Biology and Evolution* 28, 2317-2326.

MOLNAR, R.I., WITTE, H., DINKELACKER, I., VILLATE, L. & SOMMER, R.J. (2012). Tandem repeat patterns and mutation rates in microsatellites of the nematode model organism *Pristionchus pacificus*. *G3: Genes, Genomics, Genetics* 2, 1027-1034.

MUKAI, T., CHIGUSA, S. & YOSHIKAWA, I. (1964). The genetic structure of natural populations of *Drosophila melanogaster*. III. Overdominance of spontaneous mutant polygenes controlling viability in homozygous genetic background. *Genetics* 50, 711-715.

OGAWA, A. & SOMMER, R.J. (2009). Strategies to get arrested. *Science* 326, 944-945.

OGAWA, A., BENTO, G., BARTELMES, G., DIETERICH, C. & SOMMER, R.J. (2011). *Pristionchus pacificus daf-16*/FOXO regulates dauer formation but not mouth form dimorphism. *Development* 138, 1281-1284.

OPPERMAN, C.H., BIRD, D.M., WILLIAMSON, V.M., ROKHSAR, D.S., BURKE, M., COHN, J., CROMER, J., DIENER, S., GAJAN, J., GRAHAM, S. *ET AL.* (2008). Sequence and genetic map of *Meloidogyne hapla*: a compact nematode genome for plant parasitism. *Proceedings of the National Academy of Sciences of the United States of America* 105, 14802-14807.

PETERSON, K.J., COTTON, J.A., GEHLING, J.G. & PISANI, D. (2008). The Ediacaran emergence of bilaterians: congruence between the genetic and the geological fossil records. *Philosophical Transactions of the Royal Society B: Biological Sciences* 363, 1435-1443.

RAE, R., WITTE, H., RÖDELSPERGER, C. & SOMMER, R.J. (2012). The importance of being regular: *Caenorhabditis elegans* and *Pristionchus pacificus* defecation mutants are hypersusceptible to bacterial pathogens. *International Journal for Parasitology* 42, 747-753.

RAGSDALE, E., MÜLLER, M.R., ROEDELSPERGER, C. & SOMMER, R.J. (2013). A develomental switch coupled to the evolution of plasticity acts through a sulfatase. *Cell* 155, 922-933.

RÖDELSPERGER, C. & SOMMER, R.J. (2011). Computational archeology of the *Pristionchus pacificus* genome reveals evidence of horizontal gene transfer from insects. *BMC Evolutionary Biolology* 11, 239.

RÖDELSPERGER, C., STREIT, A. & SOMMER, R.J. (2013). Structure, function and evolution of the nematode genome. *eLS*. Chichester, UK, John Wiley & Sons Ltd. DOI:10.1002/9780470015902.a0024603.

RÖDELSPERGER, C., NEHER, R.A., WELLER, A., EBERHARDT, G., WITTE, H., MAYER, W., DIETERICH, C. & SOMMER, R.J. (2014). Characterization of genetic diversity in the nematode *Pristionchus pacificus* from population-scale resequencing data. *Genetics* 196, 1153-1165.

SCHLOTTERER, C. & TAUTZ, D. (1992). Slippage synthesis of simple sequence DNA. *Nucleic Acids Research* 20, 211-215.

SCHUSTER, L.N. & SOMMER, R.J. (2012). Expressional and functional variation of horizontally acquired cellulases in the nematode *Pristionchus pacificus*. *Gene* 506, 274-282.

SINHA, A., SOMMER, R.J. & DIETERICH, C. (2012a). Divergent gene expression in the conserved dauer stage of the nematodes *Pristionchus pacificus* and *Caenorhabditis elegans*. *BMC Genomics* 13, 254.

SINHA, A., RAE, R., IATSENKO, I. & SOMMER, R.J. (2012b). System wide analysis of the evolution of innate immunity in the nematode model species *Caenorhabditis elegans* and *Pristionchus pacificus*. *PLoS ONE* 7, e44255.

SINHA, A., LANGNICK, C., SOMMER, R.J. & DIEDERICH, C. (2014). Genome-wide analysis of trans-splicing in the nematode *Pristionchus pacificus* unravels conserved gene functions for germline and dauer development in divergent operons. *RNA* 20, 1386-1397.

SRINIVASAN, J. & SOMMER, R.J. (2002). From evolutionary developmental biology to genomics: towards a genome map of the free-living nematode *Pristionchus pacificus*. In: *Genome science-towards a new paradigm? Ue-*

hara Memorial Foundation Symposium 2002. Amsterdam, The Netherlands, Elsevier, pp. 101-110.

STEIN, L.D., BAO, Z., BLASIAR, D., BLUMENTHAL, T., BRENT, M.R., CHEN, N., CHINWALLA, A., CLARKE, L., CLEE, C., COGHLAN, A. *ET AL.* (2003). The genome sequence of *Caenorhabditis briggsae*: a platform for comparative genomics. *PLoS Biology* 1, e45. DOI:10.1371/journal.pbio.0000045.

STRAND, M., PROLLA, T.A., LISKAY, R.M. & PETES, T.D. (1993). Destabilization of tracts of simple repetitive DNA in yeast by mutations affecting DNA mismatch repair. *Nature* 365, 274-276.

TANAKA, H., SATO, K., SAITO, Y., YAMASHITA, T., AGOH, M., OKUNISHI, J., TACHIKAWA, E. & SUZUKI, K. (2003). Insect diapause-specific peptide from the leaf beetle has consensus with a putative iridovirus peptide. *Peptides* 24, 1327-1333.

THE *C. ELEGANS* SEQUENCING CONSORTIUM (1998). Genome sequence of the nematode *C. elegans*: a platform for investigating biology. *Science* 202, 2012-2018.

WELLER, A., RÖDELSPERGER, C., EBERHARDT, G., MOLNAR, R.I. & SOMMER, R.J. (2014). Opposing forces of A/T-biased mutations and G/C-biased gene conversions shape the genome of the nematode *Pristionchus pacificus*. *Genetics* 196, 1145-1152.

WICKER, T., SABOT, F., HUA-VAN, A., BENNETZEN, J.L., CAPY, P., CHALHOUB, B., FLAVELL, A., LEROY, P., MORGANTE, M., PANAUD, O. *ET AL.* (2007). A unified classification system for eukaryotic transposable elements. *Nature Reviews Genetics* 8, 973-982.

ZASLAVER, A., BAUGH, L.R. & STERNBERG, P.W. (2011). Metazoan operons accelerate recovery from growth-arrested states. *Cell* 145, 981-992.

ZHANG, X., ZABINSKY, R., TENG, Y., CUI, M. & HAN, M. (2011). microRNAs play critical roles in the survival and recovery of *Caenorhabditis elegans* from starvation-induced L1 diapause. *Proceedings of the National Academy of Sciences of the United States of America* 108, 17997-18002.

ZUPUNSKI, V., GUBENSEK, F. & KORDIS, D. (2001). Evolutionary dynamics and evolutionary history in the RTE clade of non-LTR retrotransposons. *Molecular Biology and Evolution* 18, 1849-1863.

Nematology Monographs & Perspectives, 2015, Vol. 11, 167-196

Chapter 7

Small-molecule signalling: encoding biological information in chemical structures

Frank C. SCHROEDER

Boyce Thompson Institute and Department of Chemistry and Chemical Biology, Cornell University, Ithaca, NY 14853, USA
schroeder@cornell.edu

Chemical information

When one thinks of chemically encoded information in a biological context, the genetic code of DNA may come to mind. Much less appreciated is the notion that, in fact, any molecule produced by a living organism – or a community of organisms – carries specific information about the state of the producer(s), simply by virtue of a molecule's chemical structure. Biogenic small molecules (BSMs) are particularly noteworthy as such 'chemical information carriers' because their chemical structures are generally much more diverse – and harder to predict and analyse – than those of biological macromolecules including DNA, RNA or proteins. More than 150 years of natural products research (Dias *et al.*, 2012) have shown that almost any chemical structure an organic chemist could imagine may conceivably exist in nature. These highly diverse structures of BSM are the result of specific cascades of chemical transformations – enzymatic or non-enzymatic, biosynthetic or catabolic – that reflect the biological state of the producing organism(s) (Meinwald, 2011). Therefore, it is not surprising that many different types of BSM have acquired signalling functions, as intracellular signalling molecules, as hormones or second messengers signalling between different cells or tissues of one organisms, or as pheromones and quorum sensing signals facilitating communication between individuals of the same or several different species (Meinwald,

2011). For many small-molecule signals, biosynthetic pathways have evolved to transduce a particular message most effectively, together with dedicated perception mechanisms, for example in the form of nuclear or membrane-bound receptor proteins that are finely tuned to respond to specific small-molecule structures, often at very low concentrations. It seems likely that nature relies so heavily on BSMs for transduction of information in large measure because of their diverse structures and often complex biosynthetic history (Meinwald, 2011). Additionally, BSMs often serve non-signalling functions related to their specific chemical properties, for example as anti-oxidants, hydrophobic protectants or adhesives.

Taken together, chemical identification of small-molecule signals along with a detailed characterisation of the associated biosynthetic cascades and perception mechanisms are of central importance for advancing organismal biology. Ultimately, the comprehensive characterisation of the entirety of the small molecules produced by an organism, its metabolome, provides a snapshot of organismal state, complementing and enhancing results from traditional transcriptomic and proteomic analyses. This chapter aims to show that the comparative analysis of metabolite structures and functions in related model organisms, specifically *Caenorhabditis elegans* and *Pristionchus pacificus*, provides unique opportunities for understanding the significance of small-molecule signalling for conserved physiological pathways.

Metabolomics for model organisms

Given the importance of BSMs as information carriers and reporters, it is striking that the metabolomes of the traditional animal model systems – *C. elegans*, *Drosophila* and mouse – have, until recently, remained largely unexplored (Schroeder, 2006; Robinette *et al.*, 2011). This deficiency can be explained, in part, by the considerable challenges associated with characterising thousands of molecules with unpredictable, yet highly diverse, chemical structures and biological roles. However, the functional characterisation of entire metabolomes may finally become feasible as a result of significant recent advances in analytical methodology, including the advent of readily available high-resolution mass spectrometry and new strategies for the processing of information-dense spectroscopic data (Forseth & Schroeder, 2011; Robinette *et al.*, 2011). Nonetheless, functional metabolomics of model organisms is, arguably, still in its infancy. Of the several thousand different metabolites one can now easily detect in the metabolomes of *C. elegans* (von Reuss *et al.*, 2012; Ludewig & Schroeder, 2013; Stupp *et al.*, 2013) or *Drosophila* (Tennessen *et al.*, 2014) using mass spectrometry, perhaps a few hundred are known and most of these represent primary metabolites that were identified many decades ago before the advent of molecular biology. However, although there was little progress in the characterisation of the metabolomes of model organisms for many decades, evidence for the existence of many additional classes of metabolites with important signalling functions continued to build (Bose *et al.*, 2012; von Reuss *et al.*, 2012).

A new beginning: small-molecule signalling in *C. elegans*

The initial analyses of the metabolome of the free-living nematode *C. elegans* was motivated by interest in the role of small molecules in regulating juvenile development. In the early 1980s, Golden & Riddle (Golden & Riddle, 1982, 1984a, b) had shown that worm-produced small molecules are required to trigger developmental arrest at the dauer stage, a long-lived and highly stress-resistant alternate life cycle stage ('dauer', from the German word for 'enduring') that corresponds to the infective juvenile stage of parasitic nematode species. While the chemical nature of these dauer-inducing small molecules (the 'dauer pheromone') still remained unknown, the laboratories of Antebi, Gems and others showed that, downstream of the dauer signal, cholesterol-derived steroids control developmental progression *via* the nuclear hormone receptor DAF-12, a homologue of vertebrate liver-X and vitamin D receptors (Antebi *et al.*, 2000; Gerisch *et al.*, 2001; Gerisch & Antebi, 2004; Gems, 2007; McCulloch & Gems, 2007).

The chemical structures of the *C. elegans* dauer pheromone were finally identified in a series of papers from several different laboratories between 2005 and 2009 (Jeong *et al.*, 2005; Butcher *et al.*, 2007; Pungaliya *et al.*, 2009). It was found that the *C. elegans* dauer pheromone consists of a mixture of glycosides, the ascarosides, which are based on the dideoxysugar ascarylose and a variety of fatty acid-derived side chains (Fig. 7.1). Soon after, it was found that ascaroside-derived small molecules of unanticipated diversity and complexity not only regulate

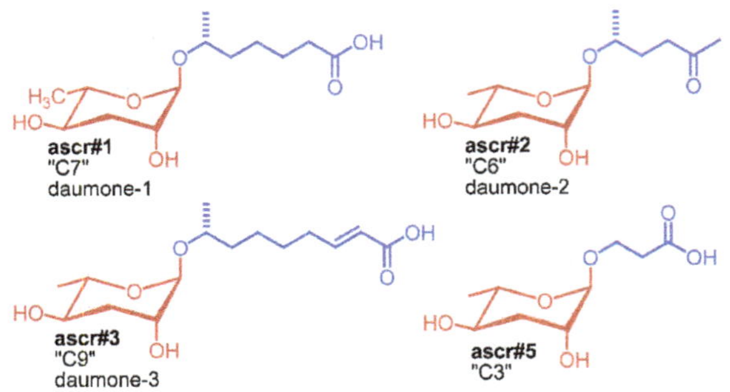

Fig. 7.1. *Major components of the* Caenorhabditis elegans *dauer pheromone.*

dauer entry and exit in *C. elegans*, but also control a wide variety of social behaviours. A series of chemical and behavioural studies revealed that nematodes employ a complex chemical 'language' that controls almost every aspect of their biology, including dauer formation, adult lifespan and stress resistance (Ludewig *et al.*, 2013), olfactory plasticity (Yamada *et al.*, 2010), dispersal (Kaplan *et al.*, 2012), avoidance (Artyukhin *et al.*, 2013), aggregation (Srinivasan *et al.*, 2012; von Reuss *et al.*, 2012), as well as sex-specific attraction and repulsion (Srinivasan *et al.*, 2008; Macosko *et al.*, 2009; Pungaliya *et al.*, 2009; Choe *et al.*, 2012a, b; Izrayelit *et al.*, 2012). Many of the ascarosides identified from *C. elegans* were found to be produced in a life stage-specific and sex-specific manner, in accordance with their often life stage- and sex-specific functions (Kaplan *et al.*, 2011; von Reuss *et al.*, 2012; Artyukhin *et al.*, 2013). This wide range of biological functions is facilitated by a great diversity of ascaroside chemical structures that include building blocks derived from amino acids, neurotransmitters, folate and other primary metabolites (Fig. 7.2), suggesting that the ascarosides' structures transduce information about the overall physiological state of the organism. Different ascarosides mediate different phenotypes and it was found that even small differences in chemical structures are often associated with strongly altered activity profiles (Izrayelit *et al.*, 2012; Srinivasan *et al.*, 2012). Additional complexity arises from synergism between different ascarosides, complex concentration-dependence, and life stage-specific effects (Srinivasan *et al.*, 2008, 2012; Pungaliya *et al.*, 2009). For example, longevity promoted by the ascaroside ascr#2 requires the sirtuin SIR-2.1 (Ludewig *et al.*, 2013), whereas dauer induction by the same compound is sirtuin-independent and instead relies on insulin/IGF and TGF-β signalling (Hu, 2007). These findings

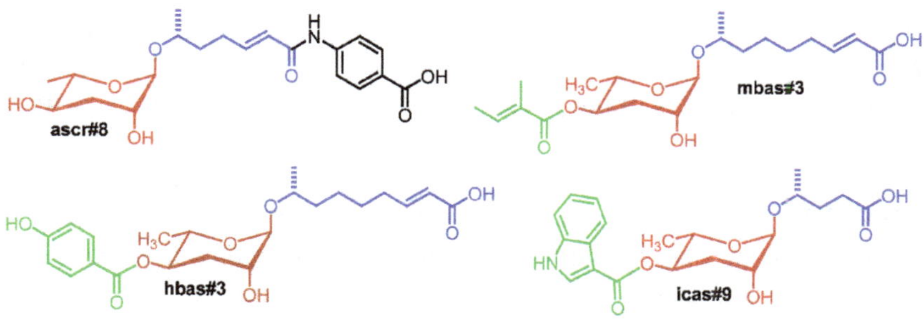

Fig. 7.2. *Examples of chemically more complex ascarosides regulating diverse behaviour in* Caenorhabditis elegans. *The biosynthesis of these ascarosides integrates building blocks from carbohydrate (red), fatty acid (blue), amino acid (green), and folate (black) metabolism.*

also demonstrate that ascarosides exert many of their functions through major conserved signalling cascades, suggesting that synthetic samples and variants of endogenously produced small molecules could be used as tools to probe specific aspects of, for example, sirtuin- or insulin signalling in *C. elegans*. Ascarosides are sensed by several types of chemosensory neurons (Srinivasan *et al.*, 2008; Kim *et al.*, 2009; Macosko *et al.*, 2009; Pungaliya *et al.*, 2009; Jang *et al.*, 2012; Srinivasan *et al.*, 2012), and their perception is mediated by several families of G protein-coupled receptors (Kim *et al.*, 2009; Greer *et al.*, 2011; McGrath *et al.*, 2011; Park *et al.*, 2012).

The rapid identification of such a large variety of ascaroside-based chemical signals in *C. elegans* was facilitated by the introduction of comparative metabolomics, a recently developed strategy that promises to revolutionise how bioactive small molecules are identified from complex biological systems (Challis, 2008; Prince & Pohnert, 2010; Forseth & Schroeder, 2011; Robinette *et al.*, 2011; von Reuss *et al.*, 2012; Izrayelit *et al.*, 2013; Mahanti *et al.*, 2014). Comparative metabolomics circumvents or reduces the need for the isolation of pure compounds *via* extensive activity-guided fractionation, usually the most challenging and time-consuming part of identifying the chemical nature of a biological signal. Instead, compound identification relies on comparing high-resolution spectroscopic data sets, for example two-dimensional NMR spectra or HPLC-MS data, from one set of metabolome samples that contain the chemical signal of interest with a second set of samples that do not contain the signal but are otherwise as similar as possible. This strategy was also employed to investigate the biosynthesis of ascarosides in *C. elegans* (Pungaliya *et al.*, 2009; Srinivasan *et al.*, 2012; von Reuss *et al.*, 2012). For example, the fatty acid-like side chains in the ascarosides were shown to be derived from peroxisomal β-oxidation of long-chain ascaroside precursors that are iteratively chain-shortened by the action of four enzymes, ACOX-1, MAOC-1, DHS-28 and DAF-22, which show a high degree of sequence similarity with corresponding enzymes in *Drosophila* and mammals (von Reuss *et al.*, 2012).

The *P. pacificus* metabolome: adventures in structure space

The discovery of the ascarosides as central mediators of *C. elegans* behaviour and life history suggested that other nematode species may use similar molecules as chemical signals. Considering that the *C. elegans* ascarosides are derived from modular assembly of building blocks of highly conserved primary metabolism, it seemed also possible that animal species from other phyla may have co-opted similar strategies for the biosynthesis of small-molecule signals. Therefore, the study of small-molecule signalling in *P. pacificus* as a satellite model to *C. elegans* has been of particular significance. Similar to *C. elegans*, harsh environmental conditions, for example food shortage, trigger developmental arrest of *P. pacificus* juveniles at a highly stress-resistant dauer stage (Ogawa *et al.*, 2009; Weller *et al.*, 2010). *Pristionchus pacificus* further exhibits a unique dimorphism of mouth development, representing an example for phenotypic plasticity in an adult metazoan (Bento *et al.*, 2010; Ogawa *et al.*, 2011). Adult worms can have either a narrow (stenostomatous) or a wide and more complex (eurystomatous) mouth opening, the latter developing in response to conditions of low food availability (see Ragsdale, Chapter 11, this volume). The two different mouth forms appear to be associated with different feeding preferences: stenostomatous worms are considered to feed primarily on bacteria, whereas the eurystomatous form is adapted for predatory behaviour toward other nematodes (Serobyan *et al.*, 2013, 2014). Previous studies had suggested that both dauer formation and mouth dimorphism are regulated by excreted small molecules that target conserved downstream signalling cascades, converging on a homologue of the *C. elegans* nuclear hormone receptor DAF-12 (Ogawa *et al.*, 2009, 2011; Bento *et al.*, 2010; Sommer & Ogawa, 2011).

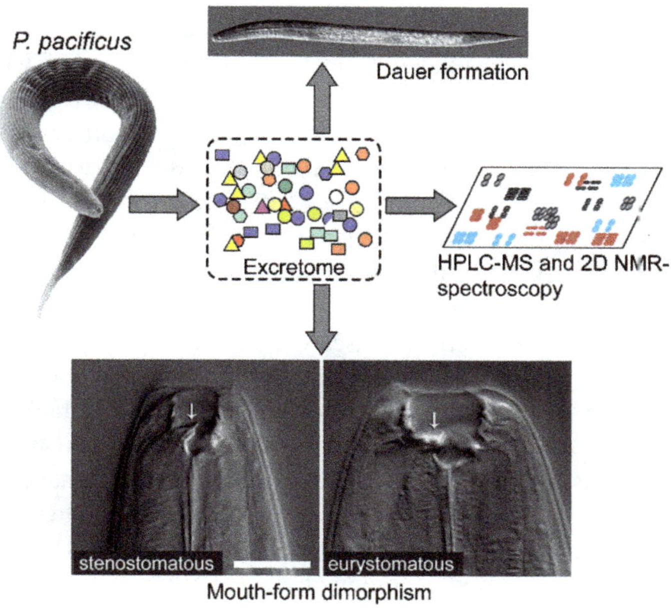

Fig. 7.3. *Biological activity of* Pristionchus pacificus *metabolite extracts. The* P. pacificus *exo-metabolome samples induce dauer arrest and affect mouth form dimorphism, promoting eurystomatous mouth development. HPLC-MS and 2D NMR-spectroscopic profiling revealed several thousand novel features. The white arrow indicates the distinctive tooth in the eurystomatous mouth form, which is absent in the stenstomatous form (Bose* et al., *2012).*

HPLC-MS and 2D NMR spectroscopic analyses of *P. pacificus* exo-metabolome samples (essentially culture supernatant, containing excreted and secreted metabolites) revealed a striking diversity of signals most of which could not be attributed to any known compounds (Fig. 7.3) (Bose *et al.*, 2012). The HPLC-MS analyses alone indicated the presence of more than 5000 unknown metabolites, in addition to signals representing familiar components of primary metabolism, such as amino acids and fatty acids. More detailed mass spectroscopic analyses suggested that many of the unknown compounds consist of building blocks of primary metabolism, reminiscent of the modular ascarosides identified previously from *C. elegans*. Subsequent 2D NMR spectroscopic analyses revealed that, similar to *C. elegans*, *P. pacificus* produces a large number of metabolites with unprecedented chemical structures based on a di-deoxysugar as the central scaffold. However, whereas *C. elegans* uses exclusively ascarylose for this purpose, the *P. pacifi-*

cus metabolites are based on two different di-deoxysugars, the familiar ascarylose and the related L-paratose, a new sugar, of which the D-enantiomer had previously been reported from bacteria (Bose *et al.*, 2012). Moreover, even though *P. pacificus*, like *C. elegans*, produces ascarosides, few of the specific ascaroside derivatives known from *C. elegans* were found in *P. pacificus*. Instead, *P. pacificus* primarily produces ascarosides whose structures differ significantly from those known from *C. elegans*. Nonetheless, the basic assembly principles underlying the biosynthesis of the *P. pacificus* paratosides and ascarosides appear to be the same as for the *C. elegans* ascarosides: modular assembly of primary metabolism-derived building blocks (Fig. 7.4). Notably, the diversity of primary metabolic pathways contributing building blocks is much greater in the case of the *P. pacificus* compounds, which incorporate building blocks derived from not only fatty acid, carbohydrate and amino acid metabolism, but also nucleoside and neurotransmitter (*e.g.*, the green-coloured phenylethanolamine unit in pasc#9; Fig. 7.4) metabolism (Bose *et al.*, 2012). Different combinations of building blocks from these pathways generate unique molecular architectures, for example dimeric ascarosides such as dasc#1, the 3′-ureido isobutyrate derivatives ubas#1 and ubas#2, and, notably, an adenosine moiety (in npar#1) which is, unlike all other nucleosides known from animals, not based on the sugar ribose, which forms the familiar five-membered ring in the structures of DNA and RNA residues, but instead incorporates the sugar xylose, which forms a six-membered ring. This xylose-based nucleoside is connected to a moiety derived from the amino acid threonine, and in this regard resembles canonical (ribo)-threonylcarbamoyl adenosine (t^6A), a highly conserved nucleoside found directly adjacent to the anticodon triplet of a subset of tRNAs (Deutsch *et al.*, 2012). The production of large quantities of a xylopyranose derivative in *P. pacificus* suggests that the conserved biosynthetic pathway for t^6A has been co-opted and modified to enable production of large quantities of a potential signalling molecule. Similarly, the 3′-ureido-isobutyrate sidechains in ubas#1 and ubas#2 probably also originate from nucleoside metabolism, specifically the conserved degradation pathway of thymine. Because of their high degree of novelty, the structures of all of the newly identified *P. pacificus* ascarosides and paratosides had to be additionally confirmed through comparison with authentic, chemically synthesised samples, demonstrating the important role of synthetic organic chemistry for the study of small-molecule signalling (Bose *et al.*, 2012).

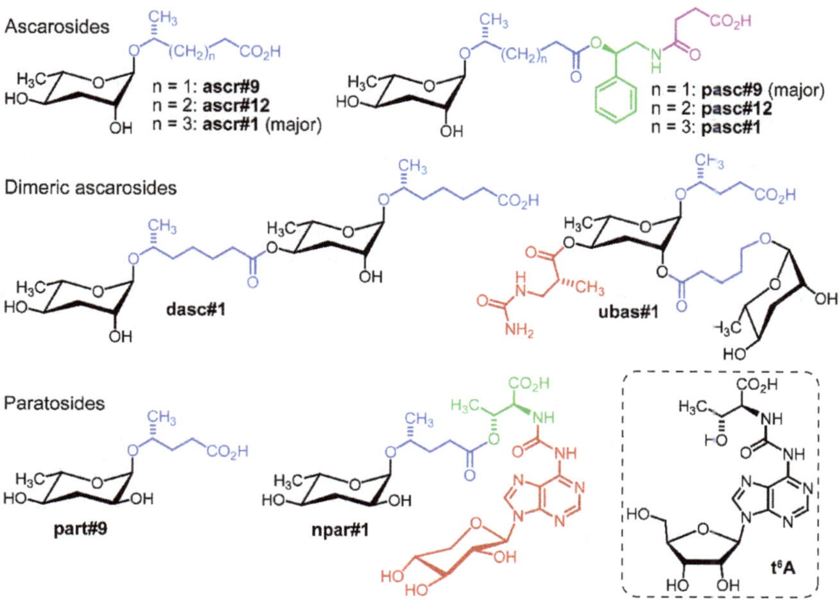

Fig. 7.4. *Chemical diversity of ascaroside and paratoside-derived metabolites in* Pristionchus pacificus. *Major components of the* P. pacificus *exo-metabolome derived from assembly of building blocks from carbohydrate (black), amino acid (green), and nucleoside (red) metabolism, as well as TCA cycle-derived succinate (magenta). Also shown is the highly conserved tRNA nucleoside, N^6-threonylcarbamoyl adenosine (t^6A), which is closely related to the paratoside npar#1 (Bose* et al., *2012).*

Specific small molecules control dauer and mouth form

To investigate the biological functions of the newly identified metabolites, synthetic samples of six major identified ascarosides and paratosides were tested in assays of dauer and mouth form dimorphisms, using two *P. pacificus* strains, RS5134 and RSB020, both of which had previously been used extensively to characterise dauer induction and mouth form dimorphism, respectively.

As expected from previous studies showing that *C. elegans* exo-metabolome samples are not active in the *P. pacificus* mouth form dimorphism and dauer assays (Ogawa *et al.*, 2009), it was found that ascr#1, a compound abundantly excreted by *C. elegans* (Jeong *et al.*, 2005), has no dauer-inducing activity in *P. pacificus*, even at very high concentrations. By contrast, the nucleoside derivative npar#1 strongly induced dauer formation in RS5134, which is similar to previously reported dauer inducing activity of unfractionated excretome in this strain (Mayer & Sommer, 2011). Additionally, weaker dauer-inducing activity was observed for part#9, whereas all other compounds tested did not induce dauer in this strain. Testing synthetic samples of the identified compounds in the mouth dimorphism assay, it was found that the dimeric compound dasc#1 strongly induces the eurystomatous mouth form. In addition, pasc#9 and npar#1 weakly induced the eurystomatous mouth form, whereas dimeric ubas#1 as well as monomeric ascr#9 and part#9 were inactive. These results showed that dauer formation and mouth form are controlled by specific members of the library of identified ascarosides and paratosides. Several of the identified compounds were inactive in both assays and thus may serve other functions, for example in regulating mating or aggregation behaviours, similar to the functions that ascarosides play for *C. elegans* behaviours; however, these possibilities have not yet been investigated.

Modular biosynthesis is selective

Following the identification of the highly unusual modular ascarosides and paratosides from *P. pacificus* it seemed necessary to clarify whether biosynthesis of these compounds is in fact directed and selective, as would be expected for potential signalling molecules. In particular, it was of concern whether the identified combinations of sugar, amino acid, lipid and nucleoside-derived building blocks are specific or merely represent examples for non-enzymatic, random oligomerisation of primary metabolites, for example, *via* ubiquitously present coenzyme-A thioesters of the incorporated fatty acid moieties. To address this question, the entire *P. pacificus* exo-metabolome was re-analysed by high resolution HPLC-MS/MS and screened for homologues or alternative combinations of the primary metabolism-derived building blocks in the identified compounds (Bose *et al.*, 2012). These analyses revealed only trace amounts of homologues and did not reveal any non-specific or seemingly random combinations of building blocks.

In fact, the analytical data indicate that assembly of the *P. pacificus* ascarosides and paratosides proceeds with extremely high selectivity (Fig. 7.4). For example, even though ascarosides with a 7-carbon sidechain (*e.g.*, ascr#1) are much more abundant than 5-carbon sidechain ascarosides (*e.g.*, ascr#9), only the 5-carbon variant is further decorated with a 3′-ureido isobutyrate substituent: there are no 7-carbon sidechain homologues of ubas#1 or ubas#2. Even more strikingly, a ω-oxygenated 5-carbon sidechain ascaroside is selectively attached to the 2-position in ubas#1, whereas all other identified compounds feature $(\omega - 1)$-oxygenated sidechains (Fig. 7.4). A similarly high level of selectivity is observed for some of the modular *C. elegans* compounds. For example, the likely folate-derived *p*-aminobenzoic acid moiety in ascr#8 (Fig. 7.2) is selectively attached to an unsaturated 7-carbon sidechain, although ascarosides based on saturated 7-carbon and unsaturated 9-carbon sidechain are much more abundant (Pungaliya *et al.*, 2009). Similarly, the indole ascarosides icas#9 (5-carbon sidechain) and icas#3 (9-carbon sidechain) are about equally abundant, although the unmodified 5-carbon ascaroside ascr#9 is orders of magnitude less abundant than the 9-carbon variant ascr#3 (Srinivasan *et al.*, 2012). Their highly selective assembly indicates that the modular ascarosides and paratosides from *P. pacificus* and *C. elegans* are the products of dedicated biosynthetic pathways, consistent with biological functions as specific signalling molecules.

Lastly, it seemed necessary to address the possibility that the identified modular *P. pacificus* metabolites are of bacterial origin or are perhaps the result of a mixed nematode/bacterial biogenesis. Analysis of the metabolome of the *E. coli* OP50 bacteria used as food for *P. pacificus* did not reveal any of the *P. pacificus* compounds. Moreover, all of the identified compounds were still found to be produced when *P. pacificus* cultures were fed with *Pseudomonas* sp. instead of *E. coli*. Direct involvement of bacterial metabolism was further excluded by growing *C. elegans* (Srinivasan *et al.*, 2012) and *P. pacificus* (Bose *et al.*, 2012) for several generations axenically, *i.e.*, under sterile conditions using a bacteria-free nutrient solution. Even under bacteria-free conditions, all of the previously identified modular ascarosides and paratosides could still be detected and thus appear to be products of nematode-autochthonus biosynthetic pathways.

Given that the building blocks of the ascarosides and paratosides identified from *P. pacificus* are derived from conserved primary metabolic pathways, it seems likely that homologues or orthologues of primary metabolic enzymes play a major role in their biosynthesis (Fig. 7.5). Given the diverse biological functions of ascarosides in *C. elegans* and *P. pacificus*, elucidation of the biosynthesis may reveal how input from conserved primary metabolism is transduced to create small-molecule

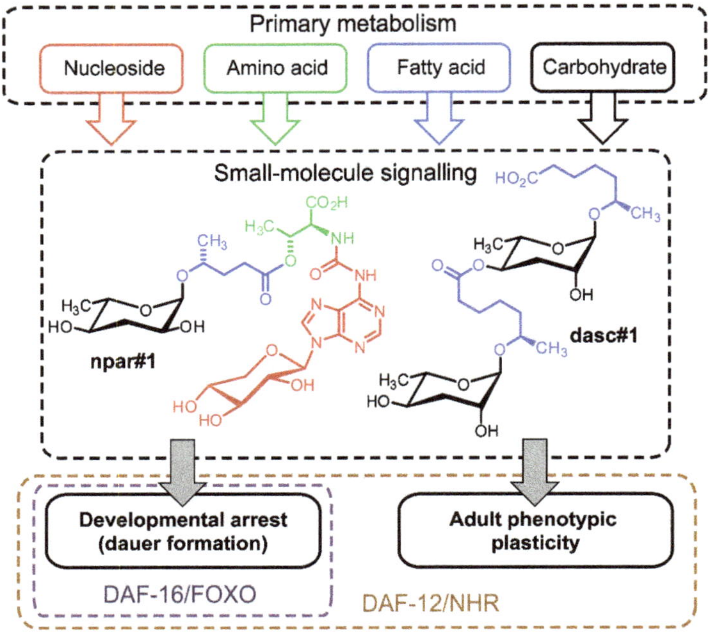

Fig. 7.5. *Small-molecule signalling in* Pristionchus pacificus. *Modular ascarosides* (e.g., *dsac#1) and paratosides* (e.g., *npar#1) connect primary metabolism to evolutionarily conserved downstream pathways, including insulin/IGF signalling* via *DAF-16/FOXO and the nuclear hormone receptor (NHR) DAF-12, a vitamin D and liver X receptor homologue (Bose* et al., *2012).*

signals that regulate development, phenotypic plasticity, and behaviour in nematode model organisms. The recent success with genome editing tools such as CRISPR in *P. pacificus* has greatly increased the feasibility and potential scope of biosynthetic studies in this model system (Witte *et al.*, 2015).

Natural variation of small-molecule biosynthesis and bioactivity

Perhaps as early as the initial discovery of the *C. elegans* dauer pheromone, it was suspected that small-molecule signals from one nematode species could affect dauer formation and other behaviours in other species (Golden & Riddle, 1982; Viney *et al.*, 2003). Since the *C. elegans* and *P. pacificus* dauer pheromones consist of non-overlapping sets of compounds, there is no evidence that the small-molecule signalling systems of these two species interact, and moreover there is no evidence that *C. elegans* and *P. pacificus* co-occur in nature. However, the frequent co-occurrence of many different *P. pacificus* strains – as many as five genetically distinct *P. pacificus* strains have been found on a single beetle – suggested the possibility of small-molecule signalling across different genotypes (Mayer & Sommer, 2011; Morgan *et al.*, 2012). Indeed, comparing the dauer pheromones of 16 different *P. pacificus* genotypes, it was found that 13 of the tested strains produce dauer pheromones that induce dauer formation more effectively in other strains ('cross-preference'), whereas only three strains produced pheromones that induced dauer most efficiently in the producing strain, *i.e.*, showed 'self preference' (Mayer & Sommer, 2011). These observations indicated that dauer signalling between different genotypes may play an important role for intraspecific competition.

Dauer pheromone cross-preference could be the result of strain-specific differences in pheromone composition or responsiveness to different pheromone components. Detailed analyses of the quantitative compositions of the dauer pheromone blends of six different strains and their responsiveness to the major components of these blends showed that both mechanisms contribute (Bose *et al.*, 2014). HPLC-MS analyses of the exo-metabolomes of the six strains revealed marked differences in composition of the ascaroside and paratoside blends. For example, the neurotransmitter-derived pasc#9 was found to vary more than six-fold between strains. Several compounds, including the ascarosides ubas#1 and ubas#2, were completely absent from the exo-metabolomes of two of the six analysed strains, for example strain RS5205 (Fig. 7.6). Because it was suspected that the absence of ubas#1 and ubas#2 from the exo-metabolomes could be due to deficiencies in pheromone *excretion*, additional analyses of the small-molecule content of the worm bodies of these strains (the 'endo-metabolome') were performed. However, ubas#1 and ubas#2 were found to be absent from the endo-metabolomes as well, suggesting that some genotypes may have lost the capacity to produce these compounds. These analyses further revealed marked differences between the endo- and exo-metabolomes of the six strains. Some of the chemically more simple ascarosides were found to be retained preferentially in the worm bodies, whereas other compounds, especially the structurally more complex modular ascarosides and most paratosides, were much more abundant in the exo-metabolomes (Fig. 7.6). These findings not only indicated that excretion of ascarosides and paratosides in *P. pacificus* is actively regulated, but

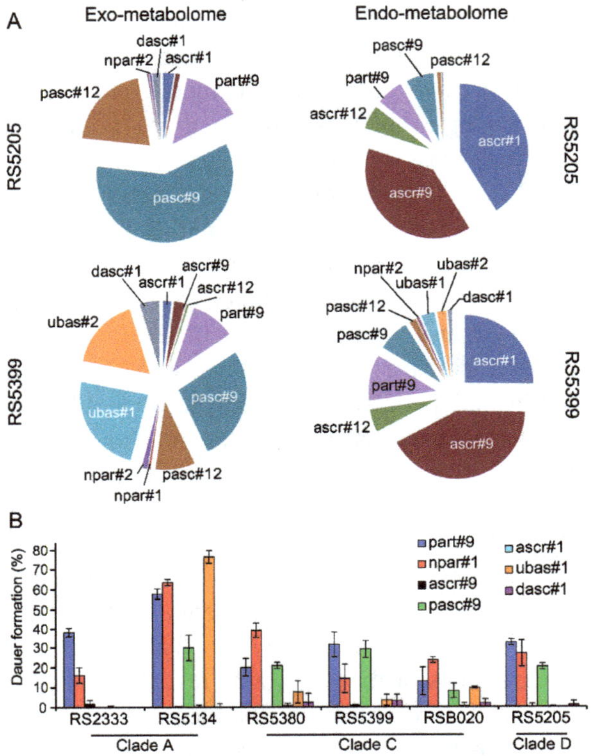

Fig. 7.6. *Natural variation of small-molecule signalling in* Pristionchus pacificus. *A: Relative abundances of ascarosides and paratosides in the exo- and endo-metabolomes of exemplary* P. pacificus *wild isolates derived from HPLC-MS analysis; B: Dauer formation of six* P. pacificus *strains in response to synthetic standards at 1 μM concentration (Bose* et al., *2014).*

also suggested that the chemically most complex, modular structures are particularly important as chemical signals. Notably, the distribution of different compounds between the exo- and endo-metabolomes is not only compound-specific, but also species-dependent. For example, entomopathogenic nematodes excrete large amounts of ascr#9 (Choe *et al.*, 2012b), whereas this compound is preferentially retained in the worm body in the case of the analysed *P. pacificus* strains. Taken together, these findings indicate that biosynthesis and excretion of ascarosides and paratosides is strongly regulated in a species- and strain-specific manner. Even sympatric strains may show very high variation in ascarosides and paratoside production.

The stark differences between the ascaroside and paratoside blends of different strains could partly explain the observed dauer pheromone cross preferences; however, given that pheromone biosynthesis is so heavily dependent on genotype, it seemed likely that pheromone response profiles may also be strain-specific. In fact, testing the dauer-inducing activity of synthetic samples of the seven generally most abundant ascarosides and paratosides (Jeong *et al.*, 2005; Bose *et al.*, 2012; Srinivasan *et al.*, 2012; von Reuss *et al.*, 2012) in the same six strains used for chemical analysis revealed strongly strain- and compound-specific variation of responsiveness (Fig. 7.6; Bose *et al.*, 2014). For example, pasc#9, a compound produced by all strains, induces dauer in five of the six strains, but is completely inactive in strain RS2333, the strain originally used to measure the dauer-inducing activity of *P. pacificus*-derived ascarosides and paratosides (Bose *et al.*, 2012). On the other hand, part#9 and npar#1, which were identified as strongly dauer-inducing in RS2333, induce dauer formation in all six strains, although to varying extent. Notably, ubas#1, a compound absent from two of the six strains, very strongly induces dauer in one strain, but is largely inactive in the other five. Detailed analysis of the dauer assay results showed that the potency of any compound in a specific strain does not correlate with the relative abundance of this compound in that strain's exo-metabolome. For example, the strongest producer of pasc#9, strain RS2333, does not respond to pasc#9, whereas of the two strongest producers of ubas#1, one strain responded very strongly to ubas#1 but the second did not respond at all.

The observation that strains may respond to compounds they do not synthesise or may synthesise compounds they do not respond to reinforced the notion that differences in dauer pheromone biosynthesis and responsiveness may play a role in intraspecific competition (Bose *et al.*, 2014). To investigate this hypothesis directly, a novel competition assay was developed using the commercially available Ussing chamber, which contains two membrane-separated compartments that are suitable for nematode liquid cultures. The membrane keeps the nematodes in the two compartments separated, but is permeable to water and small molecules and thus enables the study of dauer formation of two different *P. pacificus* strains in response to the mixture of the pheromone blends of the two strains. This setup mimics the natural environment experienced by several *P. pacificus* strains jointly colonising a decaying beetle. Competition assays exploring all three possible combinations of three

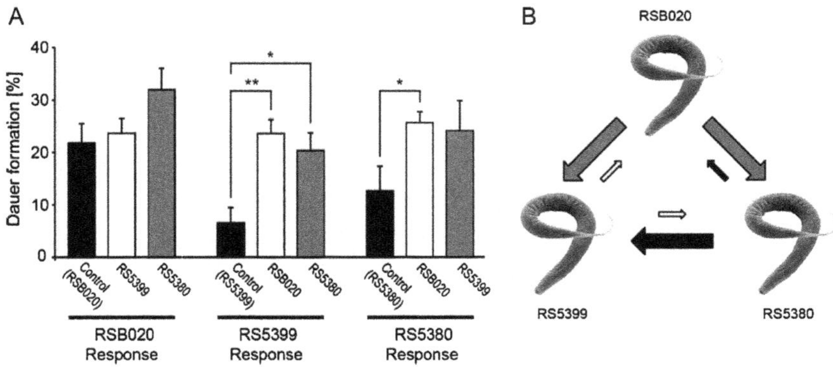

Fig. 7.7. *Intraspecific competition* via *small-molecule signalling in* Pristionchus pacificus. *A: Shown are results from a dauer pheromone competition assay with the sympatric strains RS5380, RS5399 and RSB020 grown in Ussing chambers. In control experiments, both compartments of a chamber were filled with nematodes of the same strain (control). In the competition experiments, one strain was grown in one compartment of an Ussing chamber, while a different strain was grown in the other compartment of the same chamber; B: Model for intraspecific competition among the three strains RS5380, RS5399 and RSB020 (Bose* et al., *2014).*

sympatric strains from La Réunion Island, revealed that in two combinations, dauer formation was enhanced for one of the two strains ('one sided cross-preference'), whereas in the third combination both strains showed enhanced dauer formation ('two-sided cross-preference'), relative to control experiments in which both chambers were charged with the same strain (Fig. 7.7) (Bose *et al.*, 2014).

Understanding of the role of small-molecule signalling for inter- and intra-specific competition and its evolutionary significance is still extremely limited. The analysis of pheromone biosynthesis and responsiveness in a very small set of *P. pacificus* wild isolates has demonstrated that small molecules play an important role in the interactions of different genotypes, and that the study of natural variation of small-molecule production and perception may provide important insights in the mechanisms of intraspecific competition. The large collection of *P. pacificus* isolates, compared to that of *C. elegans*, and the high genetic diversity make this species an excellent model system for studying natural variation of small-molecule biosynthesis and perception. In particular, combining comparative genomic and metabolomic approaches has the potential significantly to advance knowledge of the role of small-molecule signalling for all aspects of *P. pacificus* biology.

A conserved nuclear hormone receptor downstream of ascarosides

In *C. elegans*, the ascaroside receptors have been shown to act upstream of conserved insulin/IGF-1 and TGF-β, signalling (Hu, 2007; Ludewig & Schroeder, 2013) as well as sirtuin-dependent pathways (Ludewig *et al.*, 2013). Insulin/IGF-1 and TGF-β signalling, in turn, converge on the biosynthesis of another class of small-molecule signals: the steroidal ligands of the nuclear hormone receptor (NHR) DAF-12, one of at least 284 NHRs in *C. elegans* and orthologue of the human vitamin D (VDR) and liver-X receptors (LXR). DAF-12 functions as a ligand-dependent switch that regulates both adult lifespan and development in *C. elegans* (Antebi *et al.*, 2000; Fielenbach & Antebi, 2008; Bethke *et al.*, 2009; Kenyon, 2010; Wollam & Antebi, 2011; Wollam *et al.*, 2011, 2012; Lee & Schroeder, 2012) and has become an important model for metazoan NHR signalling because of its central role in *C. elegans* biology and close homology to functionally related mammalian NHRs (Taubert *et al.*, 2010; Hulme & Whitesides, 2011).

In the absence of its endogenous steroidal ligands, DAF-12 promotes developmental arrest at the dauer stage, whereas liganded DAF-12 allows dauer recovery and rapid maturation to reproductive adults (Antebi *et al.*, 2000; Lee & Schroeder, 2012). The decision between maturation and arrest in *C. elegans* occurs at the level of transcriptional regulation of several hormone biosynthetic genes, including DAF-9, a cytochrome p450 that had been shown to convert inactive precursor steroids into ligands that bind to DAF-12 (Gerisch *et al.*, 2001). *Via* modulation of the insulin/IGF-1 and TGF-β signalling pathways, favourable environmental conditions trigger upregulation of DAF-9 transcription and thus ligand production, whereas adverse, dauer-promoting conditions lead to DAF-9 suppression and abolishment of ligand production. In its unliganded form, DAF-12 then interact with its co-repressor DIN-1, a homologue of mammalian SHARP, resulting in repression of DAF-12 transcriptional targets, which causes developmental arrest at the dauer stage (Ludewig *et al.*, 2004). Correspondingly, loss of *daf-9* results in complete loss of ligand synthesis and constitutive dauer arrest, which can be rescued by addition of synthetic ligand.

The steroidal ligands of DAF-12 (called dafachronic acids, 'DAs'; Motola *et al.*, 2006; Gerisch *et al.*, 2007; Mahanti *et al.*, 2014) not only promote reproductive maturation, but also play an important role for adult longevity in *C. elegans*, in particular for the dramatic increase of lifespan in germline-deficient worms. As in the case of the chemical characterisation of the dauer pheromone, the use of comparative metabolomics played an important role in determining the exact structures of DAF-12 ligand structures and their biosynthetic pathways (Mahanti *et al.*, 2014). Precise knowledge of DAF-12 ligand structures and biosynthetic pathways is essential for understanding NHR function (Mangelsdorf *et al.*, 1995; Wollam & Antebi, 2011) because even small differences in ligand structures may result in dramatic changes of transcriptional activity and specificity (Brown & Slatopolsky, 2008; Singarapu *et al.*, 2011). As shown in Figure 7.8, the identified DAF-12 ligands are based on a regular 25-carbon steroid scaffold that bears a carboxy group at the end of the sidechain. Since *C. elegans* is incapable of cholesterol biosynthesis, cholesterol or a structurally similar steroid must be supplemented with the diet in order to prevent dauer arrest due to cessation of DAF-12 ligand production.

Sequencing of the genomes of diverse nematodes, including free-living, necromenic and parasitic species, has shown that the nuclear hormone receptor DAF-12 is highly conserved over many branches of the phylum Nematoda (Wang *et al.*, 2009; Sommer & Ogawa, 2011). Given that DAF-12 is strictly required for dauer formation in *C. elegans*, it was suspected that DAF-12 homologues in parasitic species may control the formation of infective juvenile stages, which in many ways resemble *C. elegans* dauers (Ogawa *et al.*, 2009; Wang *et al.*, 2009). Conservation of the DAF-12 signalling pathway was further suggested by the finding that cholesterol deprivation of *P. pacificus* increases dauer formation, as previously shown for *C. elegans* (Ogawa *et al.*, 2009). A subsequent forward genetic screen for dauer-defective mutants identified a clear homologue for *C. elegans* DAF-12 in *P. pacificus* which, considering the evolutionary distance between these two species, demonstrates very high conservation at the amino acid level in both the DNA-binding and ligand-binding domains. As in *C. elegans*, the ligand-binding domain of *Ppa*-DAF-12 was found to be essential for dauer formation, suggesting that the DAF-12 signalling cascade is *functionally* conserved, which requires that *P. pacificus* must also produce a dafachronic acid-like ligand that binds to *Ppa*-DAF-12.

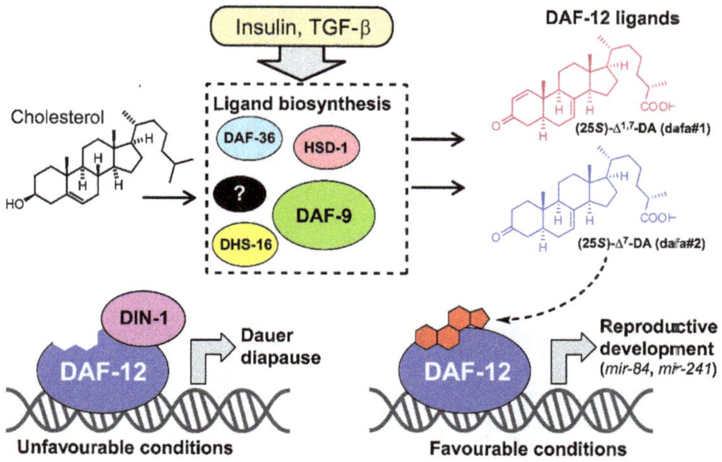

Fig. 7.8. *Steroidal ligands control development in* Caenorhabditis elegans via *the nuclear hormone receptor DAF-12. Under favourable conditions, insulin/IGF and TGF-β signalling drive biosynthesis of steroidal DAF-12-ligands. Liganded DAF-12 promotes development, in part* via *transcription of the* let-7*-family microRNAs* mir-84 *and* mir-241. *Under unfavourable conditions, ligand biosynthesis is inhibited, resulting in interaction of unliganded DAF-12 with its co-repressor DIN-1 (Mahanti* et al., *2014).*

In fact, addition of synthetic dafachronic acid was found to rescue the phenotype of several dauer constitutive *P. pacificus* mutants, closely mimicking the behaviour of dauer-constitutive *C. elegans* phenotypes that result from mutations upstream of DAF-12, *e.g.*, mutations in genes involved in dafachronic acid biosynthesis or its regulation (Ogawa *et al.*, 2009). Similarly, treatment with synthetic dafachronic acid was found completely to suppress formation of infective juveniles in the animal-parasitic nematode, *Strongyloides papillosus*, and could even redirect development to an additional free-living life cycle. Like *P. pacificus*, the *S. papillosus* genome includes a direct homologue of *Cel-daf-12*.

Although these results strongly suggested that *P. pacificus* (as well as parasitic species) produces dafachronic acid-like DAF-12 ligands, the *P. pacificus* genome does not feature a clear homologue of *Cel-daf-9*, the crucial p450 oxygenase at the end of the *C. elegans* dafachronic acid biosynthetic pathway. Furthermore, HPLC-MS-based analysis of the *P. pacificus* metabolome provided no evidence for production of any of the dafachronic acids identified from *C. elegans* (N. Bose, pers. comm.). Taken together, these findings indicate that,

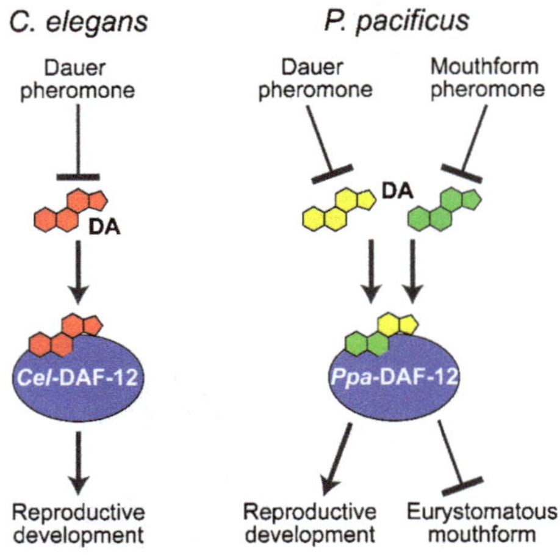

Fig. 7.9. *Schematic comparison of the roles of nuclear hormone receptor/DAF-12 signalling in* Caenorhabditis elegans *and* Pristionchus pacificus. *Upstream perception of ascarosides (*C. elegans*) or ascarosides and paratosides (*P. pacificus*) negatively regulates DAF-12 ligand (dafachronic acid; DA) biosynthesis. Binding of DA to DAF-12 promotes reproductive development in both species and has been co-opted to regulate mouth form in* P. pacificus.

although the DAF-12 signalling cascade is broadly conserved across species, the precise structures of the small-molecule ligands differ. Such structural differences are indicative of species-specific differences in biosynthetic pathways and the signalling network upstream of ligand biosynthesis. Species-specific variation of the signalling networks upstream of DAF-12 had been previously suggested by the observation that the role of TGF-β signalling for dauer formation in *C. elegans* differs markedly from its role for the formation of infective juveniles in parasitic species, *e.g.*, *Ancylostoma* spp. (Wang *et al.*, 2009). Therefore, precise characterisation of species-specific differences of DAF-12 ligand structures may provide important insights in how differences in ecology and life cycle are reflected by differences in the connections of the conserved insulin and TGF-β pathways to conserved nuclear hormone receptor signalling.

In addition to dauer formation, the *daf-12* signalling pathway also controls mouth form dimorphism in *P. pacificus* (Bento *et al.*, 2010). Mutations in *Ppa-daf-12* as well as addition of synthetic dafachronic acid suppress the eurystomatous mouth form, closely resembling the effects of *daf-12* mutation and dafachronic acid on dauer formation. Mouth form dimorphism in *P. pacificus* thus provides a remarkable example for the co-option of two conserved small-molecule signalling pathways for the regulation of phenotypic plasticity; ascaroside-based interorganismal signalling (Bose *et al.*, 2012) and the endocrine dafachronic acid/DAF-12 pathway (Bento *et al.*, 2010). The finding that mouth form dimorphism and dauer in *P. pacificus* are controlled by different sets of ascarosides suggests the intriguing possibility that these two phenotypes are controlled by different endogenous *Ppa*-DAF-12 ligands.

Dauer towers and an extremely long-chain wax ester

Lipids serve a wide variety of biological functions, as energy storage, membrane constituents, or precursors for signalling molecules such as prostaglandins or endocannabinoids. Correspondingly, lipids are chemically quite diverse, and lipid profiles often exhibit a high degree of species- and life stage-specificity. A variety of nematode-specific lipids have been reported, including unusual endocannabinoids (Izrayelit *et al.*, 2013) and other ethanolamides (Lucanic *et al.*, 2011), as well as lipids based on long-chain ascarosides (Pungaliya *et al.*, 2009). A recent study of *P. pacificus* dauer-associated behaviours revealed a highly unusual polyunsaturated wax ester, named 'nematoil', that plays an important role in a fascinating host-finding strategy (Penkov *et al.*, 2014). Clumps of up to a thousand *P. pacificus* dauers form extensive tower-like structures that reach up to 1 cm high and probably serve to increase the chance of getting picked up by a new host beetle. Formation of these 'dauer-towers' is facilitated by an oily, highly sticky secretion that only dauers produce. Chemical analysis revealed that the major component of this secretion represents a wax ester, nematoil, derived from a 30-carbon fatty acid and a 30-carbon alcohol, each including no fewer than six double bonds, for a total of 12 double bonds in the ester (Penkov *et al.*, 2014). Wax esters based on such extremely long-chained fatty acids or alcohols are extremely rare in animals and plants, and few lipids with a similarly high number of double bonds have been described. The high number of double bonds probably serves to keep nematoil liquid, as saturated wax esters of similar molecular weight stay solid at physiological temperatures. Additionally, the high molecular weight of nematoil confers extremely high viscosity, which is likely to help achieve the degree of 'stickiness' required for the formation of stable vertical

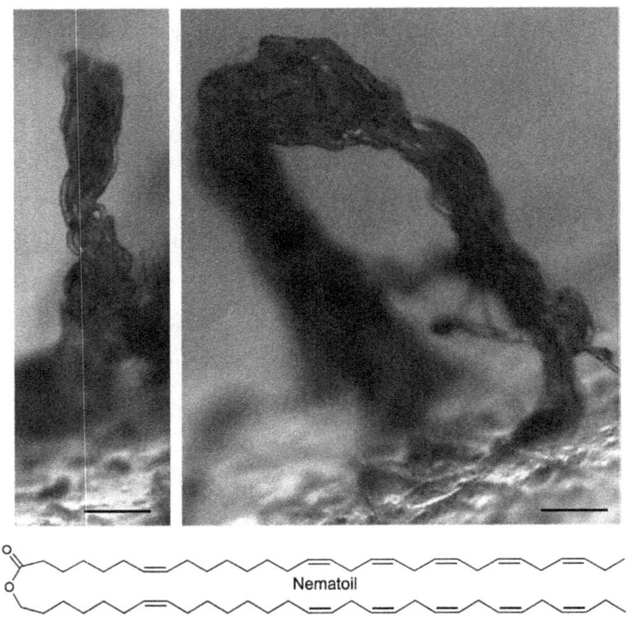

Fig. 7.10. *Dauer towers built with nematoil. Micrographs of dauer towers consisting of up to 1000* Pristionchus pacificus *dauers and chemical structure of nematoil (Penkov* et al., *2014).*

structures consisting of up to a thousand live dauers. The identification of nematoil highlights the structural and functional diversity of lipids, and further demonstrates that extremely long-chain fatty acids and their derivatives can play specific biological roles.

Conclusion

Structural and functional analyses of the *P. pacificus* metabolome have demonstrated that small molecules serve important functions relating to almost any aspect of this nematode's life history. In particular, work on *P. pacificus* has shown that the function of small molecules and their associations with specific chemical compound classes are evolutionarily conserved: in *C. elegans*, *P. pacificus* and other nematodes, glycosides derived from di-deoxysugars regulate development and phenotypic plasticity. Similarly, the dafachronic acid/NHR signalling mechanism is conserved across free-living, necromenic and parasitic nematode species. However, although associations of compound classes with types of biological roles are conserved, the exact chemical structures may vary considerably between species. For example, simple, unmodified ascarosides are the primary constituents of the *C. elegans* dauer pheromone, whereas paratosides and more complex, modular ascarosides fill this role in *P. pacificus*. Whereas the role of DAF-12 (and even biological activity of the *C. elegans*-derived dafachronic acids) is conserved, the precise structures of the DAF-12 ligands, and thus their upstream signalling network, appear to differ between species. Additionally, the ascaroside and dafachronic acid signalling pathways have been co-opted for regulating adult phenotypic plasticity in *P. pacificus*, whereas no such function is known from *C. elegans*.

Although ascarosides and paratosides appear to be nematode-specific, they exert their functions *via* conserved physiological pathways. Ascaroside perception is upstream of major physiological pathways including insulin, TGF-β, and steroid hormone signalling, which in turn control developmental progression and regulate metabolic state (Lee & Schroeder, 2012). Furthermore, it has been shown that ascaroside biosynthesis is strongly coupled to primary metabolism, *e.g.*, amino acid metabolism (Srinivasan *et al.*, 2012) and endocannabinoid biosynthesis (Izrayelit *et al.*, 2013). Further elucidation of the biosynthesis of ascarosides and paratosides will reveal how input from conserved primary metabolism is transduced to create signals that regulate development and behaviour in nematode model organisms. Moreover, since ascaroside signalling is highly conserved among nematodes, knowledge of ascaroside biosynthesis may also enable new approaches for the treatment of human nematode infections or control of parasitic nematodes in agricultural settings.

Lastly, it should be noted that the identification of the modular ascarosides and paratosides revealed entirely unexpected biosynthetic capabilities in animals. In contrast to most groups of microorganisms and plants, whose genomes have revealed a great variety of 'secondary' small-molecule biosynthetic pathways, *e.g.*, for polyketides and nonribosomal peptides (Walsh, 2007), most metazoans are not presumed to have dedicated biosynthetic machinery to generate structurally complex small molecules. As the building blocks of nematode-derived ascarosides and paratosides appear to be derived directly from conserved primary metabolism, it seems possible that similar types of modular small-molecule signals are produced by other animals, including mammals. These possibilities may inspire a comprehensive re-analysis of vertebrate small-molecule signalling pathways, taking advantage of the metabolomic approaches developed and validated using nematode model systems.

Acknowledgements

Partial support by the National Institutes of Health (GM088290) is gratefully acknowledged.

References

ANTEBI, A., YEH, W.H., TAIT, D., HEDGECOCK, E.M. & RIDDLE, D.L. (2000). *daf-12* encodes a nuclear receptor that regulates the dauer diapause and developmental age in *C. elegans*. *Genes & Development* 14, 1512-1527.

ARTYUKHIN, A.B., YIM, J.J., SRINIVASAN, J., IZRAYELIT, Y., BOSE, N., VON REUSS, S.H., JO, Y., JORDAN, J.M., BAUGH, L.R., CHEONG, M. *ET AL.* (2013). Succinylated octopamine ascarosides and a new pathway of biogenic amine metabolism in *Caenorhabditis elegans*. *Journal of Biological Chemistry* 288, 18778-18783.

BENTO, G., OGAWA, A. & SOMMER, R.J. (2010). Co-option of the hormone-signalling module dafachronic acid-DAF-12 in nematode evolution. *Nature* 466, 494-497.

BETHKE, A., FIELENBACH, N., WANG, Z., MANGELSDORF, D.J. & ANTEBI, A. (2009). Nuclear hormone receptor regulation of microRNAs controls developmental progression. *Science* 324, 95-98.

BOSE, N., OGAWA, A., VON REUSS, S.H., YIM, J.J., RAGSDALE, E.J., SOMMER, R.J. & SCHROEDER, F.C. (2012). Complex small-molecule architectures regulate phenotypic plasticity in a nematode. *Angewandte Chemie International Edition* 51, 12438-12443.

BOSE, N., MEYER, J.M., YIM, J.J., MAYER, M.G., MARKOV, G.V., OGAWA, A., SCHROEDER, F.C. & SOMMER, R.J. (2014). Natural variation in dauer pheromone production and sensing supports intraspecific competition in nematodes. *Current Biology* 24, 1536-1541.

BROWN, A.J. & SLATOPOLSKY, E. (2008). Vitamin D analogs: therapeutic applications and mechanisms for selectivity. *Molecular Aspects of Medicine* 29, 433-452.

BUTCHER, R.A., FUJITA, M., SCHROEDER, F.C. & CLARDY, J. (2007). Small-molecule pheromones that control dauer development in *Caenorhabditis elegans*. *Nature Chemical Biology* 3, 420-422.

CHALLIS, G.L. (2008). Genome mining for novel natural product discovery. *Journal of Medical Chemistry* 51, 2618-2628.

CHOE, A., CHUMAN, T., VON REUSS, S.H., DOSSEY, A.T., YIM, J.J., AJREDINI, R., KOLAWA, A.A., KAPLAN, F., ALBORN, H.T., TEAL, P.E. *ET AL.* (2012a). Sex-specific mating pheromones in the nematode *Panagrellus redivivus. Proceedings of the National Academy of Sciences of the United States of America* 109, 20949-20954.

CHOE, A., VON REUSS, S.H., KOGAN, D., GASSER, R.B., PLATZER, E.G., SCHROEDER, F.C. & STERNBERG, P.W. (2012b). Ascaroside signaling is widely conserved among nematodes. *Current Biology* 22, 772-780.

DEUTSCH, C., EL YACOUBI, B., DE CRECY-LAGARD, V. & IWATA-REUYL, D. (2012). Biosynthesis of threonylcarbamoyl adenosine (t^6A), a universal tRNA nucleoside. *Journal of Biological Chemistry* 287, 13666-13673.

DIAS, D.A., URBAN, S. & ROESSNER, U. (2012). A historical overview of natural products in drug discovery. *Metabolites* 2, 303-336.

FIELENBACH, N. & ANTEBI, A. (2008). *C. elegans* dauer formation and the molecular basis of plasticity. *Genes and Development* 22, 2149-2165.

FORSETH, R.R. & SCHROEDER, F.C. (2011). NMR-spectroscopic analysis of mixtures: from structure to function. *Current Opinion in Chemical Biology* 15, 38-47.

GEMS, D. (2007). Long-lived dwarf mice: are bile acids a longevity signal? *Aging Cell* 6, 421-423.

GERISCH, B. & ANTEBI, A. (2004). Hormonal signals produced by DAF-9/ cytochrome P450 regulate *C. elegans* dauer diapause in response to environmental cues. *Development* 131, 1765-1776.

GERISCH, B., WEITZEL, C., KOBER-EISERMANN, C., ROTTIERS, V. & ANTEBI, A. (2001). A hormonal signaling pathway influencing *C. elegans* metabolism, reproductive development, and life span. *Developmental Cell* 1, 841-851.

GERISCH, B., ROTTIERS, V., LI, D., MOTOLA, D.L., CUMMINS, C.L., LEHRACH, H., MANGELSDORF, D.J. & ANTEBI, A. (2007). A bile acid-like steroid modulates *Caenorhabditis elegans* lifespan through nuclear receptor signaling. *Proceedings of the National Academy of Sciences of the United States of America* 104, 5014-5019.

GOLDEN, J.W. & RIDDLE, D.L. (1982). A pheromone influences larval development in the nematode *Caenorhabditis elegans. Science* 218, 578-580.

GOLDEN, J.W. & RIDDLE, D.L. (1984a). A *Caenorhabditis elegans* dauer-inducing pheromone and an antagonistic component of the food supply. *Journal of Chemical Ecology* 10, 1265-1280.

GOLDEN, J.W. & RIDDLE, D.L. (1984b). The *Caenorhabditis elegans* dauer larva: developmental effects of pheromone, food, and temperature. *Developmental Biology* 102, 368-378.

GREER, E.L., MAURES, T.J., UCAR, D., HAUSWIRTH, A.G., MANCINI, E., LIM, J.P., BENAYOUN, B.A., SHI, Y. & BRUNET, A. (2011). Transgenerational epigenetic inheritance of longevity in *Caenorhabditis elegans*. *Nature* 479, 365-371.

HU, P.J. (2007). Dauer. In: The *C. elegans* Research Community (Ed.). *WormBook*. DOI:10.1895/wormbook.1.144.1.

HULME, S.E. & WHITESIDES, G.M. (2011). Chemistry and the worm: *Caenorhabditis elegans* as a platform for integrating chemical and biological research. *Angewandte Chemie International Edition* 50, 4774-4807.

IZRAYELIT, Y., SRINIVASAN, J., CAMPBELL, S.L., JO, Y., VON REUSS, S.H., GENOFF, M.C., STERNBERG, P.W. & SCHROEDER, F.C. (2012). Targeted metabolomics reveals a male pheromone and sex-specific ascaroside biosynthesis in *Caenorhabditis elegans*. *ACS Chemical Biology* 7, 1321-1325.

IZRAYELIT, Y., ROBINETTE, S.L., BOSE, N., VON REUSS, S.H. & SCHROEDER, F.C. (2013). 2D NMR-based metabolomics uncovers interactions between conserved biochemical pathways in the model organism *Caenorhabditis elegans*. *ACS Chemical Biology* 8, 314-319.

JANG, H., KIM, K., NEAL, S.J., MACOSKO, E., KIM, D., BUTCHER, R.A., ZEIGER, D.M., BARGMANN, C.I. & SENGUPTA, P. (2012). Neuromodulatory state and sex specify alternative behaviors through antagonistic synaptic pathways in *C. elegans*. *Neuron* 75, 585-592.

JEONG, P.Y., JUNG, M., YIM, Y.H., KIM, H., PARK, M., HONG, E., LEE, W., KIM, Y.H., KIM, K. & PAIK, Y.K. (2005). Chemical structure and biological activity of the *Caenorhabditis elegans* dauer-inducing pheromone. *Nature* 433, 541-545.

KAPLAN, F., SRINIVASAN, J., MAHANTI, P., AJREDINI, R., DURAK, O., NIMALENDRAN, R., STERNBERG, P.W., TEAL, P.E., SCHROEDER, F.C., EDISON, A.S. *ET AL.* (2011). Ascaroside expression in *Caenorhabditis elegans* is strongly dependent on diet and developmental stage. *PLoS ONE* 6, e17804.

KAPLAN, F., ALBORN, H.T., VON REUSS, S.H., AJREDINI, R., ALI, J.G., AKYAZI, F., STELINSKI, L.L., EDISON, A.S., SCHROEDER, F.C. & TEAL, P.E. (2012). Interspecific nematode signals regulate dispersal behavior. *PLoS ONE* 7, e38735.

KENYON, C. (2010). A pathway that links reproductive status to lifespan in *Caenorhabditis elegans*. *Annals of the New York Academy of Sciences* 1204, 156-162.

KIM, K., SATO, K., SHIBUYA, M., ZEIGER, D., BUTCHER, R., RAGAINS, J., CLARDY, J., TOUHARA, K. & SENGUPTA, P. (2009). Two chemoreceptors mediate developmental effects of dauer pheromone in *C. elegans*. *Science* 326, 994-998.

LEE, S.S. & SCHROEDER, F.C. (2012). Steroids as central regulators of organismal development and lifespan. *PLoS Biology* 10, e1001307.

LUCANIC, M., HELD, J.M., VANTIPALLI, M.C., KLANG, I.M., GRAHAM, J.B., GIBSON, B.W., LITHGOW, G.J. & GILL, M.S. (2011). N-acylethanolamine signalling mediates the effect of diet on lifespan in *Caenorhabditis elegans*. *Nature* 473, 226-229.

LUDEWIG, A.H. & SCHROEDER, F.C. (2013). Ascaroside signaling in *C. elegans*. In: The *C. elegans* Research Community (Ed.). *Wormbook*. DOI:10.1895/wormbook.1.155.1.

LUDEWIG, A.H., KOBER-EISERMANN, C., WEITZEL, C., BETHKE, A., NEUBERT, K., GERISCH, B., HUTTER, H. & ANTEBI, A. (2004). A novel nuclear receptor/coregulator complex controls *C. elegans* lipid metabolism, larval development, and aging. *Genes & Development* 18, 2120-2133.

LUDEWIG, A.H., IZRAYELIT, Y., PARK, D., MALIK, R.U., ZIMMERMANN, A., MAHANTI, P., FOX, B.W., BETHKE, A., DOERING, F., RIDDLE, D.L. *ET AL.* (2013). Pheromone sensing regulates *Caenorhabditis elegans* lifespan and stress resistance *via* the deacetylase SIR-2.1. *Proceedings of the National Academy of Sciences of the United States of America* 110, 5522-5527.

MACOSKO, E.Z., POKALA, N., FEINBERG, E.H., CHALASANI, S.H., BUTCHER, R.A., CLARDY, J. & BARGMANN, C.I. (2009). A hub-and-spoke circuit drives pheromone attraction and social behaviour in *C. elegans*. *Nature* 458, 1171-1175.

MAHANTI, P., BOSE, N., BETHKE, A., JUDKINS, J.C., WOLLAM, J., DUMAS, K.J., ZIMMERMAN, A.M., CAMPBELL, S.L., HU, P.J., ANTEBI, A. *ET AL.* (2014). Comparative metabolomics reveals endogenous ligands of DAF-12, a nuclear hormone receptor, regulating *C. elegans* development and lifespan. *Cell Metabolism* 19, 73-83.

MANGELSDORF, D.J., THUMMEL, C., BEATO, M., HERRLICH, P., SCHUTZ, G., UMESONO, K., BLUMBERG, B., KASTNER, P., MARK, M., CHAMBON, P. *ET AL.* (1995). The nuclear receptor superfamily: the second decade. *Cell* 83, 835-839.

MAYER, M.G. & SOMMER, R.J. (2011). Natural variation in *Pristionchus pacificus* dauer formation reveals cross-preference rather than self-preference of nematode dauer pheromones. *Proceedings of the Royal Society B: Biological Sciences* 278, 2784-2790.

MCCULLOCH, D. & GEMS, D. (2007). Sex-specific effects of the DAF-12 steroid receptor on aging in *Caenorhabditis elegans*. *Annals of the New York Academy of Sciences* 1119, 253-259.

MCGRATH, P.T., XU, Y., AILION, M., GARRISON, J.L., BUTCHER, R.A. & BARGMANN, C.I. (2011). Parallel evolution of domesticated *Caenorhabditis* species targets pheromone receptor genes. *Nature* 477, 321-325.

MEINWALD, J. (2011). Natural products as molecular messengers. *Journal of Natural Products* 74, 305-309.

MORGAN, K., MCGAUGHRAN, A., VILLATE, L., HERRMANN, M., WITTE, H., BARTELMES, G., ROCHAT, J. & SOMMER, R.J. (2012). Multi locus analysis of *Pristionchus pacificus* on La Réunion Island reveals an evolutionary history shaped by multiple introductions, constrained dispersal events and rare out-crossing. *Molecular Ecology* 21, 250-266.

MOTOLA, D.L., CUMMINS, C.L., ROTTIERS, V., SHARMA, K.K., LI, T., LI, Y., SUINO-POWELL, K., XU, H.E., AUCHUS, R.J., ANTEBI, A. *ET AL.* (2006). Identification of ligands for DAF-12 that govern dauer formation and reproduction in *C. elegans*. *Cell* 124, 1209-1223.

OGAWA, A., STREIT, A., ANTEBI, A. & SOMMER, R.J. (2009). A conserved endocrine mechanism controls the formation of dauer and infective larvae in nematodes. *Current Biology* 19, 67-71.

OGAWA, A., BENTO, G., BARTELMES, G., DIETERICH, C. & SOMMER, R.J. (2011). *Pristionchus pacificus daf-16* is essential for dauer formation but dispensable for mouth form dimorphism. *Development* 138, 1281-1284.

PARK, D., O'DOHERTY, I., SOMVANSHI, R.K., BETHKE, A., SCHROEDER, F.C., KUMAR, U. & RIDDLE, D.L. (2012). Interaction of structure-specific and promiscuous G-protein-coupled receptors mediates small-molecule signaling in *Caenorhabditis elegans*. *Proceedings of the National Academy of Sciences of the United States of America* 109, 9917-9922.

PENKOV, S., OGAWA, A., SCHMIDT, U., TATE, D., ZAGORIY, V., BOLAND, S., GRUNER, M., VORKEL, D., VERBAVATZ, J.-M., SOMMER, R.J. *ET AL.* (2014). A wax ester promotes collective host finding in the nematode *Pristionchus pacificus*. *Nature Chemical Biology* 10, 281-285.

PRINCE, E. & POHNERT, G. (2010). Searching for signals in the noise: metabolomics in chemical ecology. *Analytical and Bioanalytical Chemistry* 396, 193-197.

PUNGALIYA, C., SRINIVASAN, J., FOX, B.W., MALIK, R.U., LUDEWIG, A.H., STERNBERG, P.W. & SCHROEDER, F.C. (2009). A shortcut to identifying small molecule signals that regulate behavior and development in *Caenorhabditis elegans*. *Proceedings of the National Academy of Sciences of the United States of America* 106, 7708-7713.

ROBINETTE, S.L., BRUSCHWEILER, R., SCHROEDER, F.C. & EDISON, A.S. (2011). NMR in metabolomics and natural products research: two sides of the same coin. *Accounts of Chemical Research* 45, 288-297.

SCHROEDER, F.C. (2006). Small molecule signaling in *Caenorhabditis elegans*. *ACS Chemical Biology* 1, 198-200.

SEROBYAN, V., RAGSDALE, E.J., MULLER, M.R. & SOMMER, R.J. (2013). Feeding plasticity in the nematode *Pristionchus pacificus* is influenced by sex and social context and is linked to developmental speed. *Evoluton & Development* 15, 161-170.

SINGARAPU, K.K., ZHU, J., TONELLI, M., RAO, H., ASSADI-PORTER, F.M., WESTLER, W.M., DELUCA, H.F. & MARKLEY, J.L. (2011). Ligand-specific structural changes in the vitamin D receptor in solution. *Biochemistry* 50, 11025-11033.

SOMMER, R.J. & OGAWA, A. (2011). Hormone signaling and phenotypic plasticity in nematode development and evolution. *Current Biology* 21, R758-R766.

SRINIVASAN, J., KAPLAN, F., AJREDINI, R., ZACHARIAH, C., ALBORN, H.T., TEAL, P.E.A., MALIK, R.U., EDISON, A.S., STERNBERG, P.W. & SCHROEDER, F.C. (2008). A blend of small molecules regulates both mating and development in *Caenorhabditis elegans. Nature* 454, 1115-1118.

SRINIVASAN, J., VON REUSS, S.H., BOSE, N., ZASLAVER, A., MAHANTI, P., HO, M.C., O'DOHERTY, O.G., EDISON, A.S., STERNBERG, P.W. & SCHROEDER, F.C. (2012). A modular library of small molecule signals regulates social behaviors in *Caenorhabditis elegans. PLoS Biology* 10, e1001237.

STUPP, G.S., CLENDINEN, C.S., AJREDINI, R., SZEWC, M.A., GARRETT, T., MENGER, R.F., YOST, R.A., BEECHER, C. & EDISON, A.S. (2013). Isotopic ratio outlier analysis global metabolomics of *Caenorhabditis elegans. Analytical Chemistry* 85, 11858-11865.

TAUBERT, S., WARD, J.D. & YAMAMOTO, K.R. (2010). Nuclear hormone receptors in nematodes: evolution and function. *Molecular and Cellular Endocrinology* 334, 49-55.

TENNESSEN, J.M., BARRY, W.E., COX, J. & THUMMEL, C.S. (2014). Methods for studying metabolism in *Drosophila. Methods* 68, 105-115.

VINEY, M.E., GARDNER, M.P. & JACKSON, J.A. (2003). Variation in *Caenorhabditis elegans* dauer larva formation. *Development, Growth and Differentiation* 45, 389-396.

VON REUSS, S.H., BOSE, N., SRINIVASAN, J., YIM, J.J., JUDKINS, J.C., STERNBERG, P.W. & SCHROEDER, F.C. (2012). Comparative metabolomics reveals biogenesis of ascarosides, a modular library of small molecule signals in *C. elegans. Journal of the American Chemical Society* 134, 1817-1824.

WALSH, C.T. (2007). The chemical versatility of natural-product assembly lines. *Accounts of Chemical Research* 41, 4-10.

WANG, Z., ZHOU, X.E., MOTOLA, D.L., GAO, X., SUINO-POWELL, K., CONNEELY, A., OGATA, C., SHARMA, K.K., AUCHUS, R.J., LOK, J.B. *ET AL.* (2009). Identification of the nuclear receptor DAF-12 as a therapeutic target in parasitic nematodes. *Proceedings of the National Academy of Sciences of the United States of America* 106, 9138-9143.

WELLER, A.M., MAYER, W.E., RAE, R. & SOMMER, R.J. (2010). Quantitative assessment of the nematode fauna present on *Geotrupes* dung beetles reveals species-rich communities with a heterogeneous distribution. *Journal of Parasitology* 96, 525-531.

WITTE, H., MORENO, E., RÖDELSPERGER, C., KIM, J., KIM, J.-S., STREIT, A. & SOMMER, R.J. (2015). Gene inactivation using the CRISPR/Cas9 system in the nematode *Pristionchus pacificus*. *Development Genes and Evolution* 225, 55-62.

WOLLAM, J. & ANTEBI, A. (2011). Sterol regulation of metabolism, homeostasis, and development. *Annual Review of Biochemistry* 80, 885-916.

WOLLAM, J., MAGOMEDOVA, L., MAGNER, D.B., SHEN, Y., ROTTIERS, V., MOTOLA, D.L., MANGELSDORF, D.J., CUMMINS, C.L. & ANTEBI, A. (2011). The Rieske oxygenase DAF-36 functions as a cholesterol 7-desaturase in steroidogenic pathways governing longevity. *Aging Cell* 10, 879-884.

WOLLAM, J., MAGNER, D.B., MAGOMEDOVA, L., RASS, E., SHEN, Y., ROTTIERS, V., HABERMANN, B., CUMMINS, C.L. & ANTEBI, A. (2012). A novel 3-hydroxysteroid dehydrogenase that regulates reproductive development and longevity. *PLoS Biology* 10, e1001305.

YAMADA, K., HIROTSU, T., MATSUKI, M., BUTCHER, R.A., TOMIOKA, M., ISHIHARA, T., CLARDY, J., KUNITOMO, H. & IINO, Y. (2010). Olfactory plasticity is regulated by pheromonal signaling in *Caenorhabditis elegans*. *Science* 329, 1647-1650.

Nematology Monographs & Perspectives, 2015, Vol. 11, 197-219

Chapter 8

Population genetics and the La Réunion case study

Angela McGaughran [1] and Katy Morgan [2]

[1] *Bioinformatics & Phylogenomics Team, CSIRO Ecosystem Sciences, GPO Box 1700, Canberra, ACT 2601, Australia*
ang.mcgaughran@gmail.com
[2] *Computer Center, University of New Orleans, Room 200, 2000 Lakeshore Drive, New Orleans, LA 70148, USA*
kmorgan4@uno.edu

Introduction

Nematodes are an incredibly diverse and successful group of organisms. As the most abundant of the metazoans, they can be found in all habitats with a source of organic carbon, from the deep sea to hot springs and ice-covered terrestrial landscapes. Free-living nematodes play a key ecological role as decomposers and predators of micro-organisms (Neher, 2001), whilst parasitic species can have substantial impacts on the health and demography of host populations (*e.g.*, Williamson & Hussey, 1996; Brooker, 2010). Despite their great biological diversity and importance, nematodes are sorely under-represented in the population genetic literature and a wealth of questions relating to their evolutionary history remain unanswered.

Population demographic history and genetic structure has been most thoroughly addressed in parasitic nematode species that have a medical or agricultural importance. In such species, patterns of gene flow and divergence are strongly linked to the anthropogenically-associated movements of the hosts (*e.g.*, Blouin *et al.*, 1995; Wielgoss *et al.*, 2008; Gilabert & Wasmuth, 2013). However, an understanding of the demography and evolutionary history of species that differ in life history and lack such a strong anthropogenic association is largely absent. Phylogeographic studies in species such as *Caenorhabditis elegans* (*e.g.*, Sivasundar & Hey, 2003) and *C. briggsae* (Cutter *et al.*, 2006, 2010) are based on a broad geographic scale, and sparse local sampling within species prevents detailed analysis of population genetic processes. Indeed, the difficulties in sampling nematode populations are recognised as one of the factors holding back a thorough understanding of their genetic structure and dynamics (Gilabert & Wasmuth, 2013). *Pristionchus pacificus* offers the advantage of a non-anthropogenically-associated nematode species with a well-defined ecology, coupled with an ability to be sampled relatively easily on a local scale. An additional advantage of *P. pacificus* as a study system is that, like *C. elegans*, the species has been developed as a model in evolutionary biology and is easily maintained in the laboratory (Sommer, 2009). In particular, the availability of genomic tools provides great scope for studies into local adaptation and the evolution of phenotypic variation.

Although *C. elegans* continues to be a central focus of molecular, cell and developmental biology research, in practice a single strain (N2) has dominated research in this species over the past 40 years. Recent population genetic and genomic studies have begun to reverse this trend; however, the discovery that genome-wide selective sweeps have erased signals of previous evolutionary history in *C. elegans* has restricted the conclusions that can be drawn for this species (Andersen *et al.*, 2012). The inferences of population genetic structure and demographic history that we draw from well-sampled natural populations of hermaphroditic *P. pacificus* and describe in this chapter thus provide a unique example of a nematode species with a well-understood evolutionary history.

The first specific beetle host identified for *P. pacificus* was the Oriental beetle, *Exomala orientalis*, from Japan and the USA (Herrmann *et al.*, 2007). Further exploration in a biogeographic context established the presence of *P. pacificus* in locations encompassing Asia, North America, South Africa and the Mascarene Islands of the Indian Ocean (Herrmann *et al.*, 2010) (Fig. 8.1). In the latter case, *P. pacificus* was found to be in high abundance and on an array of beetle hosts on La Réunion Island (Fig. 8.2). This opened the *P. pacificus* system for population genetic and island biogeographic studies (*e.g.*, Morgan *et al.*, 2014), such that we are now in a position to integrate evo-devo (macro-evolution) with population genetics and evolutionary ecology (micro-evolution) to examine the contribution of natural variation and changing environments to the evolutionary process (Sommer, 2009). To further this aim an evolutionary field station has been established on the island of La

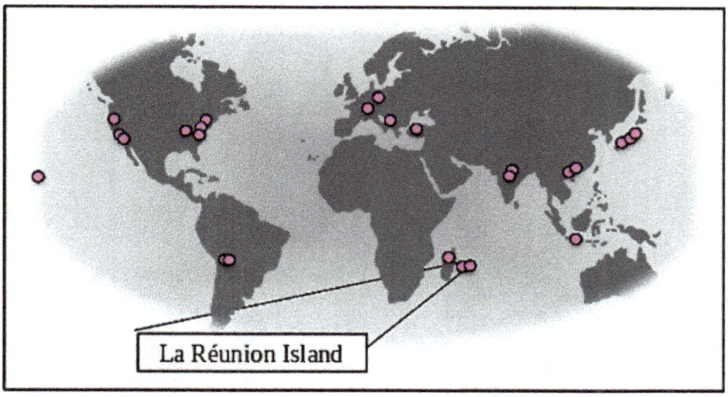

Fig. 8.1. *The cosmopolitan worldwide distribution of collected* Pristionchus pacificus *strains, and the location of La Réunion Island.*

Réunion and new samples are systematically collected across the island on an annual basis.

Considered to be a major biodiversity hotspot (Myers *et al.*, 2000; Thébaud *et al.*, 2009), La Réunion is the youngest (2-3 Ma), largest (2512 km^2), steepest (up to 3070 m a.s.l.), and most complex (both topographically and ecologically) island in the Mascarene island chain that includes neighbouring islands Mauritius and Rodriguez (Fig. 8.3). Volcanic activity, which continues to the present day, has shaped the island's rugged landscape, where short geographic distances can see dramatic altitudinal changes. Wind patterns across the island add to this diversity; climate on the north-eastern, windward side of the island is characterised by high rainfall, while the south-western, leeward side is substantially drier. Climatic variables acting upon a dynamic geological template together create a complex suite of habitat types or 'ecozones' (Strasberg *et al.*, 2005) across La Réunion (Fig. 8.3). The island thus provides an ideal setting for investigating the impacts of colonisation history, landscape, and environment on natural nematode populations.

This chapter will examine the population genetics of *P. pacificus* in its Mascarene habitat, focusing largely on La Réunion Island. We will start by describing the patterns of genetic diversity and distribution that characterise *P. pacificus* populations worldwide. Next, we will evaluate the evolutionary history of *P. pacificus* lineages on La Réunion, encompassing colonisation mechanisms, divergence estimates and demographic (population expansion and migration) properties of populations. Finally, we will focus on the role of the environment in the

Fig. 8.2. *Examples of the six most common scarab beetle hosts of* Pristionchus pacificus *collected from La Réunion Island. Photos taken by members of the Max Planck Institute.*

local population dynamics of *P. pacificus*, before discussing the scope for future work on this system. Throughout the chapter, we talk about individuals, strains and populations of *P. pacificus*. Generally speaking, individual samples are collected and processed according to the protocol of Herrmann *et al.* (2006). Briefly, freshly sacrificed carcasses of beetle hosts (see below) are monitored daily for the emergence of adult nematodes, which are then maintained in the laboratory on agar plates seeded with a bacterial food source (*Escherichia coli*). Isogenic lines are generated from each adult nematode individual by allowing reproduction and maintaining offspring in culture. Thus, the term 'strain' refers to these laboratory-maintained, isogenic lines (which can also be frozen

Fig. 8.3. *Examples illustrating the complex topographical and ecological nature of La Réunion Island, demonstrating the various habitat/ecozone types that characterise the island, from steep altitudinal gradients in arid areas (A, D) to high-altitude dry (B, C) and regenerating (C) habitats. Photos taken by A. McGaughran or K. Morgan.*

and thawed at a later date), while 'population' refers to the collection of multiple individuals (and their later isogenic lines) from the same geographic location.

Diversity and distribution

Nematode sampling across several geographic regions/ecozones and encompassing several host beetle species identified La Réunion Island as an oasis for *P. pacificus* (Herrmann *et al.*, 2010; Morgan *et al.*, 2012) (Fig. 8.4). Population genetic analyses based on both microsatellite (representative of the six *P. pacificus* chromosomes; n = 17-20) and mitochondrial markers revealed that genetic diversity in *P. pacificus* on La Réunion covers the complete known worldwide diversity of the species (Herrmann *et al.*, 2010; Morgan *et al.*, 2012). Specifically, gene diversity (H_E) at microsatellite loci among La Réunion populations

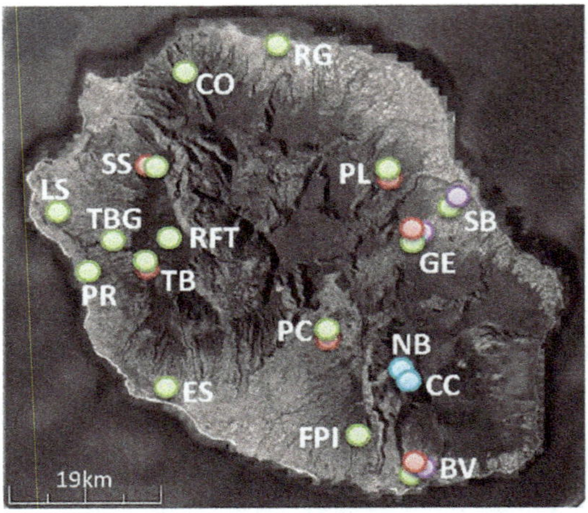

Fig. 8.4. *Diversity and distribution of the four mitochondrial lineages of* Pristionchus pacificus *sampled from La Réunion Island. Sample collection points are coloured according to lineage: purple, blue, green and red for lineages A, B, C and D, respectively. Location codes correspond to: Basse Vallée (BV), Le Cratère Commerson (CC), Colorado (CO), Etang Salé (ES), Forêt de Petite Ile (FPI), Grand Etang (GE), La Saline (LS), Nez de Bœuf-Volcano (NB), Plaine des Cafres (PC), Plaine des Lianes (PL), Petite Ravine (PR), Route Forestière des Tamarins (RFT), Roland Garros (RG), Saint Benoit (SB), Sans Souci (SS), Trois Bassins (TB), Trois Bassins Garden (TBG).*

averages approximately 0.700, with an upper limit of 0.951 (Morgan *et al.*, 2012), and the number of unique microsatellite haplotypes detected in 223 individuals in Morgan *et al.* (2012) equates to 87% of the dataset. Comparatively, mitochondrial haplotype diversity in *P. pacificus* ranges from 0.409 to 0.953, and the number of unique mitochondrial haplotypes detected in 272 individuals was 74 (27%) (Morgan *et al.*, 2012). Collectively, this amounts to the presence on La Réunion of four broad mitochondrial *P. pacificus* lineages (designated 'A', 'B', 'C' and 'D' in Herrmann *et al.*, 2010 and Morgan *et al.*, 2012), which are each separated by a high degree of genetic distance (Fig. 8.5). These lineages on La Réunion correspond globally to the following patterns: lineage A has a largely Asian distribution, including India, China and Japan, and has also been sporadically sampled in North America, Turkey, Bolivia and Hawaii; lineage B is exclusive to La Réunion Island (although this may reflect incomplete geographic sampling); lineage C is found mostly on La Réunion, but also in America, Bolivia, Montenegro and South

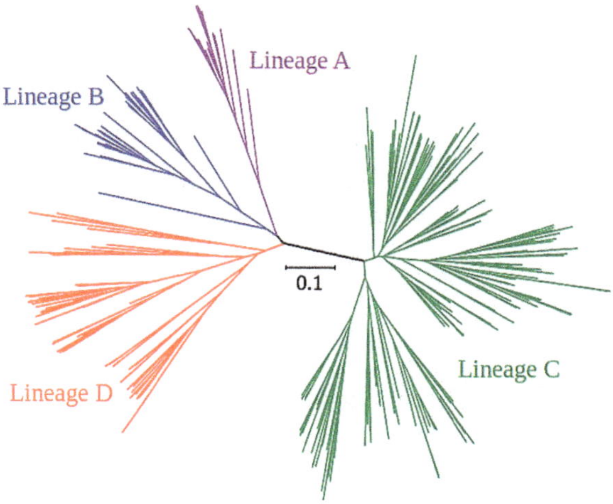

Fig. 8.5. *Neighbour-joining tree to show the four mitochondrial genetic lineages present in La Réunion* Pristionchus pacificus. *Lineages are colour-coded: purple, blue, green and red for lineages A, B, C and D, respectively.*

Africa; lineage D has been collected in South Africa and Switzerland (Herrmann *et al.*, 2010).

Analysis of the distribution of the four mitochondrial lineages on La Réunion shows that discrete differences define their geographic ranges. Generally, lineages A and D encompass predominantly eastern localities on the island, lineage B is made up almost exclusively of strains from the central volcanic plateau, and lineage C comprises largely western localities (Figs 8.4, 8.5). Genetic examination of the lineages in finer detail shows that approximately 12-15 sub-populations can be retrieved from the La Réunion microsatellite dataset when applying a clustering algorithm (Morgan *et al.*, 2012; McGaughran *et al.*, 2014). Microsatellite clusters generally correspond with sub-divided mitochondrial lineages (*e.g.*, lineage B forms two different genetic clusters that include no lineage A, C or D individuals; McGaughran *et al.*, 2014), with mitochondrial lineages A and B partitioning into one or two microsatellite genetic clusters and the remaining clusters representing vast sub-division of lineage C/D individuals (Morgan *et al.*, 2012; McGaughran *et al.*, 2014). This conforms well to recent genomic analysis, which shows an increasing pattern of complexity, such that the designation of four mitochondrial lineages A-D is most likely an over-simplification of the true genomic diversity (Rödelsperger *et al.*,

2014). Despite general patterns identified at the broader mitochondrial lineage scale, no clear relationship between nuclear/genomic genetic cluster and geographic location exists at finer scales (McGaughran *et al.*, 2014), indicating that modern geographic populations of *P. pacificus* have virtually all been compiled from multiple ancestral sources (McGaughran *et al.*, 2013).

Evolutionary history

The diversity and differentiation characterising *P. pacificus* is in stark contrast to the pattern seen in *C. elegans*, where studies have suggested that low genetic diversity and a lack of population structure are due to a high prevalence of linkage disequilibrium and selective sweeps in the recent evolutionary history of *C. elegans* (Zauner *et al.*, 2007; Rockman & Kruglyak, 2009; Anderson *et al.*, 2012). The selective sweeps in *C. elegans* are thought to have been driven by the species' anthropogenic association, and the long-distance movements of humans and agricultural produce (Andersen *et al.*, 2010, 2012). Indeed, an anthropogenic influence on population structure has been detected in several nematode species that parasitise livestock and agriculturally important plants, and often has the effect of homogenising genetic structure (Blouin *et al.*, 1995; Wielgoss *et al.*, 2008).

Species such as *C. elegans* and *P. pacificus* that reproduce primarily by hermaphroditic selfing are often characterised by strong genome-wide linkage disequilibrium, and are thus prone to selective sweeps and reduced genetic diversity. However, rare instances of out-crossing can result in the rapid generation of vast genotypic diversity because out-crossing, when followed by repeated generations of selfing, can create large arrays of novel allelic combinations across the genome (Siol *et al.*, 2008). In *P. pacificus*, out-crossing events occur with an estimated frequency of 2-20% (Morgan *et al.*, 2012), but this is not higher than the estimated range (0-22%) of out-crossing in *C. elegans* (Anderson *et al.*, 2010, and references therein). Thus, despite extensive linkage and high rates of selfing in both species, the evolutionary history of *P. pacificus* clearly differs from that of *C. elegans*.

The high diversity and strong population structure in *P. pacificus* appears more similar to phylogeographic patterns detected in the hermaphroditic species *C. briggsae*, in which geographical population structure was attributed to divergence between populations inhabiting tropical and temperate regions (Cutter *et al.*, 2010). In *P. pacificus*, population genetic patterns are probably due to historical (or ancient) divergence events within the ancestral metapopulation of this species. High genetic distances between the four major mitochondrial lineages of *P. pacificus* (>20 mutational steps/>5% uncorrected p-distance sequence divergence between lineages), and generally concordant patterns of differentiation at unlinked mitochondrial and nuclear markers together suggest a long period of isolation among lineages (Fig. 8.5). Indeed, multiple divergence dating estimates suggest that diversification occurred early in the evolutionary history of *P. pacificus* (Molnar *et al.*, 2011; McGaughran *et al.*, 2013). Divergence estimates were first made in *P. pacificus* based on a mutation accumulation (MA) line approach (Molnar *et al.*, 2011). With MA line experiments it is possible to directly assess the minimum number of new mutations that appear in a given lineage over a specified number of generations (Lynch, 2010). Such experiments are limited to a small number of model organisms including *C. elegans* (Denver *et al.*, 2000, 2004). The MA lines established in *P. pacificus* were first used to estimate a mitochondrial mutation rate of 7.6×10^{-8} ($\pm$ 2.2×10^{-8}) per site per generation, which is not statistically different from the rates derived for *C. elegans* (Denver *et al.*, 2000) and *C. briggsae* (Howe *et al.*, 2010). This mutation rate estimate was then used in multiple mitochondrial-based dating assessments. Specifically, codon-partitioned dating analyses using BEAST software (Drummond & Rambaut, 2007) for a subset of nine isolates that spanned the entire world diversity estimated the most recent common ancestor (TMRCA) of all *P. pacificus* lineages to be 1.12×10^6 (1.1-1.3×10^6) generations, while diversification estimates among lineages ranged from 280 000 to 560 000 generations (Molnar *et al.*, 2011). Further analysis of a much larger group of La Réunion-specific strains (McGaughran *et al.*, 2013), and additional follow-up work based on population genomic resequencing of the MA lines that established a nuclear mutation rate per site per generation in *P. pacificus* of 1.4-2.6×10^{-9} and a TMRCA among 47 wild isolates of 4.3×10^6 generations (Weller *et al.*, 2014), provide clear support that divergence in the *P. pacificus* metapopulation occurred early in the evolutionary history of *P. pacificus*.

This divergence, therefore, most probably preceded the emergence and colonisation of La Réunion Island. Indeed, La Réunion harbours all four mitochondrial lineages of *P. pacificus* and ecological and population genetic studies suggest that the island is too young for such diversification to have evolved *in situ*. Instead, this diversity is most probably due to repeated independent island colonisation events by *P. pacificus*. Above, we noted that no clear relationship between genetic cluster and geographic location exists at finer scales on La Réunion Island, which is a pattern indicative of multiple colonisation events from diverse source populations. To test this hypothesis in a modelling framework, an approximate Bayesian computation (ABC) approach was applied to the microsatellite and mitochondrial datasets. As well as examining the number of discrete colonisation events, this research provided insight into the order and timing of the establishment of *P. pacificus* populations on La Réunion (McGaughran *et al.*, 2013).

Using specialised ABC software (Beaumont *et al.*, 2002; Bertorelle *et al.*, 2010; Csilléry *et al.*, 2010; Guillemaud *et al.*, 2010), this work confirmed that establishment of *P. pacificus* on La Réunion occurred *via* at least four independent founding events from the (unsampled) source population (McGaughran *et al.*, 2013). Differences in both the relative timing and location of these independent colonisation episodes may explain the present differences in distribution and differentiation among the different lineages. For example, in the case of mitochondrial lineage C, early colonisation of La Réunion may have allowed for the consequent structure and widespread western distribution observed today (Fig. 8.4). Conversely, more recent independent eastern colonisation by the B and D lineages has presumably left less time for their dispersal to other geographic regions (Morgan *et al.*, 2012; McGaughran *et al.*, 2013). Lineage A is also more narrowly dispersed across its eastern distribution (Fig. 8.4). Thus, it may be that establishment in eastern regions of the island is subject to greater dispersal restriction subsequent to foundation.

Colonisation order may also have been important in terms of defining niche exclusivity across distinct La Réunion ecozones (Fig. 8.3). At Saint Benoit, for example, sympatric mitochondrial lineage A and D populations exist but appear to remain isolated. Isolation among lineages following differentially-timed foundation may have resulted in a suite of phenotypic and genotypic differences as isolated island populations diverged in adaptive traits and/or host specificity. Defining natural variation in phenotypic traits in an evolutionary context is an on-going aim of our research group (*e.g.*, Hong *et al.*, 2008; Bento *et al.*, 2010; Mayer & Sommer, 2011).

Demography

POPULATION EXPANSION

Although estimates of the evolutionary history of *P. pacificus* indicate diversification largely preceded colonisation on La Réunion Island, patterns of diversity and distribution among lineages suggest that, following colonisation, different populations and lineages have experienced further diversifying effects. As mentioned previously, discrete differences define the geographic ranges of the four mitochondrial lineages, with A and D encompassing predominantly eastern localities on the island, B consisting almost exclusively of strains from the central volcanic plateau, and C comprising largely western populations (Fig. 8.4). Further, while all lineages are characterised by a high degree of within-lineage diversity, heterogeneity and population structure is comparatively higher in mitochondrial lineage C (Morgan *et al.*, 2012).

Lineages A, C and D experience some degree of geographic overlap, and there is some evidence to support rare admixture/recombination events between these lineages. For example, admixed microsatellite haplotypes consisting of both predominantly eastern and predominantly western alleles occur in a few instances (Morgan *et al.*, 2012). However, mixing among lineages is generally rare, and lineage B individuals do not mix with the other lineages at all as their populations are almost exclusively found in isolated habitats of high altitude, and in association with a specific beetle host (Morgan *et al.*, 2012; McGaughran *et al.*, 2014). The fact that the lineages remain distinct, even in regions of sympatry, suggests that some degree of reproductive isolation may have accumulated between them. Association with different assemblages of host beetle species (Fig. 8.2) is one potential isolating mechanism, and the question of potential reproductive isolation is an issue that requires further attention.

Due to their relative isolation, the different lineages and populations are most likely subject to different demographic parameters. To test this, demographic (mismatch distribution) analyses were performed for the four mitochondrial La Réunion lineages (McGaughran *et al.*, 2013). This analysis identified signals of spatial population expansion for lineages B, C and D. Dating estimates associated with the mismatch distribution suggested that the expansion events occurred from 59 000 (lineage B) to 125 000 (lineage D) ybp. This time period may correspond to the population growth immediately succeeding foundation of the independent mitochondrial lineages (McGaughran *et al.*, 2013).

MIGRATION

Characterising demographic processes in *P. pacificus* requires knowledge about the beetle host, because the growth and movement of the nematode is necessarily dictated by the dispersal dynamics of its host (Fig. 8.2). On La Réunion, *P. pacificus* is found in association with several distinct scarab beetles that are known to have invaded the island at different times in history and in a highly species-specific manner. *Oryctes borbonicus*, an endemic La Réunion scarab beetle, shows the highest infestation rate for *P. pacificus* (Herrmann *et al.*, 2010) (Fig. 8.2). This beetle most likely invaded the island early in its history and co-evolution between *P. pacificus* and *O. borbonicus* may have resulted in the enormous radiation that encompasses many strains found in *P. pacificus* mitochondrial lineage C today. In addition to the *O. borbonicus* association, lineage C populations have undergone extensive host-switching episodes, shifting to newly invaded beetle hosts of wide habitat breadth, further accounting for the wider distribution of this lineage on La Réunion (Morgan *et al.*, 2012). By contrast, *Maladera affinis* is a beetle that is known to have invaded the island from India during the last few hundred years (Cheke & Hume, 2008) (Fig. 8.2). Consistently, *P. pacificus* strains found on *M. affinis* fall into mitochondrial lineage A, which is geographically restricted in its distribution on La Réunion. Inasmuch, beetle host utilisation may represent spurious founding events akin to the colonisation of new geographic areas in other species. Random founding events probably explain a degree of the phylogeographic patterning in *P. pacificus* as a consequence of some beetles more commonly inhabiting certain environments.

However, despite hitchhiking with beetles being a viable option for migration among geographic regions, the differentiation among populations on La Réunion indicates that gene flow is far from being a homogenising force in *P. pacificus*. Instead, rare episodes of successful dispersal are likely to have punctuated the demographic history of La Réunion lineages. Recent analyses to explore this further involved employment of the software IMA2 (Hey, 2010), which performs analyses of genetic data under the Isolation with Migration model of population divergence to calculate and date migration events between populations (Morgan *et al.*, 2014). This work extended its focus beyond La Réunion populations to encompass nearby Mauritius Island in order to examine migration both within and between islands. The analysis found evidence for high levels of bidirectional migration between La Réunion and Mauritius (Morgan *et al.*, 2014). High levels of haplotype sharing among the islands and a lack of clear genetic separation between *P. pacificus* populations on Mauritius and La Réunion suggested frequent migration despite the presence of a considerable (145 km-wide) oceanic barrier (Morgan *et al.*, 2014). This contrasts strongly with patterns in other taxa. For example, considerable genetic and phenotypic differentiation exists between La Réunion and Mauritius populations of the Mascarene grey white-eye, a small passerine bird endemic to the region (Mila *et al.*, 2010), such that the formerly 'conspecific' taxon is now considered to be two separate species (Warren *et al.*, 2006).

In addition to supporting periodic long-distance dispersal events in *P. pacificus*, dating of between-island migration events provided further information about their colonisation history. Specifically, the similar assemblage of highly divergent lineages detected on both Mauritius and La Réunion suggest that the multiple independent colonisations of La Réunion identified above (McGaughran *et al.*, 2013) were 'regional' colonisations that resulted in simultaneous establishment on both islands. Previous studies in *C. elegans* have also supported the capacity for long-distance dispersal, with a lack of clear geographical structure and frequent examples of mitochondrial haplotype sharing across continents. However, as previously mentioned, dispersal within *C. elegans* is suggested to have been strongly aided by anthropogenic associations (Sivasundar & Hey, 2003; Zauner *et al.*, 2007; Anderson *et al.*, 2012). Given that, as for *C. elegans*, *P. pacificus* is frequently found free within the soil, anthropogenically-mediated dispersal facilitated by boat, aeroplane, or even the shoes of tourists, may explain the genetic connections between La Réunion and Mauritius islands in *P. pacificus*. This may be especially important given the lack of beetle fauna overlap between the two islands, suggesting that beetle movements between islands are limited. However, the ability of *P. pacificus* to survive within the soil column independently of beetle hosts, and the capacity for self-fertilisation and thus population establishment from a single founding individual, means that isolated instances of beetle dispersal that fail to result in beetle establishment may still result in successful nematode dispersal. Thus, beetle-mediated dispersal between the islands is likely to be important, even in the absence of successful host establishment.

Within-island migration analyses were also recently performed to examine migration properties of *P. pacificus* populations on La Réunion alone (Morgan *et al.*, 2014). The program BAYESASS (Wilson & Rannala, 2003) was used to perform Bayesian inference of recent migration (*i.e.*, within the last one to three generations) in a relatively assumption-free manner. In this analysis, patterns of recent immigration were shown often to be asymmetrical (*i.e.*, greater in one direction than the other) between pairs of La Réunion populations. The estimated percentage of immigrants varied from 0.3 to 14% across all pair-wise comparisons using BAYESASS, with the total percentage of immigrants per population ranging from 10.7 to 32.2% (Morgan *et al.*, 2014). Collectively, migration analyses suggest that dispersal among Mauritius and La Réunion islands is not strongly limited, and neither is it within La Réunion Island; in relation to this, the non-panmictic nature of La Réunion populations is striking.

Examining migration further, boundary analyses were performed to make additional inferences about the connectedness of populations. Several software packages (ALLELES IN SPACE, Miller, 2005; WOMBSOFT, Crida & Manel, 2007; spatial Principal Components Analysis, Jombart *et al.*, 2010) highlighted genetic barriers that separated the north-eastern part of the island from south-western areas. Thus, it may be that environmental distinctions across the island overcome the effects of dispersal on gene flow, thereby playing a strong role in determining population subdivision in *P. pacificus*.

Environmental aspects

It is reasonable to assume that once a lineage established on La Réunion, ecological and geological factors were both important in limiting dispersal and driving differentiation among populations. The ability of *P. pacificus* to tolerate a wide variety of environments and its co-dispersal with a variety of beetle species make it a good model species for investigating the complex effects of environmental, ecological and geological factors on local adaptation and genotypic evolution. For example, isolation among lineages following differentially-timed foundation (from an already diverse gene pool; see above) may have resulted in a suite of phenotypic and genotypic differences as discrete populations diverge in adaptive traits and/or host specificity (*e.g.*, Roman & Darling, 2007; Dlugosch & Parker, 2008). La Réunion, with its diverse array of ecotypes, steep altitudinal gradients and host beetle species (Figs 8.2, 8.3), offers an ideal site for such studies.

On La Réunion, adaptive divergence has been seen in birds (Mila *et al.*, 2010) and insects (Paupy *et al.*, 2001; Morlais *et al.*, 2005), recapitulating the general trend of rapid genetic and phenotypic diversification in association with ecological shifts seen in other island systems (*e.g.*, Jordan *et al.*, 2005; Kleindorfer *et al.*, 2006; Lawton-Rauh *et al.*, 2007; Mathys & Lockwood, 2011). Defining natural variation in phenotypic traits of *P. pacificus* in an evolutionary context is an on-going aim of current research (Hong *et al.*, 2008; Mayer & Sommer, 2011). In particular, how does *P. pacificus* align with this island paradigm where niche variability can facilitate adaptive divergence in association with distinct local environments?

Some signatures in the *P. pacificus* La Réunion data suggest that local adaptation may be driving differences among populations at the broad regional scale. For example, the east/west partition of population genetic structure on La Réunion (Morgan *et al.*, 2012) could reflect a pattern of poor adaptation among eastern lineages to the arid western climate and/or better adaptation among the western-distributed mitochondrial lineage C strains. Other examples point towards additional layers of complexity. For example, lineage B strains, found exclusively on La Réunion, form a genetically distinct group that corresponds with the endemic beetle *Amneidus godefroyi*. In strict association with these beetles, lineage B strains are likely to be locally adapted to the cooler conditions that characterise high-altitude locations at which that beetle is found. The failure of lineage B to disperse both within La Réunion and from this island to Mauritius, despite putatively frequent dispersal events among the other genetic lineages (see above), is consistent with the hypothesis of dispersal limited by environmental factors. For example, the maximum altitude on Mauritius (828 m a.s.l.) is much lower than the minimum altitude of lineage B collection sites (>2000 m a.s.l.) on La Réunion; thus, the environmental conditions characterising lineage B habitats on La Réunion are not present on Mauritius. It is therefore possible that this lack of suitable environmental conditions prevented establishment and/or maintenance of a Mauritius lineage B (Morgan *et al.*, 2014).

On La Réunion, local adaptation to the distinct climates on the wet eastern and arid western sides of the island is proposed to have driven differentiation between populations of mosquitoes (Paupy *et al.*, 2001; Morlais *et al.*, 2005). Other island invertebrates have also been shown to display east/west patterns of divergence associated with environmental factors (*e.g.*, McGaughran *et al.*, 2010). To investigate environmental effects on genetic structure in more detail, recent work employed species distribution modelling approaches. This method combines information about species occurrence with climate variables to generate continuous predictions of the potential distribution of species (*e.g.*, Fielding & Bell, 1997). For *P. pacificus* on La Réunion Island, species distribution models indicated that a significant proportion of the La Réunion landscape is putatively 'excellent', 'very highly probable' or 'highly probable' habitat for *P. pacificus*. The areas of highest probabilities of occurrence largely corresponded to discrete pockets of habitat across an inner circular belt of the island, where conditions are more generally cooler and wetter, while the greater proportion of low-lying coastal areas, where conditions are hotter and drier, appear to be largely avoided by *P. pacificus* and its beetle hosts (McGaughran *et al.*, 2014).

Further work suggested that environmental factors are also important in determining the genetic structure of populations. Specifically, a series of Mantel tests detected significant associations (r values between 0.246 and 0.417; $P < 0.05$) between genetic, geographic and environmental distances among populations (McGaughran *et al.*, 2014). This correlation was higher between genetic and environmental distances than between genetic and geographic distances, providing significant support for an 'isolation by environmental distance' pattern, rather than pure isolation by distance alone. Regression and redundancy approaches supported the Mantel results, with the climatic variables, annual minimum temperature, annual precipitation and temperature seasonality explaining up to 65.4% of the total explainable genetic variance after removing variance explained by geography (McGaughran *et al.*, 2014). Collectively, environmental analyses suggest that differences among habitats in climatic variables are particularly important influences of genetic structure among La Réunion *P. pacificus*. This has important adaptive implications as populations may respond differently to local environments. Mapping of the potentially adaptive genetic variation that underlies phenotypes under selection in different habitats and linking this genomic variation with functional studies to explore the adaptive mechanisms driving present-day population structure in more detail is an ongoing aim of research in this system.

Conclusion

Using La Réunion Island as a case study, we have united ecological, genetic, developmental and population-based approaches to begin to disentangle the intricacies of the evolutionary history of *P. pacificus*. Analyses based on microsatellite, mitochondrial and genome-wide markers have shown there to be substantial genetic variation and population structure within *P. pacificus* populations in the Indian Ocean. This manifests in the presence of highly significant genetic differentiation indices and a low degree of haplotype sharing among populations. Four distinct mitochondrial genetic lineages and approximately 12-15 nuclear genetic sub-populations characterise the *P. pacificus* metapopulation on La Réunion Island. These lineages invaded the island independently over different time scales and are now defined by distinct demographic parameters. As a result, distribution and diversity patterns among the lineages differ substantially. Environmental and host-beetle variation among locations has contributed to this pattern and now complex effects of environment, ecology and geology influence local adaptation and genotypic evolution in *P. pacificus*. Elements of the life history strategy of *P. pacificus*, such as a hermaphroditic mating system with occasional out-crossing, further contribute to significant genetic differentiation among isolated nematode populations. In concert, these factors conspire strongly to set *P. pacificus* apart from *C. elegans*, with the evolutionary histories of the two species clearly differing. Combining an island system with the highly complex nature of genetic partitioning in *P. pacificus* has provided us with a unique opportunity to compare, contrast and understand the mechanisms underlying evolutionary history in nematode species.

Future directions

The studies described in this chapter have provided detailed understanding of patterns of colonisation, dispersal and gene flow within natural populations of *P. pacificus* on La Réunion Island and have revealed the influence of environmental variation on genetic structure. The next step will be to combine genomic and phenotypic data to detect patterns of local adaptation in natural populations across the island, and to identify environmental factors driving selection on specific phenotypic traits.

The differentiation of the many divergent lineages detected on the island, and the issue of potential reproductive isolation between them, also warrant further attention. These lineages have co-existed on La Réunion for a considerable time period, and three of the four mitochondrial lineages are known to have regions of sympatry. Despite the apparent opportunity for gene flow and homogenisation of genetic divergence, the lineages remain distinct, with only limited evidence for rare cases of admixture (Morgan *et al.*, 2012). This suggests a degree of reproductive isolation, which may be due to behavioural differences such as association with different host beetle species, or to a potential post-mating component. Further genomic analysis and experimental data are necessary to shed light on possible isolating mechanisms and to gain a more complete understanding of *P. pacificus* evolutionary history.

Finally, one issue that has rarely been addressed in natural populations of androdioecious nematodes is the role of males, out-crossing, and recombination in evolutionary history. For example, the impacts of sex-based dynamics on colonisation, adaptation to novel environments and population persistence are unclear. Although differences in spontaneous male production have been documented among natural isolates of *P. pacificus* (Click *et al.*, 2009), variation in out-crossing rates across lineages and under differing environmental conditions has not been characterised. Experimental evaluation of spontaneous male production rates in different environments and analysis of linkage disequilibrium and recombination signatures in *P. pacificus* will contribute to our understanding of the importance of males and out-crossing in natural populations.

References

ANDERSEN, E., GERKE, J., SHAPIRO, J., CRISSMAN, J., GHOSH, R., BLOOM, J., FÉLIX, M.-A. & KRUGLYAK, L. (2012). Chromosome-scale selective sweeps shape *Caenorhabditis elegans* genomic diversity. *Nature Genetics* 44, 285-290.

ANDERSEN, J.L., MORRAN, L.T. & PHILLIPS, P.C. (2010). Outcrossing and the maintenance of males within *C. elegans* populations. *Journal of Heredity* 101(Suppl. 1), S62-S74.

BEAUMONT, M.A., ZHANG, W. & BALDING, D.J. (2002). Approximate Bayesian computation in population genetics. *Genetics* 162, 2025-2035.

BENTO, G., OGAWA, A. & SOMMER, R.J. (2010). Co-option of the hormone-signalling module dafachronic acid-DAF-12 in nematode evolution. *Nature* 466, 494-497.

BERTORELLE, G., BENAZZO, A. & MONA, S. (2010). ABC as a flexible framework to estimate demography over space and time: some cons, many pros. *Molecular Ecology* 19, 2609-2625.

BLOUIN, M.S., YOWELL, C.A., COURTNEY, C.H. & DAME, J.B. (1995). Host movement and the genetic structure of populations of parasitic nematodes. *Genetics* 141, 1007-1014.

BROOKER, S. (2010). Estimating the global distribution and disease burden of intestinal nematode infections: adding up the numbers – a review. *International Journal for Parasitology* 40, 1137-1140.

CHEKE, A. & HUME, J. (2008). *Lost land of the Dodo: the ecological history of Mauritius, Réunion and Rodrigues*. New Haven, CT, USA, Yale University Press.

CLICK, A., SAVALIYA, C.H., KIENLE, S., HERRMANN, M. & PIRES-DASILVA, A. (2009). Natural variation of outcrossing in the hermaphroditic nematode *Pristionchus pacificus. BMC Evolutionary Biology* 9, 75.

CRIDA, A. & MANEL, S. (2007). WOMBSOFT: an R package that implements the Wombling method to identify genetic boundary. *Molecular Ecology Notes* 7, 588-591.

CSILLÉRY, K., BLUM, M.G.B., GAGGIOTTI, O.E. & FRANÇOIS, O. (2010). Approximate Bayesian Computation (ABC) in practice. *Trends in Ecology & Evolution* 25, 410-418.

CUTTER, A.D., FELIX, M.-A., BARRIÈRE, A. & CHARLESWORTH, D. (2006). Patterns of nucleotide polymorphism distinguish temperate and tropical wild isolates of *Caenorhabditis briggsae. Genetics* 173, 2021-2031.

CUTTER, A.D., YAN, W., TSVETKOV, N., SUNIL, S. & FELIX, M.-A. (2010). Molecular population genetics and phenotypic sensitivity to ethanol for a globally diverse sample of the nematode *Caenorhabditis briggsae. Molecular Ecology* 19, 798-809.

DENVER, D.R., MORRIS, K., LYNCH, M., VASSILIEVA, L.L. & THOMAS, W.K. (2000). High direct estimate of the mutation rate in the mitochondrial genome of *Caenorhabditis elegans. Science* 289, 2342-2344.

DENVER, D.R., MORRIS, K., LYNCH, M. & THOMAS, W.K. (2004). High mutation rate and predominance of insertions in the *Caenorhabditis elegans* nuclear genome. *Nature* 430, 679-682.

DLUGOSCH, K.M. & PARKER, I.M. (2008). Founding events in species invasions: genetic variation, adaptive evolution, and the role of multiple introductions. *Molecular Ecology* 17, 431-449.

DRUMMOND, A. & RAMBAUT, A. (2007). BEAST: Bayesian evolutionary analysis by sampling trees. *BMC Evolutionary Biology* 7, 214.

FIELDING, A.H. & BELL, J.F. (1997). A review of methods for the assessment of prediction errors in conservation presence/absence models. *Environmental Conservation* 24, 38-49.

GILABERT, A. & WASMUTH, J.D. (2013). Unravelling parasitic natural history using population genetics. *Trends in Parasitology* 29, 438-448.

GUILLEMAUD, T., BEAUMONT, M.A., CIOSI, M., CORNUET, J.M. & ESTOUP, A. (2010). Inferring introduction routes of invasive species using approximate Bayesian computation on microsatellite data. *Heredity* 104, 88-99.

HERRMANN, M., MAYER, W.E. & SOMMER, R.J. (2006). Nematodes of the genus *Pristionchus* are closely associated with scarab beetles and the Colorado potato beetle in Western Europe. *Zoology* 109, 96-108.

HERRMANN, M., MAYER, W., HONG, R., KIENLE, S., MINASAKI, R. & SOMMER, R.J. (2007). The nematode *Pristionchus pacificus* (Nematoda: Diplogastridae) is associated with the Oriental beetle *Exomala orientalis* (Coleoptera: Scarabaeidae) in Japan. *Zoological Science* 24, 883-889.

HERRMANN, M., KIENLE, S., ROCHAT, J., MAYER, W.E. & SOMMER, R.J. (2010). Haplotype diversity of the nematode *Pristionchus pacificus* on Réunion in the Indian Ocean suggests multiple independent invasions. *Biological Journal of the Linnean Society* 100, 170-179.

HEY, J. (2010). Isolation with migration models for more than two populations. *Molecular Biology and Evolution* 27, 905-920.

HONG, R.L., WITTE, H. & SOMMER, R.J. (2008). Natural variation in *Pristionchus pacificus* insect pheromone attraction involves the protein kinase EGL-4. *Proceedings of the National Academy of Sciences of the United States of America* 105, 7779-7784.

HOWE, D.K., BAER, C.F. & DENVER, D.R. (2010). High rate of large deletions in *Caenorhabditis briggsae* mitochondrial genome mutation processes. *Genome Biology and Evolution* 1, 29-38.

JOMBART, T., DEVILLARD, S. & BALLOUX, F. (2010). Discriminant analysis of principal components: a new method for the analysis of genetically structured populations. *BMC Genetics* 11, 94-94.

JORDAN, M.A., SNELL, H.L., SNELL, H.M. & JORDAN, W.C. (2005). Phenotypic divergence despite high levels of gene flow in Galápagos lava lizards (*Microlophus albemarlensis*). *Molecular Ecology* 14, 859-867.

KLEINDORFER, S., CHAPMAN, T.W., WINKLER, H. & SULLOWAY, F.J. (2006). Adaptive divergence in contiguous populations of Darwin's small ground finch (*Geospiza fuliginos*). *Evolutionary Ecology Research* 8, 357-372.

LAWTON-RAUH, A., ROBICHAUX, R.H. & PURUGGANAN, M.D. (2007). Diversity and divergence patterns in regulatory genes suggest differential gene flow in recently derived species of the Hawaiian silversword alliance adaptive radiation (Asteraceae). *Molecular Ecology* 16, 3995-4013.

LYNCH, M. (2010). Evolution of the mutation rate. *Trends in Genetics* 26, 345-352.

MATHYS, B.A. & LOCKWOOD, J.L. (2011). Contemporary morphological diversification of passerine birds introduced to the Hawaiian archipelago. *Proceedings of the Royal Society B: Biological Sciences* 278, 2392-2400.

MAYER, M.G. & SOMMER, R.J. (2011). Natural variation in *Pristionchus pacificus* dauer formation reveals cross-preference rather than self-preference of nematode dauer pheromones. *Proceedings of the Royal Society B: Biological Sciences* 278, 2784-2790.

MCGAUGHRAN, A., CONVEY, P., STEVENS, M.I. & CHOWN, S. (2010). Metabolic rate, genetic and microclimate variation among springtail populations from sub-Antarctic Marion Island. *Polar Biology* 33, 909-918.

MCGAUGHRAN, A., MORGAN, K. & SOMMER, R.J. (2013). Unravelling the evolutionary history of the nematode *Pristionchus pacificus*: from lineage diversification to island colonization. *Ecology and Evolution* 3, 667-675.

MCGAUGHRAN, A., MORGAN, K. & SOMMER, R.J. (2014). Environmental variables explain genetic structure in a beetle-associated nematode. *PLoS ONE* 9, e87317.

MILA, B., WARREN, B., HEEB, P. & THÉBAUD, C. (2010). The geographic scale of diversification on islands: genetic and morphological divergence at a very small spatial scale in the Mascarene grey white-eye (Aves: *Zosterops borbonicus*). *BMC Evolutionary Biology* 10, 158.

MILLER, M.P. (2005). Alleles In Space (AIS): computer software for the joint analysis of interindividual spatial and genetic information. *Journal of Heredity* 96, 722-724.

MOLNAR, R.I., BARTELMES, G., DINKELACKER, I., WITTE, H. & SOMMER, R.J. (2011). Mutation rates and intraspecific divergence of the mitochondrial genome of *Pristionchus pacificus*. *Molecular Biology and Evolution* 28, 2317-2326.

MORGAN, K., MCGAUGHRAN, A., VILLATE, L., HERRMANN, M., WITTE, H., BARTELMES, G., ROCHAT, J. & SOMMER, R.J. (2012). Multi locus analysis of *Pristionchus pacificus* on La Réunion Island reveals an evolutionary history shaped by multiple introductions, constrained dispersal events and rare out-crossing. *Molecular Ecology* 21, 250-266.

MORGAN, K., MCGAUGHRAN, A., GANESHAN, S., HERRMANN, M. & SOMMER, R.J. (2014). Landscape and oceanic barriers shape dispersal and population structure in the island nematode, *Pristionchus pacificus*. *Biological Journal of the Linnean Society* 112, 1-15.

MORLAIS, I., GIROD, R., HUNT, R., SIMARD, F. & FONTENILLE, D. (2005). Population structure of *Anopheles arabiensis* on La Réunion Island, Indian Ocean. *American Journal of Tropical Medicine and Hygiene* 73, 1077-1082.

MYERS, N., MITTERMEIER, R.A., MITTERMEIER, C.G., DA FONSECA, G.A.B. & KENT, J. (2000). Biodiversity hotspots for conservation priorities. *Nature* 403, 853-858.

NEHER, D.A. (2001). Role of nematodes in soil health and their use as indicators. *Journal of Nematology* 33, 161-168.

PAUPY, C., GIROD, R., SALVAN, M., RODHAIN, F. & FAILLOUX, A.-B. (2001). Population structure of *Aedes albopictus* from La Réunion Island (Indian Ocean) with respect to susceptibility to a dengue virus. *Heredity* 87, 273-283.

ROCKMAN, M.V. & KRUGLYAK, L. (2009). Recombinational landscape and population genomics of *Caenorhabditis elegans*. *PLoS Genetics* 5, e1000419.

RÖDELSPERGER, C., NEHER, R.A., WELLER, A.M., EBERHARDT, G., WITTE, H., MAYER, W.E., DIETERICH, C. & SOMMER, R.J. (2014). Characterization of genetic diversity in the nematode *Pristionchus pacificus* from population-scale resequencing data. *Genetics* 196, 1153-1165.

ROMAN, J. & DARLING, J.A. (2007). Paradox lost: genetic diversity and the success of aquatic invasions. *Trends in Ecology & Evolution* 22, 454-464.

SIOL, M., PROSPERI, J.M., BONNIN, I. & RONFORT, J. (2008). How multilocus genotypic pattern helps to understand the history of selfing populations: a case study in *Medicago truncatula*. *Heredity* 100, 517-525.

SIVASUNDAR, A. & HEY, J. (2003). Population genetics of *Caenorhabditis elegans*: the paradox of low polymorphism in a widespread species. *Genetics* 163, 147-157.

SOMMER, R.J. (2009). The future of evo-devo: model systems and evolutionary theory. *Nature Reviews Genetics* 10, 416-422.

STRASBERG, D., ROUGET, M., RICHARDSON, D., BARET, S., DUPONT, J. & COWLING, R.M. (2005). An assessment of habitat diversity and transformation on La Réunion Island (Mascarene Islands, Indian Ocean) as a basis for identifying broad-scale conservation priorities. *Biodiversity Conservation* 14, 3015-3032.

THÉBAUD, C., WARREN, B.H., STRASBERG, D. & CHEKE, A. (2009). Mascarene Islands, biology. In: Gillespie, R.G. & Clague, D.A. (Eds). *Encyclopedia of islands*. Oakland, CA, USA, University of California Press, pp. 612-619.

WARREN, B.H., BERMINGHAM, E., PRYS-JONES, R.P. & THÉBAUD, C. (2006). Immigration, species radiation and extinction in a highly diverse songbird lineage: white-eyes on Indian Ocean islands. *Molecular Ecology* 15, 3769-3786.

WELLER, A.M., RÖDELSPERGER, C., EBERHARDT, G., MOLNAR, R.I. & SOMMER, R.J. (2014). Opposing forces of A/T-biased mutations and G/C-biased gene conversions shape the genome of the nematode *Pristionchus pacificus*. *Genetics* 196, 1145-1152.

WIELGOSS, S., TARASCHEWSKI, H., MEYER, A. & WIRTH, T. (2008). Population structure of the parasitic nematode *Anguillicola crassus*, an invader of declining North Atlantic eel stocks. *Molecular Ecology* 17, 3478-3495.

WILLIAMSON, V.M. & HUSSEY, R.S. (1996). Nematode pathogenesis and resistance in plants. *Plant Cell* 8, 1735-1745.

WILSON, G.A. & RANNALA, B. (2003). Bayesian inference of recent migration rates using multilocus genotypes. *Genetics* 163, 1177-1191.

ZAUNER, H., MAYER, W.E., HERRMANN, M., WELLER, A., ERWIG, M. & SOMMER, R.J. (2007). Distinct patterns of genetic variation in *Pristionchus pacificus* and *Caenorhabditis elegans*, two partially selfing nematodes with cosmopolitan distribution. *Molecular Ecology* 16, 1267-1280.

Nematology Monographs & Perspectives, 2015, Vol. 11, 221-255

Chapter 9

Evo-devo and developmental systems drift: an evolving paradigm in organ formation and tissue coordination, vulva and gonad development in *Pristionchus pacificus*

David RUDEL

Department of Biology, East Carolina University, Greenville, NC 27858, USA
rudeld@ecu.edu

Introduction

I don't know the question, but sex is definitely the answer.
Woody Allen

There is nothing more fundamental to a species than the continuity of generations. As such, the vulva and the gonad that comprise the nematode hermaphrodite reproductive system have been essential to developmental and evolutionary studies. Not surprisingly, given their importance in producing the next generation, their development is highly regulated and the gonad itself is the largest organ by volume and cell number in most nematodes. The vulva and gonad must interact in a coordinated way for proper development of both organs and, in normal adult physiology, for the manufacture of gametes and progeny. Here I will describe the development and morphology of the organs and discuss their developmental genetic underpinnings in the nematode *Pristionchus pacificus*. The discussion will be viewed through the looking glass provided by the development of these organs in the model nematode, *Caenorhabditis elegans*. *Pristionchus pacificus* shares a great deal of developmental and morphological features with *C. elegans* making comparisons and developmental changes experimentally tractable (see Fig. 9.1 for comparison of reproductive systems). Developmental events in both animals are coordinated through four life stages prior to adulthood, with stages punctuated by moults. In *P. pacificus*, first- to fourth-stage 'juveniles' are labelled J1-J4 (Fig. 2.4). In *C. elegans*, first- to fourth-stage 'larvae' are labelled L1-L4. This chapter focuses upon the differences in vulva and gonad development that have been generated during the evolution of these two species, and makes several concluding arguments based upon the presentation of comparative vulva and gonad development. Among them are inferences concerning the roles of developmental systems drift, redundancy, heterochrony, pleiotropy and the co-option of regulatory pathways and cellular processes in evolution and development.

A comparative description of vulva development

The vulva is the egg-laying organ of the nematode (Fig. 9.1, open triangles and inserts). In *P. pacificus* the vulva is formed from the descendants of only three cells. The fully formed vulva connects the egg-containing uterus to the external environment by forming a channel composed of rings of tissues. This channel is opened and closed by connected sex muscles for laying eggs. Vulva development has three stages: establishment of the vulval equivalence group (VEG), induction and morphogenesis of the vulva. Vulva and gonad development are intertwined to make a fully functional reproductive system. Signals from the gonad pattern the early development of the vulva. Later, the vulva instructs the shape of the hermaphrodite gonad arms. Lastly, crosstalk between the ventral uterus and the vulva coordinate the connection between the two organs. Without this coordination, developing progeny are not laid, early progeny hatch inside the mother, and the mother is killed and consumed from the inside out. In *C. elegans*, vulva development can be subdivided into the same three stages, and work over the last three decades has provided detailed genetic and molecular insight into underlying mechanisms (Sommer, 2005; Sternberg, 2005; Podbilewicz, 2006; Gupta *et al.*, 2012), creating a paradigm for comparative and evolutionary studies.

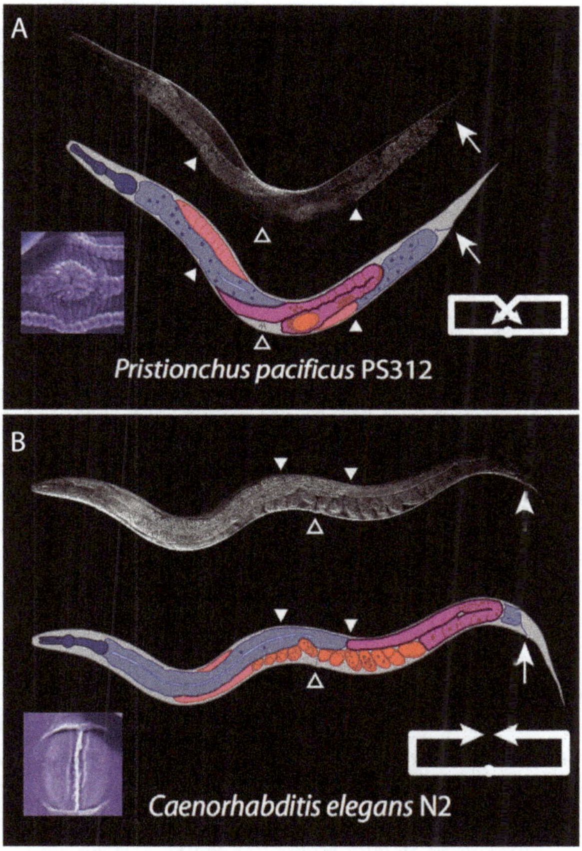

Fig. 9.1. *Summary of the differences between (A)* Pristionchus pacificus *and (B)* Caenorhabditis elegans *reproductive systems. Nematodes are approximately 1 mm in length. Nomarski photomicrographs are at the top of the panels; cartoons of the photomicrographs are shown below. Red/violet highlights gonad tissues. The anterior gonad arm is in red and is obscured by the overlying gut. The posterior arm is shown in violet. Developing eggs are shown in dark red. Blue highlights the pharynx (dark) and intestine (light). Small solid dark ovals/circles represent visible nuclei. Solid triangles represent the position of the distal tip cells (DTCs) at the end of the gonad arms, ventral for* P. pacificus *and dorsal for* C. elegans. *Inserts: The right insert is a line diagram depicting the path of gonad arm extension in the animal with the central dot representing the position of the connection to the vulva. The left inserts are scanning electron micrographs showing a surface ventral view of the adult vulva.*

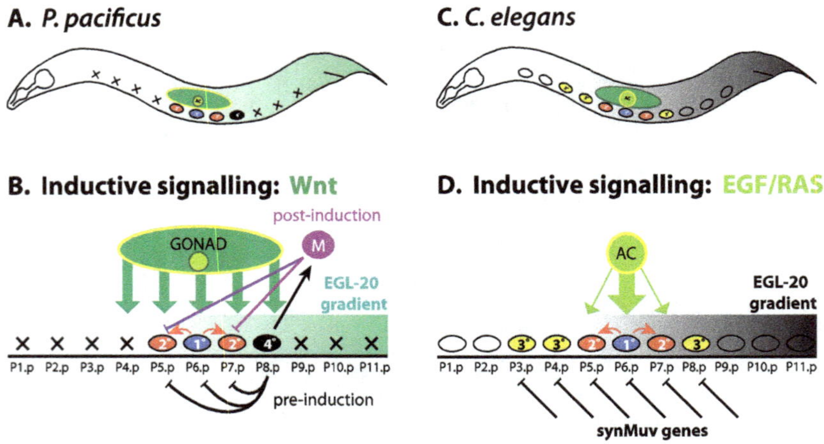

Fig. 9.2. *Changes in induction of the vulva. Anterior is towards the left. Dorsal is towards the top. A, B: Induction in the nematode* Pristionchus pacificus. *Ovals represent vulval equivalence group (VPC) cells on the ventral midline: blue, 1° cell fates, (P6.p); red, 2° cell fates, P(5,7).p; black, 4° cell fates, P8.p. The symbol '×' represents a Pn.p cell death before induction. The VEG consists of three cells; all other cells die (P(1-4,9-11).p), become part of the rectal connection (P12.p), or cannot respond to signal (P8.p). Gonad is shown as a large dark green oval; the anchor cell as a smaller yellow-green circle. Red arrows represent a molecularly uncharacterised signal reinforcing P(5,7).p 2° cell fates. The gonad is outlined in yellow to indicate it as the source of inductive Wnt signal. The Wnt signal is shown as dark green arrows. The EGL-20 inductive signal is shown as a forest-green gradient originating from the posterior. EGL-20 also establishes a basal division pattern for P5.p and P7.p, the polarity of the P7.p division pattern is reversed by a second Wnt signal not shown. P8.p acts before induction to inhibit aberrant and precocious formation of vulva tissue from the VPCs and later post-induction through the mesoblast M to inhibit 1° fates in P(5,7).p through molecularly uncharacterised pathways. C, D: As already described with the following exception. Clear ovals represent Pn.p cells outside the VEG. P(1,2,9,10).p fuse with the hypodermis before induction. Yellow ovals represent 3° cell fate; these cells are competent to be induced though are not in normal animals and fuse with the hypodermis after induction. The anchor cell is outlined in yellow to show it as the major inductive signal. The EGF/LIN-3 inductive signal is shown as a morphogen represented by yellow-green arrows emanating from the AC; P6.p receives a larger dose and P(5,7).p lesser doses. Red arrows represent a redundant Notch lateral signal, reinforcing P(5,7).p 2° cell fates. EGL-20 is shown as a black gradient from the posterior, it polarises but does not induce*

DEFINING THE EQUIVALENCE GROUP

The VEG comprises all cells capable of producing vulva tissues if induced. In all nematodes studied, the vulva develops from a subset of ventral epidermal cells termed P1.p to P12.p, or P(1-12).p in an anteroposterior order (Fig. 9.2A, C). In *P. pacificus*, the vulva is formed from the vulva precursor cells (VPCs), P(5-7).p (Figs 9.2A, B; 9.3), and they comprise the equivalence group (Sommer & Sternberg, 1996). P6.p forms the centre of the vulva, the primary (1°) cell fate, and P(5,7).p the outer portions of the vulva, the secondary (2°) cell fate. Removal of P6.p and P8.p by cell ablation using a laser results in P6.p's 1° fate being adopted by either P5.p or P7.p while the remaining cell adopts a 2° fate (Sommer, 1997; Jungblut & Sommer, 2000). P8.p is not competent to form the vulva and serves a unique organising function (Fig. 9.2B), termed the quaternary (4°) cell fate (Jungblut & Sommer, 2000). Cells outside the equivalence group, P(1-4).p and P(9-11).p, undergo programmed cell death (PCD)/apoptosis prior to vulva induction (Fig. 9.2A, B), eliminating them from consideration as members of the VEG (Sommer & Sternberg, 1996). Lastly, P12.p forms part of the connection between the epidermis and rectum.

How are the borders of the VEG established in *P. pacificus*? The Hox gene *lin-39* is expressed in P(5-8).p, and this expression domain is essential to prevent the apoptotic deaths of P(5-8).p. Loss of LIN-39 results in P(5-8).p undergoing apoptosis (Eizinger & Sommer, 1997; Sommer *et al.*, 1998). A HAIRY/GROUCHO transcription module controls the anterior border of the *lin-39* gene expression domain in the Pn.p cells (Schlager *et al.*, 2006). Loss-of-function mutations in either *Ppa-hairy* or *Ppa-groucho* result in expansion of LIN-39 into P(3,4).p and increases in the VEG as both P3.p and P4.p survive and are competent to form the vulva upon induction. In comparison to the anterior border, *Ppa*-PAX-3, a second homeodomain protein, regulates the posterior border. Intriguingly, *Ppa*-LIN-39 probably acts directly through regulating the transcriptional expression of *Ppa*-PAX-3 to preserve P(5-8).p (Yi & Sommer, 2007). Like LIN-39, loss-of-

---

the vulva. There are no 4° fates and no inhibitory signals from P8.p. Inhibitory arrows represent a general signal involving the synMuv genes from the hypodermis to prevent aberrant and precocious vulva tissue differentiation in the VPCs pre-induction.

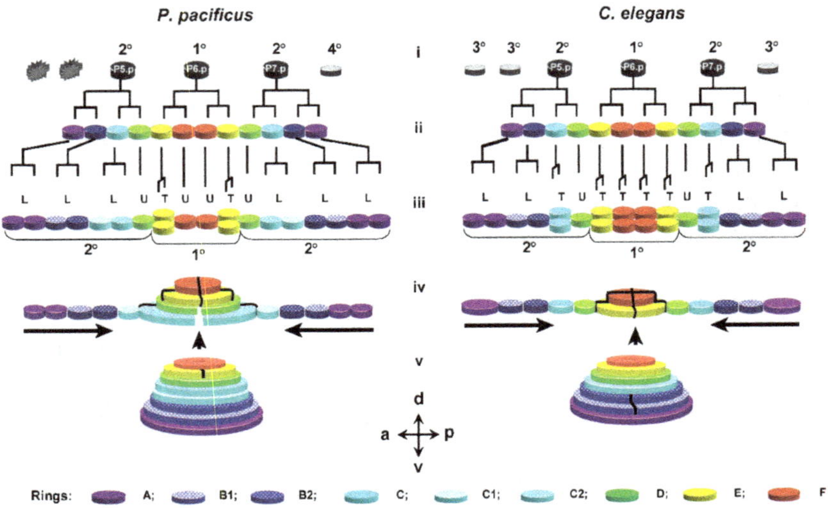

Fig. 9.3. *Stages of vulva morphogenesis in* Pristionchus pacificus *(left) and* Caenorhabditis elegans *(right). Anterior/posterior and dorsal/ventral are as indicated. Vul(A-F) cell fates are colour-coded and represented as oval cells and correlated vulval rings. Line diagrams represent cell divisions within a cell lineage. L denotes a longitudinal division, T a transverse division, and U an undivided cell.* i*) The vulval equivalence groups: P(5-7).p in* P. pacificus *and P(3-8).p in* C. elegans. *Jagged ovals represent apoptotic deaths (*P. pacificus*). Smaller grey ovals represent cell fates that ultimately fuse with the hypodermis, 3° cell fates divide once and 4° cell fates do not divide prior to fusion;* ii*) The 12-cell stage where vul(A-F) lineages have been generated;* iii*) The vul cells undergo differential divisions between species to give rise to the vulval rings. Cells descended from P6.p, vul(E,F) lineages, form the central rings and cells from P(5,7).p, vul(A-D) lineages, form the outer rings. Differences in the division patterns of the vul cells give rise to more or fewer vulval rings and change vulval morphology.* Pristionchus pacificus *has more longitudinal divisions and has eight vulval rings.* Caenorhabditis elegans *has fewer longitudinal divisions and therefore seven rings;* iv*) The arrows represent migration of the vul(A-F) lineages in towards the centre of the vulva. Migration commences in* P. pacificus *prior to completed vul cell divisions. In* C. elegans *migration commences following completed vul cell divisions. As mirrored counter parts meet, they form rings that move into the interior of the worm, perhaps elevated by the formation of the next more ventral ring creating an invagination. During vulva invagination cells within rings fuse in a defined pattern many making fully syncytial toroids;* v*) Vulva structure at the end of*

function mutations in *Ppa-pax-3* also result in the PCD of the VPCs; however, in addition, P(9-11).p, which normally die, survive. LIN-39 is not expressed in P(9-11).p; thus, regulation of the survival of central Pn.p cells and induction of apoptosis of posterior Pn.p cells involve differential genetic circuitry involving *Ppa*-PAX-3 (Yi & Sommer, 2007).

The VEG of *C. elegans* is highly similar to that of *P. pacificus* and characterised by an expression domain of *lin-39* to distinguish it from the other Pn.p cells (Clandinin *et al.*, 1997). However, there are four notable differences. First, the *C. elegans* equivalence group consists of P(3-8).p, although in wild-type animals the VPCs P(5-7).p still form the vulva (Fig. 9.2C, D); the remaining un-induced cells adopt a tertiary (3°) hypodermal cell fate and fuse with the epidermis in a wild-type animal (Sulston & White, 1980; Sternberg & Horvitz, 1986). As in *P. pacificus*, the remaining cells in the equivalence group can replace experimentally laser-ablated VPCs (Sulston & White, 1980; Sommer & Sternberg, 1996). Second, cells outside the VEG do not undergo apoptosis; rather they fuse with other cells of the hypodermis (Sommer & Sternberg, 1996). Third, unlike *P. pacificus*, the borders of the VEG are determined by other means. Mutant analysis indicates that *Cel-pax-3* is not involved in defining the posterior border, as loss of PAX-3 expression has no effect on the VEG (Yi & Sommer, 2007). Additionally, in *P. pacificus*, HAIRY and GROUCHO work together *via* direct protein interaction to restrict the anterior border of *Ppa*-LIN-39 expression (Schlager *et al.*, 2006). This interaction requires a GROUCHO-interacting domain in the HAIRY transcription factor; *C. elegans* does not contain a HAIRY orthologue, and no other HAIRY-family basic helix-loop-helix (bHLH) DNA-binding protein in the *C. elegans* genome encodes a protein with a GROUCHO-interacting domain (Schlager *et al.*, 2006). This interacting module is simply not there in *C. elegans*; however, in a strangely parallel fashion, the HAIRY-family bHLH protein LIN-22 regulates the more anteriorly expressed HOX gene *mab-5* (Clandinin *et al.*, 1997; Wrischnik & Kenyon, 1997; Schlager *et al.*, 2006). In turn, MAB-5 expression acts to determine the anterior limit of *lin-39* expression and

ring formation and prior to vulva eversion. Upon maturation the connection is made with the uterus and the internal vulval structures collapse together to fill the invagination and form a tight pore/slit. Figure modified from Kolotuev & Podbilewicz (2008).

the VEG in *C. elegans*. Fourth and last, P8.p is a *bona fide* VPC with 3° fate capable of producing vulva tissue and does not have a known organising function as in *P. pacificus* (Sulston & White, 1980).

VULVA INDUCTION

The induction of the vulva involves at least three signalling centres (Fig. 9.2B). First, P(5-7).p receive a continuous inductive signal from the entire developing somatic gonad to form the vulva (Sigrist & Sommer, 1999). Second, the VPCs receive a second signal from the tail of the animal (Tian *et al.*, 2008; Wang & Sommer, 2011). Third, P8.p sends signals to P5.p and P7.p to restrict them to 2° fates (Jungblut & Sommer, 2000). Full induction proceeds over an extended period of time. Based upon proximity to the centre of the gonad, the central cell P6.p potentially receives the largest dose of signal, adopts a 1° cell fate, and forms the centre of the vulva with a unique cell division pattern resulting in six descendants. In addition to being restricted by inhibitory signalling from P8.p, P5.p and P7.p presumably receive a lesser dose of signal from the gonad and, as a result of potentially redundant mechanisms, adopt 2° cell fates. They form the periphery of the vulva, adopting mirror image cell lineages, each producing seven descendants. Once induced, it is likely that additional signalling from P6.p acts in a redundant way with the inductive signal and inhibitory signal from P8.p to promote the adoption of 2° cell fates by P5.p and P7.p, and reinforce the 2°-1°-2° pre-pattern of the vulva (Fig. 9.2B, red arrows).

Interruptions in the canonical WNT pathway result in a lack of induction (Zheng *et al.*, 2005; Tian *et al.*, 2008). In this pathway, diffusible secreted Wnt ligands are putatively sent from the gonad to the vulva. The ligand binds and interacts with an LRP/Frizzled receptor complex on the surface of the VPCs. Activated signalling inhibits formation of a cytoplasmic destruction complex that targets the β-catenin transcriptional co-activator for degradation. Freed from repression, β-catenin transports into the nucleus to bind a TCF-family DNA-binding protein. On its own, TCF often acts as a transcriptional repressor that resides upon the promoter of target genes; when bound by β-catenin, TCF activates transcription of those same genes instead (Cadigan, 2012). The *P. pacificus* WNT cell-signalling pathway is extremely redundant, with the *P. pacificus* genome encoding at least five diffusible WNT ligands, four Frizzled-like cell surface receptors and a single Ryk-like receptor (Tian *et al.*, 2008). Individual loss-of-function mutations in a ligand or a receptor either result in no phenotype or a reduced induction of the vulva; only the single β-catenin null mutation or triple mutants involving ligand/receptor combinations result in total loss of vulva induction (Tian *et al.*, 2008). The WNT ligands MOM-2 and LIN-44 have been implicated as putative signals from the anchor cell and the somatic gonad, respectively, and EGL-20 as a putative signal from the tail of the animal (Tian *et al.*, 2008; Wang & Sommer, 2011). Strikingly, the Frizzled-like receptor LIN-17 and the RYK-like receptor LIN-18 have been implicated as having antagonistic roles in vulva induction, with LIN-18 promoting and LIN-17 inhibiting induction (Tian *et al.*, 2008; Wang & Sommer, 2011). In contrast to most other systems, Frizzled-like *Ppa-lin-17* mutants have a multivulva phenotype that is opposite of mutations in Wnt-ligands, indicating an antagonistic role of *Ppa*-LIN-17. Evidence suggests that LIN-18 is inactive in the absence of a ligand due to binding of an inhibitor to SH3-binding domain motifs (SBDMs) in the intracellular portion of the receptor (Wang & Sommer, 2011). LIN-17 putatively acts to sequester the EGL-20 signal to prevent activation of the LIN-18 receptor, probably in tandem with another Frizzled-like receptor yet to be identified functionally (Wang & Sommer, 2011). Consistently, *Ppa-egl-20*/Wnt and *Ppa-lin-17*/Frizzled are co-expressed in the posterior tail of the animal (Wang & Sommer, 2011). Intriguingly, in *P. pacificus*, PAX-3 is involved in the induction of the vulva downstream of WNT signalling, as well as earlier during specification of the posterior VEG border (Yi & Sommer, 2007).

In *C. elegans*, the VPCs receive an EGF signal that acts through a RAS/MAPK signal cascade from a single cell of the gonad (Fig. 9.2D), the anchor cell (AC), at a specific time for induction (Kimble, 1981; Aroian *et al.*, 1990; Han & Sternberg, 1990). No second source of inductive signal has been identified. The AC sits directly above P6.p and the EGF signal putatively acts in a graded manner, signalling P6.p more than P5.p and P7.p, resulting in the initial cue for the 2°-1°-2° pattern of the vulva (Katz *et al.*, 1995). In contrast to *P. pacificus*, *C. elegans* LIN-39 has a second role post VEG specification and is required for induction of the vulva downstream of RAS/MAPK signalling; PAX-3 is not involved (Katz *et al.*, 1995; Clandinin *et al.*, 1997; Maloof & Kenyon, 1998; Sommer *et al.*, 1998; Yi & Sommer, 2007). The role of EGF signalling in *P. pacificus* remains unresolved. In *C. elegans*, following initial specification of the 1° cell fate by EGF/RAS, P6.p sends a second redundant signal to its neighbours, P5.p and P7.p, using the Notch pathway to reinforce their adoption of 2° cell fates (Sulston & White, 1980; Sternberg, 1988; Sternberg & Horvitz, 1989; Huang *et al.*, 1994). Thus the Notch pathway initiates a positive feedback loop laterally to inhibit 1° fates in P(5,7).p and further lock in the 2°-1°-2° pattern (Fig. 9.2D, red arrows). Although P6.p also promotes 2° cell fates in its neighbours in *P. pacificus*, the role of the Notch pathway in *P. pacificus* remains unresolved, yet recent analyses of multiple *P. pacificus* wild-isolates have implicated changes in the expression of the Notch-associated DSL-like ligand APX-1 in cryptic variation of vulva specification (Kienle & Sommer, 2013). These variations in vulva specification seem to trace back to alteration in a *cis*-acting HAIRY-binding site in the *Ppa-apx-1* promoter (Kienle & Sommer, 2013). Likewise, the WNT pathway shows minor induction defects in some *C. elegans* strains (Braendle & Félix, 2008; Milloz *et al.*, 2008), though RAS/MAPK signalling is the major inductive pathway; however, WNT signalling also has additional roles in forming and patterning the VEG. In *C. elegans*, the Wnt pathway is involved in establishing the equivalence group, with Wnt signalling required to maintain LIN-39 expression in the VPCs (Eisenmann *et al.*, 1998). This signalling has yet to be demonstrated in *P. pacificus*. Wnt signalling also helps to establish a basal cell division pattern for the 2° cells lineages and a second Wnt signal reverses the polarity of this basal pattern in P7.p to generate a symmetrical vulva in *C. elegans* (Inoue *et al.*, 2004; Zheng *et al.*, 2005). Certainly the role of WNT signalling in vulva symmetry has been retained in *P. pacificus* (Zheng *et al.*, 2005). Of note, no antagonistic/inhibitory role for the *Cel*-LIN-17 Frizzled-like receptor has been reported in WNT signalling, and the *Cel*-LIN-18 protein does not have the SDBMs present in *Ppa*-LIN-18 (Wang & Sommer, 2011). Thus, despite a preserved 2°-1°-2° pre-pattern, and surprisingly similar adult structures, the molecular mechanism of vulva specification is remarkably diverged. These changes are likely to involve the evolution of protein expression domains by *cis*-acting DNA regulatory elements and the evolution of small protein domains to co-opt and recruit additional existing molecular players and signalling pathways.

Perhaps the single most striking difference between vulva induction in *P. pacificus* and *C. elegans* is the role of P8.p (Fig. 9.2B). In the current working model for vulva patterning in *P. pacificus*, P8.p is not competent to respond to the inductive signal but has two organising functions

(Jungblut & Sommer, 2000). First, P8.p inhibits early ectopic vulva formation in the absence of gonad signalling, whereas in *C. elegans* the synMuv genes, synthetic multivulva genes including *lin-15*, are involved in signalling from the epidermis to inhibit vulva induction in the absence of inductive signal (Fig. 9.2D) (Ferguson *et al.*, 1987; Sternberg, 1988; Huang *et al.*, 1994). The synMuv genes have been shown to regulate vulva induction by chromatin remodelling by forming a rather complex and highly redundant system of two pathways (Sternberg, 2005; Cui *et al.*, 2006). By contrast, LIN-15 and synMuv function in *P. pacificus* remain unknown. Second, P8.p inhibits the 1° cell fate in P5.p and P7.p, hence the requirement to ablate P8.p in addition to P6.p (Fig. 9.2B). This inhibition requires mediation by the M cell. The M cell gives rise to the sex muscles used to open and close the vulva for laying eggs (Jungblut & Sommer, 2000; Photos *et al.*, 2006). No analogous organising centre like this has been found in rhabditids. However, the interplay of this signalling with vulva inductive signals, the molecular players involved, and their *modus operandi* remain poorly understood.

Vulva morphogenesis

The *P. pacificus* vulva is a round contractile pore-like opening (Fig. 9.1A, insert). By contrast, the *C. elegans* vulva is more slit-like in appearance (Fig. 9.1B, insert). For both animals, the vulva is formed along the anterior-posterior axis upon the ventral midline and proceeds in steps. First, six vulva cell lineages, vulA, vulB, vulC, vulD, vulE, and vulF, *i.e.*, vul(A-F), are produced in a mirrored pattern at the position of the future vulva (Fig. 9.3, i, ii) (Sommer & Sternberg, 1996; Kolotuev & Podbilewicz, 2004). These lineages undergo a morphogenetic programme starting at the centre of the vulva where paired mirrored cells migrate towards the centre of the group to meet and form rings (Fig. 9.3, iv, arrows). As one ring forms it invaginates into the body of the worm as subsequent cells reach their mirrored counterparts to form the following ring. The process repeats until the full complement of cells are used and all rings are made (Kolotuev & Podbilewicz, 2004). During this process, cells that comprise a given ring fuse in an invariant manner to form syncytial multinucleated cells and sometimes fully syncytial toroid rings (Fig. 9.3, v) (Kolotuev & Podbilewicz, 2004). Ultimately, the uterus forms connections with the interior-most ring and sex muscle cells connect with the forming channel to produce the working structure.

In *P. pacificus*, the migration of cells towards the centre of the developing vulva, the formation of the vulva rings and the fusion of cells proceed as the last vulva divisions are being completed (Kolotuev & Podbilewicz, 2004). In *C. elegans*, morphogenetic programmes follow the completed cell divisions (Sharma-Kishore *et al.*, 1999). Increasing evidence suggests that heterochronic shifts among developmental programmes is a common mechanism for the advent of morphological and functional novelty in organ formation (Smith, 2003; Spicer *et al.*, 2011; Tills *et al.*, 2011).

Cell division pattern dictates vulva morphology. In *P. pacificus*, the vulva is formed from the descendants of P(5-7).p. These three cells undergo a characteristic set of cell divisions to produce eight rings that form the structure of the pore (Sulston & Horvitz, 1977; Sharma-Kishore *et al.*, 1999). The descendants of P6.p give rise to a symmetrical lineage and form the two central-most rings of the vulva, and the descendants of P5.p and P7.p give rise to respective mirror image lineages that form the outer rings of the vulva (Fig. 9.3). For all lineages the first two divisions result in four granddaughters each, *i.e.*, 12 cells in total (Sommer & Sternberg, 1996). Taken in pairs from the outside to the inside of the group, these 12 cells are labelled vul(A-F) and pre-pattern six of the eight rings (Fig. 9.3, ii). These 12 cells either can undergo no additional division, divide transversely along the left-right axis of the animal, or divide longitudinally along the anterior-posterior axis (Fig. 9.3, iii) (Sommer & Sternberg, 1996). Undivided cells or cells that divide transversely produce a single vulval ring. Generally, longitudinal division results in two rings (Fig. 9.3, compare steps iii and v). Three longitudinal divisions occur in the vulA, vulB and vulC cells at the anterior and posterior periphery. The vulB and C divisions result in two rings each, whereas the vulA daughters, which also have longitudinal divisions, are an exception and give rise to only a single ring (Kolotuev & Podbilewicz, 2004).

Caenorhabditis elegans has only seven rings, not eight, forming the vulva. This is due to differences in the axis of division of the vulC cell compared to *P. pacificus* (Fig. 9.3, iii). Initially, as in *P. pacificus*, after two rounds of division P(5-7).p give rise to 12 cells linearly arranged along the anterior-posterior axis on the ventral midline that pre-pattern the vulval rings. In contrast to *P. pacificus*, only two longitudinal divisions occur at the anterior and posterior periphery, one each in the vulA and vulB lineages; vulC does not divide longitudinally but transversely (Sulston & Horvitz, 1977). This difference generates only seven total rings (Fig. 9.3, v), with vulB descendants giving rise to two rings as in *P. pacificus*, but vulC giving rise to a single ring unlike in *P. pacificus*. These differences led to the hypothesis that changes in the axis of cell division of vul cells alter the numbers of rings and vulva morphology in nematodes; longitudinal divisions appear to be more or less indicative of additional or fewer vulval rings. The ring hypothesis, first noted in comparisons between *P. pacificus* and *C. elegans*, has garnered additional support through the comparison of mutations in *C. elegans* that alter the division pattern of the vul(A-F) cells and result in corresponding changes in the number of rings (Kolotuev & Podbilewicz, 2008). This correlation was further expanded in the examination of the Pn.p cell divisions and vulva formation of additional nematode species (Kolotuev & Podbilewicz, 2008).

A comparative description of gonadogenesis

The gonad is formed from the descendants of only four founding cells. Signalling from the vulva influences the adult gross morphology of the *P. pacificus* hermaphrodite gonad. The gonad is a complex organ composed of both somatic and germline tissues (Fig. 9.4). The soma acts to pattern the underlying germ line to produce viable gametes and functional offspring in a highly regulated fashion. The gonad must coordinate its spatial and temporal development with the vulva to form a proper connection for expulsion of embryos and, thus, the gonad instructs vulva development as outlined in part 1. The germ line and the somatic gonad have been a longstanding exemplar for signalling between tissues, tissue patterning, translational regulation, and organogenesis (Hubbard & Greenstein, 2005; Kimble & Crittenden, 2005; Strome, 2005; Ellis & Schedl, 2007).

GONAD DEVELOPMENT, ANATOMY AND PHYSIOLOGY

Early gonad development in *P. pacificus* is similar to that in *C. elegans*. Both the hermaphrodite and male *P. pacificus* gonad starts out as a four-cell primordium consisting of two somatic founder cells, Z1 and Z4, and two germline founder cells, Z2 and Z3 (Fig. 9.5A) (Kimble & Hirsh, 1979; Félix *et al.*, 1999; Rudel *et al.*, 2005). These cells are arranged roughly linearly along the anterior-posterior

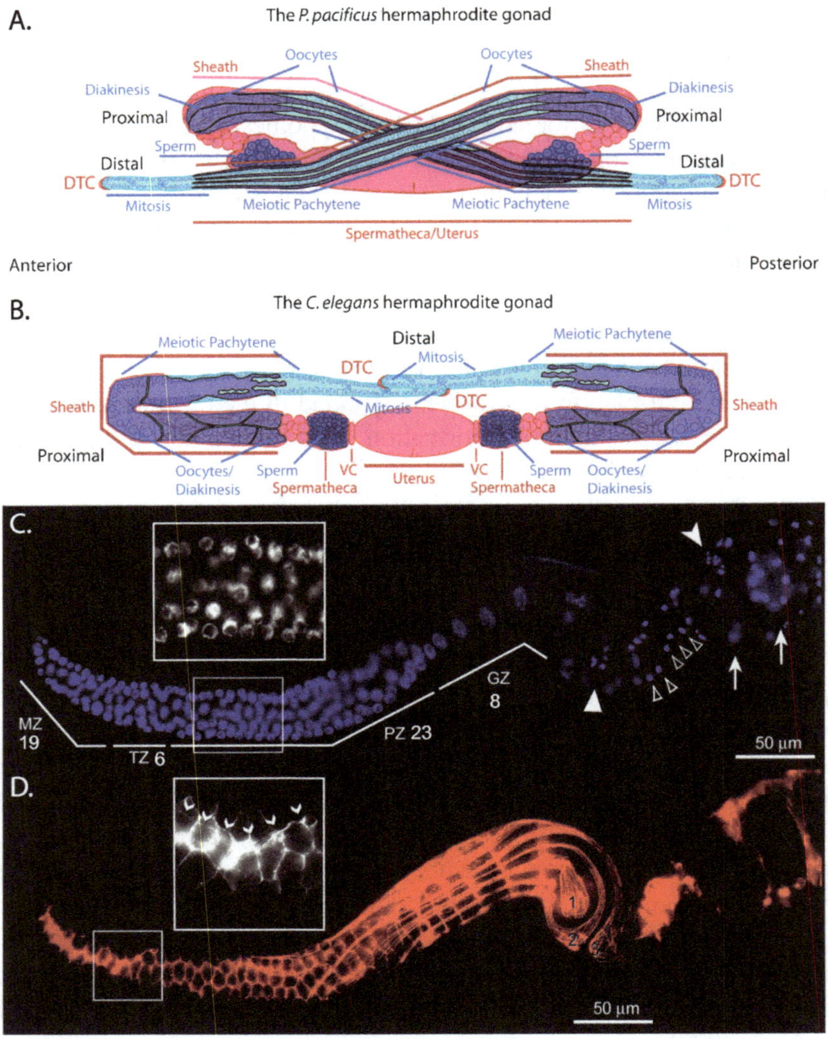

Fig. 9.4. *The adult hermaphrodite gonad. A, B: Cartoons of the* Pristionchus pacificus *and* Caenorhabditis elegans *hermaphrodite gonads. Anterior is towards the left, dorsal towards the top. Somatic tissues are shown in red. Germ line is shown in blue. Gonad arms are patterned along a distal-proximal axis, the distal tip cells (DTCs) sit at the distal pinnacles of the axes and the proximal ends terminate at the vulva shown as a slit in the uterus. The principal somatic tissues given in order from distal to proximal location along this axis are DTC, sheath, spermatheca and uterus. The sheath is contractile; actin myosin filaments are depicted as thin red lines within individual sheath cells. The underlying germ cells are also patterned along this axis in zones: the mitotic zone*

containing germline stem cells, the transition zone from mitosis to meiosis, the pachytene zone of meiosis, and the gamete zone containing first developing sperm and later developing oocytes. The mature germ line is a syncytial tissue with incomplete germ cell division until the most proximal oocytes. A few sperm are found at the proximal base of sheath with the majority stored in the spermatheca. Oocytes are fertilised upon reaching the sperm. Once fertilised the eggshell and barriers to other sperm form. Ovulation or a contraction of the sheath moves eggs into the spermatheca and uterus. Sperm occasionally get pushed into the uterus but migrate back into the spermatheca following an attractive cue. Nematode sperm are amoeboid. A: P. pacificus *pretzel-shaped gonad. The sheath is arranged in a roughly circular pattern around the proximal edge of the gamete zone. The sheath cells extend process(es) down the distal-proximal axis forms a non-continuous tube encasing the germ line (overlap purple). This allows nutrients from the intestine to travel directly to the germ line. Individual sheath cells touch multiple zones, putatively requiring substantial organisation of signalling along and within a sheath cell for instruction of the gonad; B:* C. elegans *gonad with U-shaped arms. The sheath is arranged as pairs of cells along the distal proximal axis, individual cells overlying specific regions of germline zones. The sheath forms a continuous sheet covering the developing gametes requiring nutrients to transport through the sheath cells to the germ line for proper nutrition.* Caenorhabditis elegans *has a contractile syncytial toroid-shaped valve cell between the spermatheca and uterus to aid in separating the spermatheca and uterus; this is absent in* P. pacificus*; C: DAPI stained young adult hermaphrodite* P. pacificus *gonad arm: MZ, mitotic zone, contains nuclei that have diffuse uniform DNA staining of circular nuclei. TZ, transition zone, contains sickle-like nuclear staining (see insert). PZ, pachytene zone, contain DNA condensing into thick chromosomal strands. GZ, gamete zone, contains differentiating gametes with differing cytoplasmic morphology and DNA is further condensing into pachytene chromosomes; proximal oocytes contain six very tightly condensed bivalent chromosomes, solid triangle. Open triangles indicate the nuclei of the spermathecal corridor. Solid arrowheads indicate sperm nuclei, sperm nuclei are small very condensed dots. Arrows indicate the position of fertilised eggs. Maternal meiosis completes following fertilisation. Numbers represent the average number of nuclei along a single line going from distal to proximal within a zone; D: Phalloidin staining of actin in the same gonad arm. Photomicrograph is taken at a more peripheral focal plane to see the sheath cells. The cell-body position of three sheath cells are shown and numbered. Specific sheath cells have characteristic locations and cellular morphologies. With the exception of the second sheath cell shown, which extends two, the remaining sheath cells extend a single process down the distal-proximal axis. Processes do not cross and may regulate each other's progression down the axis. Processes continue to grow down the arm until one reaches the distal tip cell. Insert shows distal sheath cell processes as indicated by small arrowheads.*

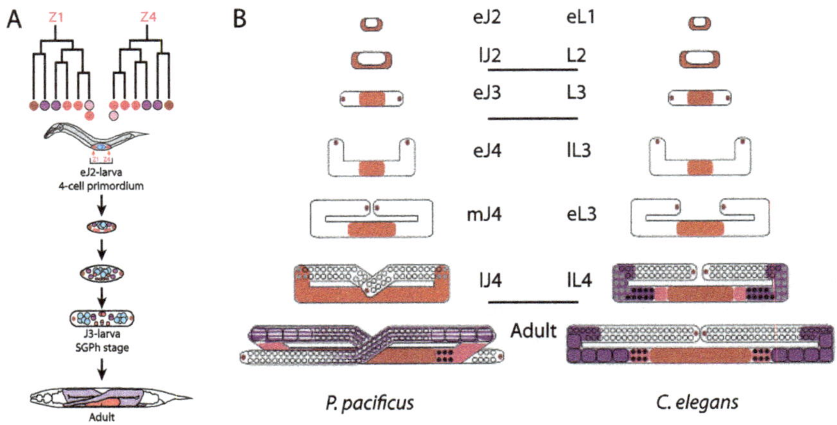

Fig. 9.5. *Development of the hermaphrodite gonad. Anterior towards the left; dorsal towards the top. A: Top. Cell lineage diagram for the somatic gonad precursors Z1 and Z4 to the hermaphrodite somatic gonad primordium stage. In terminal cells: Dark red, DTC. Violet/purple, sheath/spermathecal precursor. Red, uterine precursor. Pink, anchor cell. Bottom diagrams show nuclear position during key developmental stages. Nuclei and adult somatic tissues are coloured as indicated above; B: Diagram comparing migration, somatic differentiation and germline differentiation between* Pristionchus pacificus *(left) and* Caenorhabditis elegans *(right). Tissue colours are as shown in (A) with the exception that spermatheca is shown in dark pink. Stages are labelled as follows: J for juvenile (*P. pacificus*); L for larva (*C. elegans*); e/m/l, early/mid/late. Juvenile/larval stages are punctuated by cuticle moults. Dark circles, sperm. Grey circles, spermatocytes. Clear circles, undifferentiated germ cells. Rounded blocks, oocytes. Note the following differences:* i*)* P. pacificus *DTCs turn dorsally at a more central position;* ii*)* P. pacificus *has a novel ventral migration;* iii*)* P. pacificus *has not completed somatic differentiation before adulthood;* iv*)* P. pacificus *has no differentiated gametes before adulthood.*

axis above the ventral midline of the epidermis, Z1 to Z4, with the somatic precursors sandwiching the germline precursors. Z1 and Z2 reside slightly to the right of the ventral midline with Z3 and Z4 to the left. In both hermaphrodites and males, Z1 and Z4 subsequently undergo a reproducible pattern of divisions and cell rearrangements to produce the somatic gonad primordium (SGP) that establishes a pre-pattern of the adult gonad morphology (Fig. 9.5A) (Kimble & Hirsh, 1979; Félix *et al.*, 1999). By contrast, the germline precursors appear to divide with no obvious polarity to the direction of division and the ultimate placement of germline daughter cells. In hermaphrodites the SGP is symmetric; at both the anterior and posterior end there resides a single distal tip cell (DTC) and the remaining somatic founder cells reside in a roughly mirror image midway between the DTCs, thereby dividing the germ cells into two populations. The hermaphrodite DTCs migrate in a well-characterised pattern and their movements are correlated with developmental stage and ultimately give rise to two rotationally-symmetrical gonad arms on the right and left sides of the body (Fig. 9.5B) (Hirsh *et al.*, 1976; Hedgecock *et al.*, 1987; Rudel *et al.*, 2005). While the male gonad has not been studied in detail in *P. pacificus*, it is known from *C. elegans* that the male SGP rearranges such that the two DTCs reside at the posterior end and the remainder of the somatic founder cells at the anterior end (Kimble & Hirsh, 1979). At the anterior end of the male SGP, a capping linker cell leads migration and formation of the single male gonad arm. In both sexes the gonad arms develop as tubes of patterned somatic tissues that encase, instruct and nurture the developing syncytial germ cells in the centre of the arm (Fig. 9.4). The somatic tube and germ line are patterned along a proximal-distal axis, proximal defined as the end of the arm closest to the uterus and vulva in hermaphrodites and the cloaca for males (Hirsh *et al.*, 1976; Kimble & Hirsh, 1979; Rudel *et al.*, 2005). Developing gametes are produced in an assembly line fashion (Fig. 9.4C). Germline stem cells at the distal end undergo mitosis. As germ cells move proximally they enter meiosis, which progresses as they move further still. Following meiosis germ cells begin to differentiate terminally into three fates; sperm, oocytes, or PCDs. Approximately the first 100-200 (in *C. elegans* even 300) gametes produced are spermatozoa. Subsequently, there is a switch and later gametes are uniformly oocytes. PCDs are probably germ cells that either served a nursing function or have genomic or physiological damage. Later development of specific tissues is well characterised for the *P. pacificus* hermaphrodite, but remains unresolved for the male.

Later development of the adult *P. pacificus* hermaphrodite gonad exhibits a number of developmental and physiological differences in comparison with *C. elegans* (Fig. 9.4, compare A and B) (Rudel *et al.*, 2005). Two very intriguing differences include a change in the gross morphology of the gonad, in part due to altered DTC migrations and changes in the regulation of germline patterning. Experiments have revealed some of the cellular and genetic differences underpinning these changes and these are described in later sections. In addition to these there are several other differences. First, the timing of somatic differentiation with respect to life stage as determined by cuticle moults is altered (Fig. 9.5B, compare differentiated tissues at the late J4/L4 stage and in adults) (Rudel *et al.*, 2005). Both *P. pacificus* and *C. elegans* go through four juvenile stages prior to adulthood. In comparison with *C. elegans*, *P. pacificus* somatic tissues are retarded in their differentiation and do not fully differentiate until entry into adulthood. Second, though not surprising given the dependence of the germ line on the soma, the production of gametes and the earliest progeny is also retarded in *P. pacificus* (Fig. 9.5B, compare differentiated tissues at the late J4/L4 stage and in adults) (Rudel *et al.*, 2005). The somatic gonad is instructive in adult germline patterning, putatively necessitating this delay. Third, while the tissues that comprise the somatic gonad are patterned along the proximal-distal axis in a similar order (uterus, spermatheca, sheath, and DTC), the number of cells and their arrangement within tissues vary greatly (Fig. 9.4, compare A and B) (Rudel *et al.*, 2005). An example of this variation is the observations that the spermatheca, spermathecal corridor and sheath contain five, ten and eight cells in *P. pacificus*, whilst in *C. elegans* they are composed of 16, four and ten cells, respectively. Additionally, *P. pacificus* hermaphrodite gonad arms lack valve cells; in *C. elegans* valve cells are multinucleated toroid-shaped cells that constrict the gonad arm between the spermatheca and uterus.

The difference in tissue construction has implications for the physiology of the tissue. In *P. pacificus*, the sheath forms a ball and tendon joint with the spermathecal corridor (Fig. 9.4D) (Rudel *et al.*, 2005). Sheath cells are arranged as long fingers in a circle around the periphery of the gonad arm circumference. These fingers stretch down along the proximal-distal axis of the gonad arm. Actin/myosin bundles are also aligned along this axis internal to the sheath cells. Based upon this morphology and observed intercellular connections, it seems likely that these fingers grasp the underlying oocytes and contract to pull the oocytes along. When oocytes hit the back wall of the ball and socket joint they are forced down into the spermatheca (Rudel *et al.*, 2005). Following fertilisation, eggs are quickly laid. The uterus of healthy young *P. pacificus* hermaphrodites typically contains one to four eggs at a given time. In *C. elegans*, the sheath forms a smooth transition into the spermatheca; there is no sharp join or drastic change in the direction of oocyte movement during ovulation (Hirsh *et al.*, 1976; Hall *et al.*, 1999). The sheath is arranged as pairs of cells along the proximal-distal axis (Strome, 1986; Hall *et al.*, 1999). These cells make an enclosed sheet that envelops the proximal portion of the germ line containing the developing gametes. Within the sheath cells are actin/myosin bundles, but not structured along any given access (Hall *et al.*, 1999; Rudel *et al.*, 2005). The sheath contracts like squeezing a tube of toothpaste and the oocytes move along in a straightforward progression. *Caenorhabditis elegans* retains its eggs *in utero* for a longer time, hence there are a larger number of eggs and the embryos are more developed when laid (Fig. 9.1, compare A and B).

Regulation of gonad arm shape and gross gonad morphology

The two-armed gonad morphology of the *P. pacificus* hermaphrodite resembles a 'squashed pretzel'-like shape (Fig. 9.4A) (Rudel *et al.*, 2005). During the J3 stage, the hermaphrodite DTCs migrate away from the anterior-posterior centre of the animal along the ventral midline, one towards the head and one towards the tail. During the early J4 stage, the arms turn to migrate dorsally, one along the left body wall and one along the right body wall. Upon reaching the dorsal side, both DTCs migrate back towards the anterior-posterior centre of the animal. As the DTCs approach the centre, they leave the dorsal body wall and begin to migrate back towards the ventral side crossing over the developing vulva (Rudel *et al.*, 2005, 2008). Upon reaching the ventral side, the DTCs continue to migrate along and reside at the body wall (see Fig. 9.5B, left side, for staged gonad arm extensions).

Ablation of the vulva results in the failure of the gonad arms to migrate back towards the ventral side during late J4 (Rudel *et al.*, 2008). This results in U-shaped arms where the DTCs migrate along and reside at the dorsal body wall upon reaching young adulthood. Genetic analysis has implicated a canonical Wnt signalling pathway in the crosstalk between the developing vulva and the DTCs, the developing vulva being a putative source of Wnt ligands, MOM-2, CWN-2 and LIN-44, and the DTCs expressing the downstream transcriptional co-activator BAR-1/β-catenin in the nucleus (Rudel *et al.*, 2008). It has been proposed that the Wnt signal results in transcriptional activation of guidance molecules resulting in the ventral migration. Based upon what is known from *C. elegans*, the Netrin pathway has been proposed as a target of this signalling. Netrin, a secreted diffusible protein, is thought to be expressed in ventral cells forming a ventral to dorsal expression gradient. Two families of structurally unrelated cell surface receptors, the UNC-5 and UNC-40/DCC families, are expressed on the surface of migrating cells and instruct cell migrations either towards or away from Netrin sources (Hedgecock *et al.*, 1990; Ziel & Sherwood, 2010; Ogura *et al.*, 2012). In *C. elegans*, the Netrin pathway strongly influences postembryonic dorsal and ventral cell migrations, including the ventral to dorsal migration of the DTCs at the L3 stage (Hedgecock *et al.*, 1990). Talk between the vulva and the DTCs is not the only signalling that affects the terminal ventral migration; there also appears to be cross talk between gonadal arms. Early cell ablation of a single distal tip cell impairs formation of one arm. In these animals, the remaining arm often fails to make a terminal dorsal to ventral migration.

Development of the *C. elegans* hermaphrodite gonad is highly similar with three exceptions (Fig. 9.5B, right side). First, there is a slight alteration in the timing of the individual migrations with respect to larval stages (Hedgecock *et al.*, 1987; Rudel *et al.*, 2005). Second, the *C. elegans* hermaphrodite gonad arms migrate from the ventral body wall to the dorsal body wall farther towards the head and tail of the animal, respectively (Rudel *et al.*, 2005). Lastly and most notably, there are no terminal dorsal to ventral migrations, which leads to two U-shaped gonad arms (Hirsh *et al.*, 1976; Hedgecock *et al.*, 1987). Cladistic analysis of rhabditids (including *C. elegans*), diplogastrids (including *P. pacificus*) and basal nematodes indicates that the ventral migration could be a unifying novelty specific to diplogastrids (Rudel *et al.*, 2008). This would imply transcriptional co-option of an existing ancestral Wnt-Netrin genetic cassette, given the current hypothesis from *P. pacificus* developmental studies. Isolates of *P. pacificus* vary in their terminal morphologies. In the laboratory strain PS312, approximately 70% of gonad arms make the terminal ventral migration; the remaining 30% remain dorsal. In other strains, such as PS1843, only 23% of gonad arms make the terminal ventral migration. In yet other strains, like JU726, the gonad arms migrate from dorsal to ventral 97% of the time. These disparate phenotypes from genetically distinct populations highlight the potential for quantitative trait loci analysis to identify causative regulatory changes involved in modulating, and perhaps in the evolution of, these novelties.

Somatic instruction of the germ line

The *P. pacificus* somatic gonad has a highly instructive role in developmental regulation of the germ line. First, in addition to the loss of the vulva, cell ablation of Z1 and Z4, the somatic gonad precursor cells, results in either the loss of the germ line or in germline tumours (Rudel *et al.*, 2005). Like gonad arm migration, whether the majorities of primordial germ cells (Z2 and Z3) die, give rise to a small group of germ cells or result in a large tumour depends upon the wild isolate in question. It appears in *P. pacificus* that the soma acts to check unregulated germ cell mitosis. In *Caenorhabditis* strains tested to date, ablation of Z1 and Z4 results in Z2 and Z3 either dying or persisting without further divisions (Kimble, 1981). Germ cell tumours from such cell ablations have yet to be reported. Second, ablation of the DTCs after birth in *P. pacificus* results in a failure of the gonad arms to extend and in all germ cells exiting mitosis and differentiating as sperm. Hence, the DTCs play a crucial role in providing a niche to nurture and maintain the germline stem cell population (Rudel *et al.*, 2005) and *C. elegans* DTCs perform the same function (Kimble, 1981). Third, ablation of the *P. pacificus* sheath/spermatheca precursor cells results in relatively full-sized gonad arms with smooth bends; that is, the ball and socket joint is missing and gametes can be pushed in an unimpeded manner into the uterus. Additionally, sheath/spermatheca cell-ablated animals are sterile and lack embryos *in utero*. The distal mitotic and transition zones leading into meiosis are reduced in these animals, whereas the pachytene region of meiosis is expanded to nearly twice the normal length. The sheath-ablated arms produce sperm and oocytes; however, the oocytes have abnormal nuclear morphologies and lack diplotene chromosomes. Instead of condensing into diplotene chromosomes, the chromatin in the oocytes become more diffuse than in earlier stages of meiosis and gamete development (Rudel *et al.*, 2005). Ablation of the majority of the sheath/spermatheca in *C. elegans* results in smaller gonad arms with a substantial reduction in the mitotic zone, cells stuck in pachytene, only a few sperm and no oocytes (McCarter *et al.*, 1997). Cell ablation of half the sheath/spermathecal cell precursors most commonly results in full-sized arms with reduced mitotic and transition zones, an extended pachytene region and the production of oocytes and sperm. In a fraction of these ablated arms, however, the chromosomes of the oocytes do not condense and undergo endomitotic replication. This results in large nuclei with massive amounts of diffuse DNA. Also in a fraction of the arms, the germ line is feminised producing only oocytes but no sperm (McCarter *et al.*, 1997). Intriguingly, the results of ablating the sheaths in *P. pacificus* and *C. elegans* are analogous at the gross level of the size of major germline zones; however, differing nuclear morphology suggests that the points of regulation that are disrupted within the zones are likely to be different (Rudel *et al.*, 2005). Also, in *P. pacificus*, each sheath cell may extend down the length of all zones. By contrast, in *C. elegans*, sheath cells are arranged in pairs along the length of the zones and studies have suggested that individual pairs can specifically instruct individual steps of gamete production in particular zones during the development of the underlying germ line (Killian & Hubbard, 2005). Thus, soma instruction of the germ line must involve a different cellular organisation of signalling molecules based on the nature of the sheath in *P. pacificus*; that is, every *P. pacificus* sheath cell can potentially contact all zones.

Conclusions

Taken together, the knowledge gained from studies of vulva and gonad development in *P. pacificus* and *C. elegans* leads to many surprising conclusions: *i*) developmental systems drift may have resulted in many signalling differences seen between the two species; *ii*) it seems likely that functional redundancy in the developmental genetic circuitry is an essential prerequisite for change; *iii*) vulva induction using the Wnt pathway in *P. pacificus* may be ancestral and thus vulva induction in *C. elegans* derived; *iv*) in *P. pacificus*, the cross-talk between the two organs and the rest of the worm that properly patterns both adult organs is Wnt signalling; *v*) pleiotropic use of Wnt signalling may require the evolution of changes in developmental timing; *vi*) alterations in development involve co-opting molecular genetic pathways both at the protein level, the level of transcriptional activation, and the level of translational activation; *vii*) changes in development, not surprisingly, result ultimately from signalling that feeds into cell autonomous behaviours such as PCD, axis of cell division, and cell migration; and *viii*) these studies and their findings required the use of an unbiased forward genetic approach and could not have been accomplished as efficiently using a reverse genetic approach.

Developmental systems drift is the observation that similar morphologies can have dramatically different underlying genetic circuitry. Wnt signalling in *P. pacificus* is an example of this. Coordinated development of the reproductive system in *P. pacificus* is about Wnt signalling; in this system not only does the Wnt pathway retain functions in common with *C. elegans* but has additional functions unknown from *C. elegans*. A Wnt signal from the gonad and posterior of the animal induces and patterns the vulva. Wnt signalling establishes a basal 2° cell-fate division pattern. Another Wnt signal reverses the polarity of this division pattern in P7.p to establish the mirror image vul(A-F) cell lineages. A Wnt signal from the vulva to the somatic gonad instructs the movements of the DTC and facilitates the overall shape of the hermaphrodite gonad. Finally, if *P. pacificus* follows what is known from *C. elegans*, Wnt signalling could play a role in polarising and establishing the symmetry of the hermaphrodite somatic gonad primordium. Not only are there new interactions like the gonadal and posterior induction of the vulva by Wnt signals and instruction of gonad arm migration by the vulva, but also these new Wnt signalling events often involve co-option of additional genetic and protein interactions.

Vulva development between *P. pacificus* and *C. elegans* is highly diverged. The EGF/RAS pathway is the principal pathway involved in vulva induction in *C. elegans*. The Notch pathway plays a redundant reinforcement role. Curiously, Wnt signalling in other *Caenorhabditis* strains/species has also been shown to have an effect on proper vulva induction. Given the essential nature of the vulva for reproductive continuity, it is not surprising that multiple regulatory pathways act redundantly to ensure the terminal morphology and physiological function. It seems likely that the differences exhibited in vulva patterning between *P. pacificus* and *C. elegans* are probably due to developmental system drift acting upon these pathways (True & Haag, 2001), which leads to dramatically altered molecular genetic control of what, on the surface, is a highly conserved morphology at the level of cells, organs and physiological function. That is, as long as one pathway is acting to enforce proper patterning, other redundant pathways are free to accumulate changes through genetic drift. Over time this could have led to the highly diverged developmental programmes noted.

While the somatic gonad displays many obvious morphological differences, for example, gonad arm extension and sheath morphology, the patterning of the hermaphrodite germ line into mitotic, transition, meiotic and gamete zones is strikingly conserved. Additionally, the progression of germ cell cytology and nuclear morphology as illustrated by staining of DNA is remarkably similar between *P. pacificus* and *C. elegans*. Despite this morphological similarity, several observations suggest the patterning of the germ line among species is, similar to vulval patterning, also likely to involve substantial developmental systems drift. I will consider the data from *P. pacificus* first. One line of evidence is that there is a dramatically altered morphology of the overlying sheath (McCarter *et al.*, 1997, 1999; Killian & Hubbard, 2004, 2005; Rudel *et al.*, 2005), which probably necessitates innovations and alterations in the instruction from the sheath to the germ line. Unlike *C. elegans* where individual pairs of sheath cells overlie specific zonal regions, *P. pacificus* sheath cells contact multiple, and potentially all, zones. The second line of evidence concerns the phenotypes resulting from interfering with soma germline interactions *via* cell ablation. While ablation of the sheath in *P. pacificus* results in germline defects at the level of zones that is reminiscent of the correlating ablations in *C. elegans*, the resulting aberrant nuclear morphologies within the germ cells exhibit differences (McCarter *et al.*, 1997; Rudel *et al.*, 2005).

One line of evidence supporting the potential for a high level of developmental systems drift between *P. pacificus* and *C. elegans* is the nature of hermaphroditism across phyla. Hermaphrodites are largely females that make a few sperm; cladistics analysis of nematodes suggest hermaphroditism has arisen across lineages independently many times and that the hermaphrodite germline morphology is convergent in these nematodes (Kiontke *et al.*, 2004, 2007). The decision between specifying an oocyte or a sperm is largely controlled by complexes of translational regulatory proteins and their target mRNAs (Ellis & Schedl, 2007). Data from *C. elegans* and its close relatives indicate that known regulators of *C. elegans* hermaphrodite germ cell fates show dramatically altered roles in germline sex determination among species; *i.e.*, the roles of *fog-2*, *tra-2*, *gld-1*, and *puf* family genes change. Some of these proteins are not involved in patterning the germ line at all in other species, *i.e.*, *C. elegans* FOG-2 (Nayak *et al.*, 2005; Guo *et al.*, 2009) and FBF1/2 (Liu *et al.*, 2012); they are missing altogether in other species. Others have germline roles but opposite to those known from *C. elegans*, *e.g.*, *C. briggsae gld-1* (Beadell *et al.*, 2011). Often, even if there is the possibility that genetic interactions are conserved, for example, the repression of *tra-2* mRNA by GLD-1 (Jan *et al.*, 1997; Haag & Kimble, 2000), these regulators are not working through the same molecular interactions. For example, the *C. briggsae* and *C. remanei tra-2* mRNA lacks elements in its 3′UTR required for robust GLD-1 binding of the mRNA in *C. elegans* (Goodwin *et al.*, 1993; Beadell *et al.*, 2011). Together, the genetic and molecular data suggest many independent co-options of disparate regulators leading to hermaphroditism in very closely related species. Often, more distantly related translational regulators in the same protein families are used. In part, this may be because translational regulation in the germ line and early embryo is the primary patterning mechanism; thus, these families of translational regulators are hanging around and available for co-option. As a result, their use does not reflect conservation of biological function, but conservation of biochemical role, *i.e.*, RNA binding.

One last argument for the likelihood of a highly diverged regulation of even the most basic germline decision, mitosis *vs* meiosis, is predicated on what we have learned from vulva development between *P. pacificus* and *C. elegans*. The role of translational regulators of the switch between mitotic germline stem cells and meiosis of gametes is likely to be more pliable than generally acknowledged. This may seem a surprising statement given the essential nature of the regulation of a stem cell pool and the absolute requirement to generate haploid gametes, especially as what little data exist from comparative studies within the *Caenorhabditis* group suggest conservation of at least GLD-1 function in the mitosis/meiosis decision (Beadell *et al.*, 2011). But, if one considers the lessons learned from the specification of the nematode vulva, this may not be too unreasonable in retrospect. Such an essential tissue as the germ line must, for such an important cell fate distinction, have many redundant pathways to ensure proper patterning. In point of fact, multiple parallel pathways regulate the mitosis/meiosis boundary in *C. elegans* (Kimble & Crittenden, 2005, 2007; Crittenden *et al.*, 2006). This high level of redundancy may leave pathways open to developmental systems drift over time; thus, even in this most critical decision, the germ line, in fact, may be one of the tissues most amenable to molecular if not morphological change.

The continuous induction of the vulva seen in *P. pacificus* using Wnt signals from a large expression domain may be less derived than the single-cell short-duration event that patterns the *C. elegans* vulva. Comparative cell ablation studies reveal that continuous or two-step vulva induction by the gonad is common in nematodes (Félix &

Sternberg, 1997, 1998; Sigrist & Sommer, 1999; Félix *et al.*, 2000), somewhat similar to *P. pacificus*. By contrast, single AC induction is limited to *C. elegans* and some close relatives. All this in conjunction with notable vulva induction defects in Wnt mutant backgrounds in some *C. elegans* strains argues that a combinatorial control of vulval induction using broader Wnt expression domains may be ancestral in comparison to EGF/RAS induction. Many traits of the Wnt signalling cascades may contribute to the Wnt's amenability for this type of vulval induction. Among them is that there are multiple Wnt ligands and receptors, and they can act in concert both cooperatively and antagonistically in *P. pacificus*. There are also both canonical and non-canonical Wnt signalling pathways in nematodes (Eisenmann, 2005; Archbold *et al.*, 2012; Cadigan, 2012; Sawa, 2012; Gomez-Orte *et al.*, 2013). These alternative Wnt signalling cascades often act in combination with additional signalling pathways in animal development to control fates.

A second line of evidence also suggests that the Wnt pathway has an ancestral role in ventral hypodermal inductions in nematodes. This evidence comes from male hook specification in *C. elegans*. The male hook is a sclerotised spike of cuticle formed by the ventral hypodermis. During coitus the male rubs the ventral part of his tail against the hermaphrodite in a characteristic search pattern, and the hook catches onto the vulval lips to open the vulva facilitating sperm transfer and mating. The hook is specified from a hook competence group (HCG) similar to the VEG in hermaphrodites. The male hook is formed from the descendants of three ventral epidermal cells, P(9-11).p. The hook is an inherently polarised structure and not symmetric. P11.p is induced to form a 1° hook fate and P10.p and P9.p have 2° and 3° fates respectively; this gives rise to an asymmetric 3-2-1 pattern (Emmons, 2005). Upon cell ablation of a HCG member other members are competent to replace higher fates. In contrast to the VEG induction in *C. elegans*, a Wnt signal through the LIN-17 Frizzled-like receptor is the primary signal; EGF inductive signal is also observed but it is only observed when Wnt signalling is compromised (Yu *et al.*, 2009). This makes EGF a redundant secondary signal, an observation similar though opposite to that seen for Wnt and EGF signalling in the vulva of some *C. elegans* strains. Also similar to vulval induction once the 1° HCG fate is induced, P11.p signals through Notch to promote the 2° cell fate in P10.p. Thus, from *P. pacificus* to *C. elegans*, Wnt has likely retained an ancestral role in ventral epidermis patterning. Given this scenario, in the evolution of

C. elegans, redundant pathways allowed developmental drift to produce a novel mechanism of anchor cell induction of the vulva using EGF.

Many of these Wnt developmental signalling events in the vulva and gonad occur disparately in space and, maybe more importantly, disparately in developmental time. Perhaps these alterations in developmental timing underscore the permissive nature of heterochronic changes in the evolution of developmental signalling events. It could be that differences in the timing of gonad and vulva development in *P. pacificus* in comparison to *C. elegans* reflect compensatory timing changes both within the development of individual organs and between gonad and vulva development in *P. pacificus*. In other words, these timing differences may play roles in separating multiple Wnt signalling events, allowing this pathway to be used in such a pleiotropic manner. Separating these individual events in developmental time avoids confusion from potential cross signalling from concurrent developmental events.

Change in developmental patterning involves alterations in the interactions of members of molecular signalling pathways. There is a long-standing debate over whether selection is more likely to act upon the principal workers that carry out most cellular jobs (the coding sequence of the proteins themselves) or the instructions for proper expression of the molecular players (*cis*-acting genomic regulatory elements). It seems clear from the differences seen between *P. pacificus* and *C. elegans* that the answer is that both are common. Size restriction of the VEG and regulation of vulva induction both involve the loss and gain of protein interactions. In the specification of the VEG, *P. pacificus* requires both a HAIRY homologue with a GROUCHO interacting domain and GROUCHO. In *C. elegans*, a HAIRY family member with a GROUCHO interacting domain is absent and the VEG is larger. HAIRY is a highly conserved protein throughout metazoans, and the lack of HAIRY in *C. elegans* has been the source of some speculation. Given the lack of knowledge from additional nematodes at key nodes in the nematode phylogeny, it is impossible to determine whether the loss of HAIRY was concurrent with, or followed, the loss of the protein interaction domain. It is an interesting question and begs acknowledgement of the importance of protein interaction domains as agents of evolutionary variation. More striking, and more transparent, is the antagonistic role of *Ppa*-LIN-17/Frizzled and the gain of the SBDM motif by *Ppa*-LIN-18 involved in *P. pacificus* vulva induction. It seems likely that the gain of this single domain brought disparate molecular pathways together and led to altered mechanisms for patterning vulva formation. Several working hypotheses also suggest the importance of the evolution of transcriptional regulation. For example, it seems plausible that co-option of a Netrin 'cell guidance cassette' could have led to changes in DTC migration during gonadogenesis *via* altered regulation of Netrin receptor expression in DTCs upon Wnt signalling. Likewise, variability in vulva specification in *P. pacificus* has been traced to a *cis*-acting element in a gene encoding a Notch ligand. Given the earlier discussion of translational regulators involved in germline sex determination, it seems appropriate that evolution of regulatory elements in the untranslated regions of mRNAs should be added to the evolution of protein domains and transcriptional *cis*-acting elements as a mechanism for change as well.

Ultimately, changes in molecular and genetic architecture result in readouts in the cells themselves. It is, in fact, this last step of morphogenesis that truly gives rise to morphological and functional change. Such key changes are in basic cellular processes like PCD (defining the VEG), cell division polarity (defining the number of vulva rings), and cell migration (the shape of the gonad arm). Signalling often seems to feed into cellular physiological cassettes. These cassettes appear to be complete bundled molecular packages that, when turned on, carry out a given function. As such it seems likely that the regulations of molecular players early in the initiation of these cassettes are likely points of dramatic evolutionary change. Repeated studies in nematodes and other animal phyla reiterate the ease with which these cassettes are gained and lost. Yet an explanation for their apparently evolutionary flexibility during animal development has yet to be fully pursued.

Comparative work on the nematode reproductive system, particularly the vulva, is amongst the most comprehensive evo-devo studies in invertebrates, together with insect segmentation, for example. This is largely due to the adoption of an unbiased forward genetic approach using traditional mutagenesis to isolate mutants exhibiting vulva phenotypes. These screens were subsequently followed up with reverse genetic, molecular and biochemical approaches to flesh out pathways and protein interactions. In these days of readily available and obtainable genomes and easy, almost universal, reverse genetic approaches, many systems rely upon knocking out players known from established systems. As these studies have shown, this is an inadequate approach as extensive developmental drift ensured the principle players in the biology are different. And yet with all these molecular details now known, the evolutionary and ecology reasons (the WHY questions) still remain unanswered: why do developmental systems drift so much?

References

ARCHBOLD, H.C., YANG, Y.X., CHEN, L. & CADIGAN, K.M. (2012). How do they do Wnt they do?: regulation of transcription by the Wnt/beta-catenin pathway. *Acta Physiologica* 204, 74-109.

AROIAN, R.V., KOGA, M., MENDEL, J.E., OHSHIMA, Y. & STERNBERG, P.W. (1990). The *let-23* gene necessary for *Caenorhabditis elegans* vulval induction encodes a tyrosine kinase of the EGF receptor subfamily. *Nature* 348, 693-699.

BEADELL, A.V., LIU, Q., JOHNSON, D.M. & HAAG, E.S. (2011). Independent recruitments of a translational regulator in the evolution of self-fertile nematodes. *Proceedings of the National Academy of Sciences of the United States of America* 108, 19672-19677.

BRAENDLE, C. & FÉLIX, M.-A. (2008). Plasticity and errors of a robust developmental system in different environments. *Developmental Cell* 15, 714-724.

CADIGAN, K.M. (2012). TCFs and Wnt/beta-catenin signaling: more than one way to throw the switch. *Current Topics in Developmental Biology* 98, 1-34.

CLANDININ, T.R., KATZ, W.S. & STERNBERG, P.W. (1997). *Caenorhabditis elegans* HOM-C genes regulate the response of vulval precursor cells to inductive signal. *Developmental Biology* 182, 150-161.

CRITTENDEN, S.L., LEONHARD, K.A., BYRD, D.T. & KIMBLE, J. (2006). Cellular analyses of the mitotic region in the *Caenorhabditis elegans* adult germ line. *Molecular Biology of the Cell* 17, 3051-3061.

CUI, M., CHEN, J., MYERS, T.R., HWANG, B.J., STERNBERG, P.W., GREENWALD, I. & HAN, M. (2006). SynMuv genes redundantly inhibit *lin-3/EGF* expression to prevent inappropriate vulval induction in *C. elegans*. *Developmental Cell* 10, 667-672.

EISENMANN, D.M. (2005). Wnt signaling. In: The *C. elegans* Research Community (Ed.). *WormBook*. DOI:10.1895/wormbook.1.7.1.

EISENMANN, D.M., MALOOF, J.N., SIMSKE, J.S., KENYON, C. & KIM, S.K. (1998). The beta-catenin homolog BAR-1 and LET-60 Ras coordinately regulate the Hox gene *lin-39* during *Caenorhabditis elegans* vulval development. *Development* 125, 3667-3680.

EIZINGER, A. & SOMMER, R.J. (1997). The homeotic gene *lin-39* and the evolution of nematode epidermal cell fates. *Science* 278, 452-455.

ELLIS, R. & SCHEDL, T. (2007). Sex determination in the germ line. In: The *C. elegans* Research Community (Ed.). *WormBook*. DOI: 10.1895/wormbook.1.82.2.

EMMONS, S.W. (2005). Male development. In: The *C. elegans* Research Community (Ed.). *WormBook*. DOI:10.1895/wormbook.1.33.1.

FÉLIX, M.-A. & STERNBERG, P.W. (1997). Two nested gonadal inductions of the vulva in nematodes. *Development* 124, 253-259.

FÉLIX, M.-A. & STERNBERG, P.W. (1998). A gonad-derived survival signal for vulval precursor cells in two nematode species. *Current Biology* 8, 287-290.

FÉLIX, M.-A., HILL, R.J., SCHWARZ, H., STERNBERG, P.W., SUDHAUS, W. & SOMMER, R.J. (1999). *Pristionchus pacificus*, a nematode with only three juvenile stages, displays major heterochronic changes relative to *Caenorhabditis elegans*. *Proceedings of the Royal Society B: Biological Sciences* 266, 1617-1621.

FÉLIX, M.-A., DE LEY, P., SOMMER, R.J., FRISSE, L., NADLER, S.A., THOMAS, W.K., VANFLETEREN, J. & STERNBERG, P.W. (2000). Evolution of vulva development in the Cephalobina (Nematoda). *Developmental Biology* 221, 68-86.

FERGUSON, E.L., STERNBERG, P.W. & HORVITZ, H.R. (1987). A genetic pathway for the specification of the vulval cell lineages of *Caenorhabditis elegans*. *Nature* 326, 259-267.

GOMEZ-ORTE, E., SAENZ-NARCISO, B., MORENO, S. & CABELLO, J. (2013). Multiple functions of the noncanonical Wnt pathway. *Trends in Genetics* 29, 545-553.

GOODWIN, E.B., OKKEMA, P.G., EVANS, T.C. & KIMBLE, J. (1993). Translational regulation of *tra-2* by its 3′ untranslated region controls sexual identity in *C. elegans*. *Cell* 75, 329-339.

GUO, Y., LANG, S. & ELLIS, R.E. (2009). Independent recruitment of F box genes to regulate hermaphrodite development during nematode evolution. *Current Biology* 19, 1853-1860.

GUPTA, B.P., HANNA-ROSE, W. & STERNBERG, P.W. (2012). Morphogenesis of the vulva and the vulval-uterine connection. In: The *C. elegans* Research Community (Ed.). *WormBook*. DOI:10.1895/wormbook.1.152.1.

HAAG, E.S. & KIMBLE, J. (2000). Regulatory elements required for development of *Caenorhabditis elegans* hermaphrodites are conserved in the *tra-2* homologue of *C. remanei*, a male/female sister species. *Genetics* 155, 105-116.

HALL, D.H., WINFREY, V.P., BLAEUER, G., HOFFMAN, L.H., FURUTA, T., ROSE, K.L., HOBERT, O. & GREENSTEIN, D. (1999). Ultrastructural features of the adult hermaphrodite gonad of *Caenorhabditis elegans*:

relations between the germ line and soma. *Developmental Biology* 212, 101-123.

HAN, M. & STERNBERG, P.W. (1990). *let-60*, a gene that specifies cell fates during *C. elegans* vulval induction, encodes a ras protein. *Cell* 63, 921-931.

HEDGECOCK, E.M., CULOTTI, J.G., HALL, D.H. & STERN, B.D. (1987). Genetics of cell and axon migrations in *Caenorhabditis elegans Development* 100, 365-382.

HEDGECOCK, E.M., CULOTTI, J.G. & HALL, D.H. (1990). The *unc-5*, *unc-6*, and *unc-40* genes guide circumferential migrations of pioneer axons and mesodermal cells on the epidermis in *C. elegans*. *Neuron* 4, 61-85.

HIRSH, D., OPPENHEIM, D. & KLASS, M. (1976). Development of the reproductive system of *Caenorhabditis elegans*. *Developmental Biology* 49, 200-219.

HUANG, L.S., TZOU, P. & STERNBERG, P.W. (1994). The *lin-15* locus encodes two negative regulators of *Caenorhabditis elegans* vulval development. *Molecular Biology of the Cell* 5, 395-411.

HUBBARD, E.J. & GREENSTEIN, D. (2005). Introduction to the germ line. In: The *C. elegans* Research Community (Ed.). *WormBook*. DOI: 10.1895/wormbook.1.18.1.

INOUE, T., OZ, H.S., WILAND, D., GHARIB, S., DESHPANDE, R., HILL, R.J., KATZ, W.S. & STERNBERG, P.W. (2004). *C. elegans* LIN-18 is a Ryk ortholog and functions in parallel to LIN-17/Frizzled in Wnt signaling. *Cell* 118, 795-806.

JAN, E., YOON, J.W., WALTERHOUSE, D., IANNACCONE, P. & GOODWIN, E.B. (1997). Conservation of the *C. elegans tra-2* 3′UTR translational control. *The EMBO Journal* 16, 6301-6313.

JUNGBLUT, B. & SOMMER, R.J. (2000). Novel cell-cell interactions during vulva development in *Pristionchus pacificus*. *Development* 127, 3295-3303.

KATZ, W.S., HILL, R.J., CLANDININ, T.R. & STERNBERG, P.W. (1995). Different levels of the *C. elegans* growth factor LIN-3 promote distinct vulval precursor fates. *Cell* 82, 297-307.

KIENLE, S. & SOMMER, R.J. (2013). Cryptic variation in vulva development by *cis*-regulatory evolution of a HAIRY-binding site. *Nature Communications* 4, 1714.

KILLIAN, D.J. & HUBBARD, E.J. (2004). *C. elegans pro-1* activity is required for soma/germline interactions that influence proliferation and differentiation in the germ line. *Development* 131, 1267-1278.

KILLIAN, D.J. & HUBBARD, E.J. (2005). *Caenorhabditis elegans* germline patterning requires coordinated development of the somatic gonadal sheath and the germ line. *Developmental Biology* 279, 322-335.

KIMBLE, J. (1981). Alterations in cell lineage following laser ablation of cells in the somatic gonad of *Caenorhabditis elegans*. *Developmental Biology* 87, 286-300.

KIMBLE, J. & CRITTENDEN, S.L. (2005). Germline proliferation and its control. In: The *C. elegans* Research Community (Ed.). *WormBook*. DOI: 10.1895/wormbook.1.13.1.

KIMBLE, J. & CRITTENDEN, S.L. (2007). Controls of germline stem cells, entry into meiosis, and the sperm/oocyte decision in *Caenorhabditis elegans*. *Annual Review of Cell and Developmental Biology* 23, 405-433.

KIMBLE, J. & HIRSH, D. (1979). The postembryonic cell lineages of the hermaphrodite and male gonads in *Caenorhabditis elegans*. *Developmental Biology* 70, 396-417.

KIONTKE, K., GAVIN, N.P., RAYNES, Y., ROEHRIG, C., PIANO, F. & FITCH, D.H.A. (2004). *Caenorhabditis* phylogeny predicts convergence of hermaphroditism and extensive intron loss. *Proceedings of the National Academy of Sciences of the United States of America* 101, 9003-9008.

KIONTKE, K., BARRIERE, A., KOLOTUEV, I., PODBILEWICZ, B., SOMMER, R., FITCH, D. & FÉLIX, M.-A. (2007). Trends, stasis, and drift in the evolution of nematode vulva development. *Current Biology* 17, 1925-1937.

KOLOTUEV, I. & PODBILEWICZ, B. (2004). *Pristionchus pacificus* vulva formation: polarized division, cell migration, cell fusion, and evolution of invagination. *Developmental Biology* 266, 322-333.

KOLOTUEV, I. & PODBILEWICZ, B. (2008). Changing of the cell division axes drives vulva evolution in nematodes. *Developmental Biology* 313, 142-154.

LIU, Q., STUMPF, C., THOMAS, C., WICKENS, M. & HAAG, E.S. (2012). Context-dependent function of a conserved translational regulatory module. *Development* 139, 1509-1521.

MALOOF, J.N. & KENYON, C. (1998). The Hox gene *lin-39* is required during *C. elegans* vulval induction to select the outcome of Ras signaling. *Development* 125, 181-190.

MCCARTER, J., BARTLETT, B., DANG, T. & SCHEDL, T. (1997). Soma-germ cell interactions in *Caenorhabditis elegans*: multiple events of hermaphrodite germline development require the somatic sheath and spermathecal lineages. *Developmental Biology* 181, 121-143.

MCCARTER, J., BARTLETT, B., DANG, T. & SCHEDL, T. (1999). On the control of oocyte meiotic maturation and ovulation in *Caenorhabditis elegans*. *Developmental Biology* 205, 111-128.

MILLOZ, J., DUVEAU, F., NUEZ, I. & FÉLIX, M.-A. (2008). Intraspecific evolution of the intercellular signaling network underlying a robust developmental system. *Genes and Development* 22, 3064-3075.

NAYAK, S., GOREE, J. & SCHEDL, T. (2005). *fog-2* and the evolution of self-fertile hermaphroditism in *Caenorhabditis*. *PLoS Biology* 3, e6.

OGURA, K., ASAKURA, T. & GOSHIMA, Y. (2012). Localization mechanisms of the axon guidance molecule UNC-6/Netrin and its receptors, UNC-5 and UNC-40, in *Caenorhabditis elegans*. *Development, Growth & Differentiation* 54, 390-397.

PHOTOS, A., GUTIERREZ, A. & SOMMER, R.J. (2006). *sem-4/spalt* and *egl-17*/FGF have a conserved role in sex myoblast specification and migration in *P. pacificus* and *C. elegans*. *Developmental Biology* 293, 142-153.

PODBILEWICZ, B. (2006). Cell fusion. In: The *C. elegans* Research Community (Ed.). *WormBook*. DOI:10.1895/wormbook.1.52.1.

RUDEL, D., RIEBESELL, M. & SOMMER, R.J. (2005). Gonadogenesis in *Pristionchus pacificus* and organ evolution: development, adult morphology and cell-cell interactions in the hermaphrodite gonad. *Developmental Biology* 277, 200-221.

RUDEL, D., TIAN, H. & SOMMER, R.J. (2008). Wnt signaling in *Pristionchus pacificus* gonadal arm extension and the evolution of organ shape. *Proceedings of the National Academy of Sciences of the United States of America* 105, 10826-10831.

SAWA, H. (2012). Control of cell polarity and asymmetric division in *C. elegans*. *Current Topics in Developmental Biology* 101, 55-76.

SCHLAGER, B., ROSELER, W., ZHENG, M., GUTIERREZ, A. & SOMMER, R.J. (2006). HAIRY-like transcription factors and the evolution of the nematode vulva equivalence group. *Current Biology* 16, 1386-1394.

SHARMA-KISHORE, R., WHITE, J.G., SOUTHGATE, E. & PODBILEWICZ, B. (1999). Formation of the vulva in *Caenorhabditis elegans*: a paradigm for organogenesis. *Development* 126, 691-699.

SIGRIST, C.B. & SOMMER, R.J. (1999). Vulva formation in *Pristionchus pacificus* relies on continuous gonadal induction. *Development Genes and Evolution* 209, 451-459.

SMITH, K.K. (2003). Time's arrow: heterochrony and the evolution of development. *International Journal of Developmental Biology* 47, 613-621.

SOMMER, R.J. (1997). Evolutionary changes of developmental mechanisms in the absence of cell lineage alterations during vulva formation in the Diplogastridae (Nematoda). *Development* 124, 243-251.

SOMMER, R.J. (2005). Evolution of development in nematodes related to *C. elegans*. In: The *C. elegans* Research Community (Ed.). *WormBook*. DOI:10.1895/wormbook.1.46.1.

SOMMER, R.J. & STERNBERG, P.W. (1996). Apoptosis and change of competence limit the size of the vulva equivalence group in *Pristionchus pacificus*: a genetic analysis. *Current Biology* 6, 52-59.

SOMMER, R.J., EIZINGER, A., LEE, K.-Z., JUNGBLUT, B., BUBECK, A. & SCHLAK, I. (1998). The *Pristionchus* HOX gene *Ppa-lin-39* inhibits programmed cell death to specify the vulva equivalence group and is not required during vulval induction. *Development* 125, 3865-3873.

SPICER, J.I., RUNDLE, S.D. & TILLS, O. (2011). Studying the altered timing of physiological events during development: it's about time... or is it? *Respiratory Physiology & Neurobiology* 178, 3-12.

STERNBERG, P.W. (1988). Lateral inhibition during vulval induction in *Caenorhabditis elegans*. *Nature* 335, 551-554.

STERNBERG, P.W. (2005). Vulval development. In: The *C. elegans* Research Community (Ed.). *WormBook*. DOI:10.1895/wormbook.1.6.1.

STERNBERG, P.W. & HORVITZ, H.R. (1986). Pattern formation during vulval development in *C. elegans*. *Cell* 44, 761-772.

STERNBERG, P.W. & HORVITZ, H.R. (1989). The combined action of two intercellular signaling pathways specifies three cell fates during vulval induction in *C. elegans*. *Cell* 58, 679-693.

STROME, S. (1986). Fluorescence visualization of the distribution of microfilaments in gonads and early embryos of the nematode *Caenorhabditis elegans*. *Journal of Cell Biology* 103, 2241-2252.

STROME, S. (2005). Specification of the germ line. In: The *C. elegans* Research Community (Ed.). *WormBook*. DOI:10.1895/wormbook.1.9.1.

SULSTON, J.E. & HORVITZ, H.R. (1977). Post-embryonic cell lineages of the nematode, *Caenorhabditis elegans*. *Developmental Biology* 56, 110-156.

SULSTON, J.E. & WHITE, J.G. (1980). Regulation and cell autonomy during postembryonic development of *Caenorhabditis elegans*. *Developmental Biology* 78, 577-597.

TIAN, H., SCHLAGER, B., XIAO, H. & SOMMER, R.J. (2008). Wnt signaling induces vulva development in the nematode *Pristionchus pacificus*. *Current Biology* 18, 142-146.

TILLS, O., RUNDLE, S.D., SALINGER, M., HAUN, T., PFENNINGER, M. & SPICER, J.I. (2011). A genetic basis for intraspecific differences in developmental timing? *Evolution & Development* 13, 542-548.

TRUE, J.R. & HAAG, E.S. (2001). Developmental system drift and flexibility in evolutionary trajectories. *Evolution & Development* 3, 109-119.

WANG, X. & SOMMER, R.J. (2011). Antagonism of LIN-17/Frizzled and LIN-18/Ryk in nematode vulva induction reveals evolutionary alterations in core developmental pathways. *PLoS Biology* 9, e1001110.

WRISCHNIK, L.A. & KENYON, C.J. (1997). The role of *lin-22*, a *hairy/enhancer of split* homolog, in patterning the peripheral nervous system of *C. elegans*. *Development* 124, 2875-2888.

YI, B. & SOMMER, R.J. (2007). The *pax-3* gene is involved in vulva formation in *Pristionchus pacificus* and is a target of the Hox gene *lin-39*. *Development* 134, 3111-3119.

YU, H., SEAH, A., HERMAN, M.A., FERGUSON, E.L., HORVITZ, H.R. & STERNBERG, P.W. (2009). Wnt and EGF pathways act together to induce *C. elegans* male hook development. *Developmental Biology* 327, 419-432.

ZHENG, M., MESSERSCHMIDT, D., JUNGBLUT, B. & SOMMER, R.J. (2005). Conservation and diversification of Wnt signaling function during the evolution of nematode vulva development. *Nature Genetics* 37, 300-304.

ZIEL, J.W. & SHERWOOD, D.R. (2010). Roles for netrin signaling outside of axon guidance: a view from the worm. *Developmental Dynamics* 239, 1296-1305.

Nematology Monographs & Perspectives, 2015, Vol. 11, 257-299

Chapter 10

Dauer formation and dauer-specific behaviours in *Pristionchus pacificus*

Akira OGAWA [1] and Federico BROWN [2]

[1] *Laboratory for Developmental Dynamics, RIKEN Quantitative Biology Center, 2-2-3 Minatojima-minamimachi, Chuo-ku, Kobe, 650-0047, Japan*
akira.ogawa@riken.jp

[2] *Departamento de Zoologia, Instituto de Biociências Universidade de São Paulo, Rua do Matão, Travessa 14, No. 101 Cidade Universitária, São Paulo, SP 05508-090, Brazil*
fdbrown@usp.br

Introduction

DAUER FORMATION AS PREVALENT SURVIVAL STRATEGIES IN NEMATODES

To survive adverse environmental conditions, many nematodes enter into the dauer stage, which is specialised for enduring various environmental stresses (Grant & Viney, 2011). Dauer juveniles were initially described in species closely associated with beetles about a century ago (Maupas, 1899; Fuchs, 1915). After these initial descriptions, dauer and dauer-like juveniles have been described in many other species, including distantly related, free-living and parasitic nematodes (Lee, 2002). The decision of whether to go into the dauer stage or to continue reproductive development is made based on environmental cues. A harsh environment induces the development of the dauer stage, usually as an alternative to the third-stage juvenile (J3), although the dauer stage is the fourth-stage juvenile in the endoparasitic nematode *Bursaphelenchus xylophilus* (Mota & Vieira, 2008; Perry & Moens, 2011). In species where the J3 can develop into the dauer, the second-stage juvenile (J2) either continues development into a dauer J3 specialised for survival and dis-

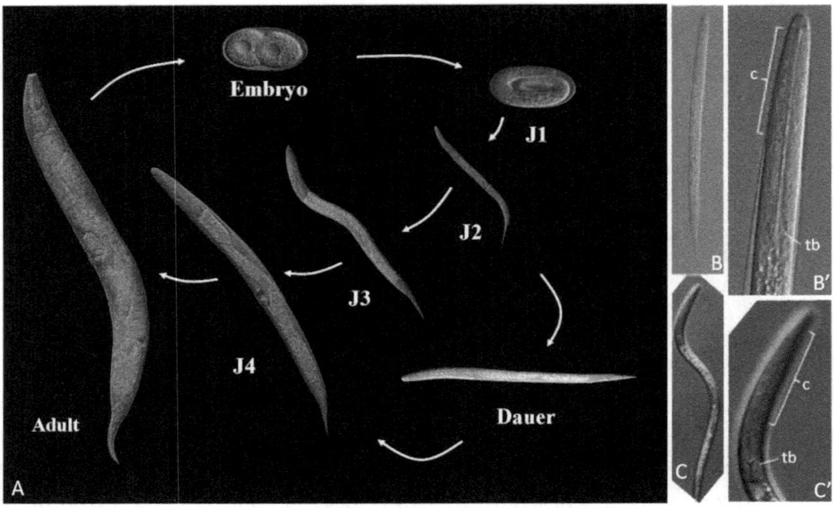

Fig. 10.1. *Environmentally regulated development of an alternative developmental stage in nematodes: the dauer. A: Life cycle of* Pristionchus pacificus. *Apart from the dauer diapause there are two other developmental arrests (*i.e., *first-stage juvenile and reproductive diapause) during the life cycle; B:* P. pacificus *dauer in a typical straight posture; B′: The inset shows the sealing of the buccal cavity and shrinkage of pharyngeal structures; C:* C. elegans *dauer; C′: The inset shows the sealed buccal cavity and a conspicuous grinder due to shrinkage of the terminal bulb. Abbreviations: c, corpus; tb, terminal bulb.*

persal, or in favourable conditions continues development into the J3 (Fig. 10.1). When dauer juveniles are formed, development remains arrested until the environment improves. Once they find a suitable environment for reproduction, they exit the dauer stage and resume development to the sexually mature adult. Facultative dauer formation in nematodes enables a 'boom-and-bust' lifestyle whereby worms proliferate as much as possible when plenty of food is available and form arrested dauer juveniles after food is depleted.

To resist harsh environmental conditions such as starvation, anoxia and high temperature, dauer juveniles have specialised morphological and physiology features, *e.g.*, they have a thick cuticle and can survive for long periods without feeding. Dauer juveniles also show specialisation for dispersal strategies, mainly in behaviour (Ishibashi, 2002). Specialised behaviours of dauer juveniles increase chances of finding and attaching onto their hosts as means of transportation to find a new food source or, in other cases, for trophic associations such as necromeny or parasitism.

Because many nematode species are found predominantly in the dauer stage in the wild, many adaptations are expected to occur in the dauer according to the ecological niche that the nematode species occupies. Furthermore, the facultative dauer formation under unfavourable environmental conditions is the best-studied example of phenotypic plasticity in nematodes, where a single genome can produce multiple phenotypes. With phenotypic plasticity, organisms can express highly specialised features adapted to extreme environmental conditions, without affecting the life in moderate environments. Therefore, phenotypic plasticity is proposed to be one of the most important facilitators of morphological and life-history evolution (West-Eberhard, 2003). Nematode dauer formation may serve as an excellent model for studying relationships between developmental plasticity and evolution. In this chapter, we compare mechanistic studies of dauer formation in *Pristionchus pacificus* to other studies mostly done in *Caenorhabditis elegans*. Studies of dauer juveniles and dauer-specific behaviours in other nematode species, including the infective juvenile of parasitic nematodes, allow us to discuss the implications of this important stage for nematode evolution.

Studies on *C. elegans* dauer formation

The model organism *C. elegans* has provided a wealth of knowledge about the genetic mechanisms regulating the entry into the dauer stage; the decision of whether to go into the dauer stage is based on inputs from sensory neurons (Bargmann & Horvitz, 1991). Therefore, mutations affecting the structural and functional integrity of these neurons often result in abnormal dauer formation (Albert *et al.*, 1981; Starich *et al.*, 1995). Among the environmental cues that regulate dauer formation, population density cues have been extensively studied with genetic and biochemical approaches. In 1982, Golden and Riddle suggested that *C. elegans* constitutively secretes hydrophilic compounds that act as dauer pheromones (Golden & Riddle, 1982). The increase of pheromone concentration caused by constitutive secretion by individuals and high population density, triggers dauer formation. Dauer-inducing activity secreted into the culture medium of *C. elegans* was attributed to a class of compounds called ascarosides (Jeong *et al.*, 2005; Butcher *et al.*, 2007, 2008; Srinivasan *et al.*, 2008; see Schroeder, Chapter 7, this volume).

Ascarosides are glycosides that contain a dideoxyhexose (ascarylose) as the sugar moiety (Fig. 7.1). The aglycon residue (side chain) can have diverse structures and the dauer-inducing activity depends strongly on the structure of the side chain. In addition to dauer induction, ascarosides are known to act as pheromones involved in social and dispersal behaviours (*e.g.*, mating pheromone in *C. elegans*) or as structural components in other species (Jezyk & Fairbairn, 1967; Srinivasan *et al.*, 2008; Kaplan *et al.*, 2012). Two enzymes of the peroxisomal β-oxidation pathway (Butcher *et al.*, 2009) are known to be involved in the biosynthesis of dauer pheromone. These enzymes play a role in the synthesis of the fatty aglycones by shortening their long-chain precursors. However, identification of the enzymes involved in other steps of the pheromone synthesis, *e.g.*, ascarylose synthesis and conjugation of the sugar and non-sugar moieties, remains elusive. So far, there are several G-protein coupled receptors belonging to three distinct families that are proposed to act as dauer pheromone receptors (Kim *et al.*, 2009; McGrath *et al.*, 2011; Park *et al.*, 2012). These receptors are differentially expressed in the amphid neurons, regulate different traits and respond to different sets of ascarosides, suggesting an unexpected complexity of the ascaroside signalling system in *C. elegans*. In addition, *P. pacificus* utilises dauer pheromones that are chemically distinct from *C. elegans* (see Fig. 7.2). Therefore, the nematode dauer pheromone induction system appears to evolve relatively rapidly.

Environmental cues indicating harsh environmental conditions are perceived by sensory neurons, which initiate a cascade of neuroendocrine events involving both inter- and intra-cellular signalling events. The signalling pathways involved in the neuroendocrine signalling include the TGF-β and insulin/IGF pathways (Georgi *et al.*, 1990; Estevez *et al.*, 1993; Gottlieb & Ruvkun, 1994; Morris *et al.*, 1996; Ren *et al.*, 1996; Kimura *et al.*, 1997). Mutations in the genes in these signalling pathways result in either the 'dauer formation defective' (Daf-d) phenotype of mutants that do not enter the dauer stage even under harsh conditions, or the 'dauer formation constitutive' (Daf-c) phenotype of animals that constitutively form dauers (Georgi *et al.*, 1990; Estevez *et al.*, 1993; Gottlieb & Ruvkun, 1994; Morris *et al.*, 1996; Ren *et al.*, 1996; Kimura *et al.*, 1997). Under favourable conditions, ligands for the TGF-β and insulin/IGF pathways are expressed in sensory neurons that suppress dauer

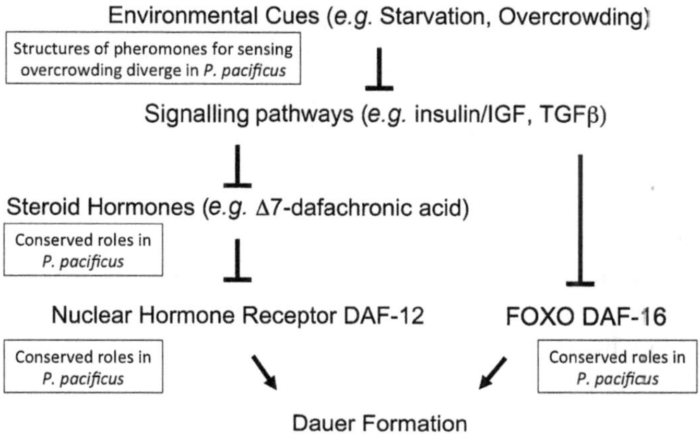

Fig. 10.2. *Regulatory pathway for* Caenorhabditis elegans *dauer formation. Current knowledge for corresponding molecular components in* Pristionchus pacificus *dauer formation is indicated in boxes.*

formation, thus promoting reproductive development (Fig. 10.2). Under harsh conditions, expression of these ligands is repressed, specifying the dauer fate (Ren *et al.*, 1996; Li *et al.*, 2003). In the TGF-β pathway, several kinase-linked transmembrane receptors and SMAD transcription factors are involved in signal transduction (Georgi *et al.*, 1990; Estevez *et al.*, 1993; Ren *et al.*, 1996; Ogg *et al.*, 1997; Patterson *et al.*, 1997; Riddle & Albert, 1997).

Conversely, in the insulin/IGF pathway, the DAF-2 insulin/IGF receptor transmits the signal through a cascade of phosphorylation events, which culminates in regulation of the FOXO transcription factor DAF-16 (Fig. 10.2) (Gottlieb & Ruvkun, 1994; Kimura *et al.*, 1997; Ogg *et al.*, 1997; Ogg & Ruvkun, 1998; Paradis & Ruvkun, 1998; Paradis *et al.*, 1999; Lee *et al.*, 2001; Wolkow *et al.*, 2002; Li *et al.*, 2003). Downregulation of insulin signalling and downstream kinases leads to nuclear translocation of unphosphorylated DAF-16, which is required for dauer formation (Lee *et al.*, 2001; Lin *et al.*, 2001; Hertweck *et al.*, 2004). DAF-16 not only is required for the normal regulation of dauer formation, but also is essential for dauer morphogenesis (Vowels & Thomas, 1992; Ogg *et al.*, 1997; Matyash *et al.*, 2004). Placing *daf-16* animals in strong dauer-inducing conditions results in morphologically aberrant dauer juveniles. By contrast, mutations in the DAF-2 receptor result in a Daf-c phenotype with extended lifespan of adults and enhanced stress-resistance (Friedman & Johnson, 1988; Kenyon *et al.*, 1993; Dorman *et al.*, 1995). The Daf-c, extended longevity and stress resistance phenotype caused by mutations in the DAF-2 receptor and other components of the pathway can be suppressed by *daf-16* (Dorman *et al.*, 1995; Ogg *et al.*, 1997). An interesting approach is to determine which aspect of the phenotype, *i.e.*, dauer formation, extended longevity, or stress-resistance, represents the ancestral role of insulin/IGF signalling in nematodes.

Steroid hormone dafachronic acid (DA) and its nuclear hormone receptor DAF-12 are two key downstream targets of the above-mentioned signalling pathways that regulate dauer formation (Figs 7.8, p. 185; 7.9, p. 186; 10.2, p. 261). Steroid hormone regulation of dauer formation was first postulated in genetic studies by Antebi and colleagues (Antebi *et al.*, 2000; Gerisch *et al.*, 2001) and later confirmed by Mangelsdorf and colleagues in their identification of two ligands, Δ-4 and Δ-7 DA (Motola *et al.*, 2006). Later Schroeder and colleagues showed that among these ligands Δ-7 DA represents an endogenous ligand, and identified several other endogenous ligands that include Δ-1, 7 DA (Mahanti *et al.*, 2014). These ligands are bile acid-like molecules containing a 3-keto sterol backbone and a carboxyl group (Fig. 7.8). Nuclear hormone receptor DAF-12 is their receptor (Antebi *et al.*, 2000). Like other nuclear hormone receptors, DAF-12 has a C4-type zinc-finger DNA-binding domain and a ligand-binding domain. Binding of DA to the ligand-binding domain of DAF-12 directly regulates transcriptional activity. Loss-of-function mutations in *daf-12* lead to a strong Daf-d phenotype, and can suppress daf-c mutations in TGF-β, insulin and other signalling pathways (Vowels & Thomas, 1992; Thomas *et al.*, 1993; Antebi *et al.*, 1998, 2000). Administration of DA ligands also strongly suppresses the daf-c mutants (Motola *et al.*, 2006; Martin *et al.*, 2008; Sharma *et al.*, 2009), suggesting that DA inhibits the dauer fate specification *via* DAF-12. DAF-12 is widely expressed in multiple tissues of *C. elegans* juveniles and probably mediates dauer transition at the level of individual cells (Antebi *et al.*, 2000).

Nematodes are sterol auxotrophs and synthesis of DA ligands requires supply of external sterols from food sources (Matyash *et al.*, 2004). DA synthesis depends on several steroidogenic enzymes and steroid-binding proteins (Jia *et al.*, 2002; Gerisch & Antebi, 2004; Li *et al.*, 2004; Mak & Ruvkun, 2004; Motola *et al.*, 2006; Rottiers *et al.*, 2006; Patel *et al.*, 2008; Dumas *et al.*, 2010; Mahanti *et al.*, 2014). The best characterised of these is DAF-9, a cytochrome P450 oxygenase (Mak & Ruvkun, 2004). *daf-9* is one of the earliest Daf mutations obtained in early genetic screens (Albert & Riddle, 1988) and results in the constitutive formation of morphologically abnormal dauer juveniles. It was shown biochemically that DAF-9 mediates successive mono-oxygenation of the C-26 position in the sterol backbone, thereby introducing a carboxyl group (Motola *et al.*, 2006). Among other enzymes, Rieske-like oxygenase DAF-36, orthologous to insect Neverland (*nvd*) that converts cholesterol to 7-dehydrocholesterol (Yoshiyama-Yanagawa *et al.*, 2011) during ecdysone biosynthesis, has been proposed as catalyst for other steps of DA synthesis (Rottiers *et al.*, 2006; Patel *et al.*, 2008; Dumas *et al.*, 2010). However, the precise biochemical functions of these enzymes have not been determined.

In summary, the studies presented above suggest that *C. elegans* dauer formation is regulated by a cascade of neuroendocrine events that involve multiple signalling pathways. These two-decade long molecular studies embody a detailed and basic framework for comparative studies in dauer formation (Fig. 10.2). Environmental cues that indicate unfavourable conditions are perceived by several sensory neurons in the head (amphid neurons). These neurons transmit the signal through TGF-β and insulin/IGF pathways, which, in turn, decrease the humoral concentration of the steroid hormone DA by suppressing steroidogenic enzymes required for DA synthesis. Finally, ligand-unbound DAF-12 specify the dauer fate of individual tissues. Some of the morphogenetic processes involved in dauer formation might be mediated directly by DAF-16, which is activated by the down-regulation of the upstream insulin/IGF-signalling pathway.

Studies on *P. pacificus* dauer formation

General features of *P. pacificus* dauer juveniles

In the wild, *P. pacificus* dauers are found in close association with scarab beetles due to their necromenic lifestyle (see Ragsdale *et al.*, Chapter 4, this volume). As with many other free-living nematodes, *P. pacificus* enters into the dauer stage as an alternative J3 when the environmental conditions are harsh. Environmental cues include population density, as indicated by the concentration of secreted pheromone, or availability of food. If bacterial food on culture plates is killed with antibiotics or heat, dauer formation is enhanced, suggesting that live bacteria secrete a chemical(s) that serves as 'food signal' that antagonises dauer formation (Golden & Riddle, 1982, 1984). High temperatures (27°C) induce dauer formation in *C. elegans*, whereas *P. pacificus* does not show the same response. In fact, *P. pacificus* dauers, but not other stages, can survive and recover from dauer when maintained at low (8°C) or high (30°C) temperatures, and can live for 1 year without feeding, suggesting a greater temperature tolerance than *C. elegans* (Mayer & Sommer, 2011). Morphologically, *P. pacificus* dauer juveniles have a thick cuticle and a closed mouth (Fig. 10.1). A pair of amphid openings in the head develops in the dauer stage but its adaptive or functional significance is not well understood. During dauer entry, and after moulting of J2, dauers undergo radial shrinkage and at the same time secrete a massive amount of lipids from the cuticle. Because these secreted substances are sticky, dauer juveniles can attach to each other, often forming 'dauer towers' that will be described below.

P. pacificus dauer pheromone

Since the early days of nematode pheromone research (Golden & Riddle, 1982) it has been suggested that dauer pheromone evolved relatively rapidly, although the precise molecular nature of pheromone diversity remained elusive. Recent findings of multiple *C. elegans* dauer pheromone compounds provide a platform for comparative studies. Supernatant of cultured *P. pacificus* contains dauer-inducing activity on conspecifics but not on *C. elegans*. Similarly, *C. elegans* dauer pheromone extracts do not induce dauer formation in *P. pacificus*, suggesting that *P. pacificus* uses distinct chemical cues for sensing population density (Ogawa *et al.*, 2009). *Pristionchus pacificus* shows tremendous natural variation in pheromone production and responsiveness (Mayer & Sommer, 2011; Bose *et al.*, 2014). While natural variation is also observed for *C. elegans* (Mahanti *et al.*, 2014), the extent of variation in *P. pacificus* is substantially larger. For example, in most investigated strains it was seen that an isolated dauer pheromone induces the highest number of dauers in individuals of other *P. pacificus* haplotypes, a phenomenon that was called 'cross-preference', in contrast to the expected but only rarely observed 'self-preference' (Mayer & Sommer, 2011). More recent follow-up studies of this phenomenon are described below.

NMR analysis revealed several glycosides secreted in *P. pacificus* culture supernatant (Fig. 7.2, p. 171) (Bose *et al.*, 2012; see also Schroeder,

Chapter 7, this volume). These compounds include ascarosides with an ascarylose moiety, which is also incorporated in all the *C. elegans* pheromone compounds so far identified. Another class of the glycosides is paratosides with a paratose moiety. Paratose differs from ascarylose only in the stereochemistry of one of the hydroxyl groups attached to the sugar backbone. By contrast, structures of the side chains show high structural complexity. Unlike the side chains of *C. elegans* pheromone compounds that are predominantly derived from hydroxylated fatty acids and the like, the side chains of *P. pacificus* pheromone compounds are derived from various primary metabolites that include short chain fatty acids, threonine, adenosine, succinate, xylose and phenylethanolamine. Such complex structural features illustrate that the diversity of molecular structures have been generated by the 'combinatorial chemistry' of primary metabolites in the course of pheromone evolution to enable intricate communication between conspecifics.

By synthesising artificial pheromone compounds, the dauer-inducing activity was largely attributed to the paratosides, whereas some of the ascarosides also showed activity (Bose *et al.*, 2012). Recently, these glycosides were shown to regulate adult mouth form dimorphism (see Ragsdale, Chapter 11, this volume). Strikingly, the response of the mouth form to the pheromone compounds shows a distinct profile from dauer formation (Fig. 10.3). In contrast to dauer formation, where paratosides play predominant roles, mouth form dimorphism is regulated by mixture of three classes of glycosides, *i.e.*, ascarosides, diascarosides and paratosides. Such differences suggest that the repertoire of pheromone compounds generates not only the species specificity of dauer induction, but also the specificity among the traits in the same species that are regulated by the same class of chemical compounds.

Most recent studies looked at the chemistry of dauer pheromones in the context of the natural variation between strains of *P. pacificus* (Bose *et al.*, 2014) (see Schroeder, Chapter 7, this volume). Chemical investigations of the exo- and endometabolome revealed substantial variation in pheromone production but also pheromone sensing among six natural isolates of *P. pacificus*. For example, some strains respond to small molecules they do not synthesise themselves and others do not respond to some of their own small molecules (Figs 7.4, p. 175; 7.6, p. 180) (Bose *et al.*, 2014). These surprising findings would be consistent with intraspecific competition between *P. pacificus* strains in the wild. Indeed, using a novel experimental assay, intraspecific

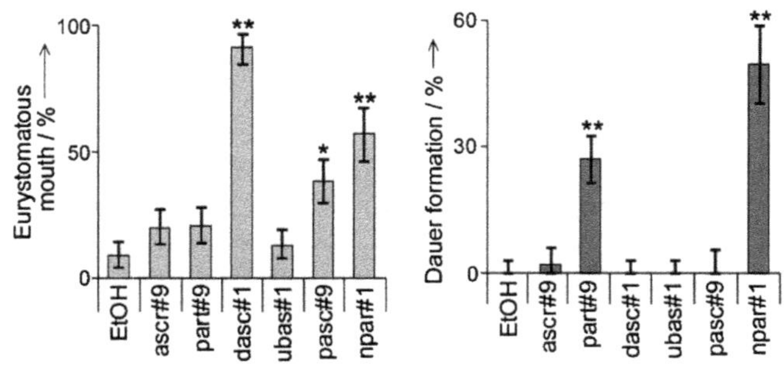

Fig. 10.3. *Response profiles to six glycosides, ascr#9 (an ascaroside also secreted by* Caenorhabditis elegans*), pasc#9 (ascaroside found only in* Pristionchus pacificus *culture media), par#9 (a paratoside), npar#1 (a paratoside with nucleoside moiety), dasc#1 (a diascaroside), ubas#1 (a diascaroside with an ureido isobutyrate moiety) secreted in* P. pacificus *culture media, for mouth form dimorphism (left) and dauer formation (right). Modified from Bose* et al. *(2012).*

competition between three sympatric strains from La Réunion Island has been observed (Fig. 7.7, p. 182) (Bose *et al.*, 2014). So-called 'cheater' strains drive other haplotypes into dauer early in order to feed themselves longer on the limited food source. Simultaneously, so-called 'escaper' strains go into early dauer and are thought to specialise for survival and dispersal (Bose *et al.*, 2014). While the molecular mechanism of this astonishing behaviour is currently unknown, this observation adds a new type of interaction that might be of crucial importance for nematode evolution and ecology. These findings can serve as testimony for the need for integrative studies of development, ecology and evolution.

Still, many unanswered questions remain on chemical communication in nematodes and, therefore, evolutionary studies are promising for the field. Future research questions include identification of enzymes involved in the synthesis of pheromone compounds, and receptors that transmit the pheromone signal across different taxa. Also, co-option and modification of these pheromone compounds and of their receptors in other developmentally plastic traits, *e.g.*, the mouth form dimorphism of *P. pacificus* or social behaviours of *C. elegans* (Srinivasan *et al.*, 2008; Macosko *et al.*, 2009), include areas of promising future research. Addressing some of these questions will eventually lead to the understanding of how ecology and genomes interact, and how pheromone compounds induce plastic traits, thus shaping the complexity and diversity of chemical communication observed in nematodes today.

REGULATORY MECHANISMS OF *P. PACIFICUS* DAUER FORMATION

Multiple Daf-d and Daf-c strains have been obtained by chemical mutagenesis in *P. pacificus* (Ogawa *et al.*, 2009). Although molecular identity of most of the mutations remains unknown, there are two genes identified so far that are responsible for the Daf-d phenotypes. The first is nuclear hormone receptor DAF-12 that was identified as a conserved regulator of dauer formation (Ogawa *et al.*, 2009). Loss-of-function mutations in *Ppa-daf-12* resulted in a strong Daf-d phenotype and DAF-12 ligands, Δ4- and Δ7-DA, strongly rescued *P. pacificus* daf-c mutants, indicating that the DAF-12/DA endocrine module is conserved in *P. pacificus* (Ogawa *et al.*, 2009). The conservation of the DA/DAF-12 endocrine signalling module between *C. elegans* and *P. pacificus* dauer formation has encouraged similar studies in parasitic nematodes. In fact it was shown that Δ7-DA, but not Δ4-DA, is able to block infective juvenile (IJ) formation in both the homogonic and the heterogonic cycle of the animal-parasitic nematode, *Strongyloides papillosus* (Ogawa *et al.*, 2009). Later, Wang *et al.* (2009) made similar observations with Δ7-DA in the related species *Strongyloides stercoralis*, a human parasite, and in the hookworm *Ancylostoma caninum*. These findings strongly support a common origin of dauer and IJ formation as related pathways of phenotypic plasticity identify DA and DAF-12 regulation. These may serve as potential pharmacological targets in therapies against parasitic nematodes.

The second locus identified that is essential for dauer formation in *P. pacificus* is *Ppa-daf-16*/FOXO (Ogawa *et al.*, 2011). Mutations in *Ppa-daf-16* have a Daf-d phenotype similar to *Ppa-daf-12*. However, *Ppa-daf-16* is also required for dauer morphogenesis, and *Ppa-daf-16* mutants arrest as partial dauer juveniles after dauer induction under certain conditions, *e.g.*, with supplementation of lophenol and depletion of other sterol compounds. Thus, DAF-12 and DAF-16 represent two conserved transcriptional modules for the regulation of dauer formation in *P. pacificus* and *C. elegans*, indicating strong molecular conservation of phenotypic plasticity. However, it should be noted that the identification of the signalling pathways involved in *P. pacificus* dauer regulation, and thus the degree of evolutionary conservation of the signalling pathways controlling dauer formation, await future analysis. Based on findings in other developmental processes, the assumption that

IGF and TGF-β signalling are the key pathways in *P. pacificus* just as they are in *C. elegans* should be made with caution (Tian *et al.*, 2008).

The conserved DAF-12/DA endocrine module has been co-opted for regulating evolutionarily novel mouth form dimorphism in *P. pacificus* and close relatives (see Ragsdale, Chapter 11, this volume). Another well studied system of co-option is ecdysone signalling that primarily regulates moulting in insects but also various other developmental traits (Nijhout, 2003). Co-option of conserved endocrine modules circumvents the need for inventing *de novo* switch mechanisms for regulating novel phenotypic plasticity, thus facilitating phenotypic evolution. Given the deep conservation of DAF-12/DA, it is interesting to explore other phenotypically plastic traits in *P. pacificus*, and to investigate whether the same endocrine modules regulate similar traits across nematode species.

Lipid secretion of *P. pacificus* dauers

One particular feature of *P. pacificus* dauer juveniles is the secretion of an oil on the surface of their cuticle, an evolutionary novelty restricted to the diplogastrids (Penkov *et al.*, 2014). Because of secreted lipids, dauer juveniles of *P. pacificus* float and stay at the surface of water, and tend to stick to each other. In other words, secreted lipids provide dauers with a waterproof cuticle and serve as glue for 'dauer towers' that they form as a collective host-finding behaviour (Fig. 10.4A) (see below). When J2 are placed in dauer-inducing conditions, they start to accumulate lipids as lipid droplets in the hypodermal tissue. The size and number of these lipid droplets increase as they become closer to the J2-dauer moult and secretion of lipids begins after moulting. During maturation of dauer juveniles the body shrinks radially, rendering the body shape thinner. Shrinkage of the body begins at the anterior and extends to the posterior, and lipids appear to be squeezed out from lipid droplets beneath the cuticle where the shrinkage occurs (Fig. 10.4B, C). These observations suggest that physical force exerted by the radial shrinkage might be involved in the lipid secretion.

Biochemical analysis revealed that *P. pacificus* dauer juveniles secrete several species of lipid molecules (Penkov *et al.*, 2014). Using NMR and total synthesis techniques, the most abundant secreted lipid was a wax-ester derived from very long chain fatty acid and alcohol and was named 'nematoil' (Fig. 10.4D). The fatty acid chains are highly unsaturated, containing 12 double bonds in total. Such a structure makes the molecule

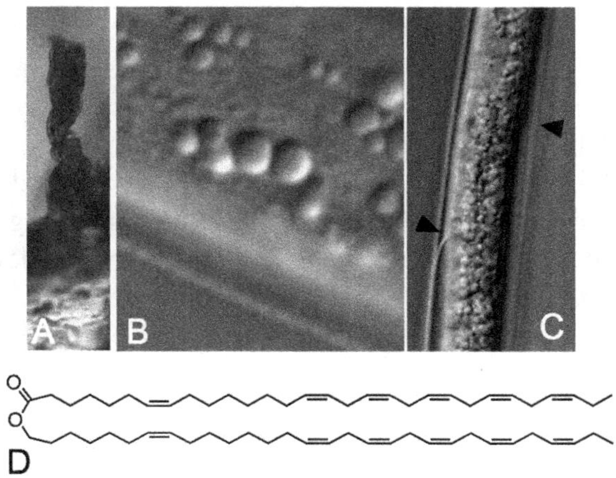

Fig. 10.4. *Lipids in* Pristionchus pacificus *facilitate dauer tower aggregates. A: Dauer towers are formed by sticky* P. pacificus *dauer juveniles; B: Lipids beneath the cuticle of* P. pacificus *dauer juvenile; C:* P. pacificus *dauer juvenile secreting lipids on its surface. The juvenile is undergoing radial shrinkage from the anterior part (bottom) to the posterior (top). Two secreted lipid fronts are indicated (arrowheads); D: Nematoil structure.*

highly hydrophobic, thus protecting dauer juveniles from desiccation, while keeping the covering material in a liquid form that gives a glue-like property required for dauer tower formation. Nematoil starts to appear in the thin-layer chromatography (TLC) of total lipid extracts around the J2-dauer moult, suggesting that production of the wax ester and the lipid secretions occur concomitantly. However, It is currently unknown if nematoil has a direct role with lipid secretion.

By chemical mutagenesis, several mutants were isolated that show defective lipid secretion (Penkov *et al.*, 2014). One of these wax secretion-defective (wsd) mutants failed to secrete surface wax, retaining lipids beneath the cuticle, and was found to lack the ability to form dauer towers. Molecular identity of the mutation is currently under investigation.

Dauer behaviours in *P. pacificus*

Nematodes of the genus *Pristionchus* have a necromenic association with scarab beetles (Osche, 1956; Sudhaus, 2008; Dieterich & Sommer, 2009; Ogawa *et al.*, 2009), in which arrested dauer stage nematodes invade the insect and wait for the host to die to resume development by feeding on growing microorganisms on the carcass (see Ragsdale *et al.*, Chapter 4, this volume). *Pristionchus pacificus* and related species are not parasitic (Herrmann *et al.*, 2006; Weller *et al.*, 2010) but display host-finding behaviours. In contrast to adult stages (Hong & Sommer, 2006; Okumura *et al.*, 2013), *P. pacificus* dauers do not engage actively in cruising behaviours but employ ambushing behaviours (Brown *et al.*, 2011). Dauer juveniles constitute the developmental stage for propagation in *P. pacificus* because nematodes isolated from live beetles in the wild are exclusively dauers (Herrmann *et al.*, 2010; Weller *et al.*, 2010). By contrast, the dauer stage in *Caenorhabditis* species is apparently important, but not necessary, for propagation. Wild populations of *C. elegans*, generally found in the soil or decaying organic matter, and *C. briggsae*, which occasionally associate with mites, isopods, springtails, flies, spiders, earwigs, beetles, ants, snails and slugs (Baird *et al.*, 1994; Petersen *et al.*, 2014), have been found in all stages of development in vegetal debris, suggesting that reproduction in wild populations can bypass the dauer stage and can occur in the absence of an animal intermediate host (Barriere & Félix, 2005; Félix & Duveau, 2012).

Exploring the genetic and neural mechanisms that control dauer-specific behaviours is necessary to understand how developmental and metabolic pathways affect specific behavioural adaptations. Mutations in *daf-2*, the gene encoding the insulin-like growth factor 1 receptor (IGF-1R), can cause adult *C. elegans* worms to behave like dauers (Gems *et al.*, 1998). DAF-2 is involved in the metabolic pathway that regulates dauer entry and aging. Downregulation of *daf-2* leads to long-lived worms, and has therefore been named the 'Grim Reaper' gene by the ageing-research scientist C. Kenyon (Kenyon, 2006, 2010). After dauer exit, *daf-2* adult mutants show temperature-dependent dauer-like behaviours, including reduced pharyngeal pumping, impaired and uncoordinated movement, coiling behaviour, and frequent adoption of a kinked posture similar to that seen in *C. elegans* dauer juveniles (Gems *et al.*, 1998). How is metabolic pathway disruption in adult worms related to dauer-like behaviours? It can be speculated that behaviours are complex phenotypes and may indirectly be affected by misregulation of several metabolic pathway genes. As an alternative, metabolic and developmental genes may have been co-opted to regulate behaviour. Below, we review what is known for each of the dauer-specific nematode behaviours in more detail.

Still behaviours during diapause

Dauer diapause in nematodes is characterised by a highly resistant and developmentally-arrested juvenile, which not only undergoes morphological and metabolic change, as discussed earlier, but also undergoes important behavioural modifications. Studies in *C. elegans* have focused on the developmental regulation of dauer entry or exit; however, less is known about the specific behaviours of dauer juveniles. In laboratory culture with NGM plates, *C. elegans* dauers show characteristic sharp coiling and kinked postures that are not observed during other stages, except in some mutant strains (Gems *et al.*, 1998). As occurs for dormancy states in other animals, dauers crawl less actively and often remain motionless (Table 10.1), although movement can be stimulated in these juveniles by either reduced or increased levels of dopamine (Gaglia & Kenyon, 2009). Developmental arrest during the dauer diapause stage is reported to last up to 6 months in *C. elegans*, and up to 1 year in *P. pacificus* under laboratory conditions (Mayer & Sommer, 2011). Inactivity during the dauer stage was also reported at the cellular level, as quantifications of RNA transcripts in dauer juveniles in both *P. pacificus* and *C. elegans* showed reduced transcriptional activity of RNA pol II that resulted in a 20-fold lower overall transcript levels in the dauer compared with other stages (Dalley & Golomb, 1992; Sinha *et al.*, 2012). Low metabolism and reduced locomotion of dauer juveniles guarantee prolonged survival rates that enhance survival over extended periods of harsh environmental conditions. Therefore, specific behavioural adaptations occur during the diapause stage of nematodes that may be related in function to dormancy states of other animals.

Resumption of feeding and mouth dimorphism

As dauers develop, the feeding apparatus undergoes two important changes: shrinkage of pharyngeal structures and sealing of the buccal cavity by a cuticle plug (Fig. 10.1). Both morphological changes occur as pharyngeal pumping ceases. *Pristionchus pacificus* pharyngeal contractions occur in the corpus region of the pharynx (Kroetz *et al.*, 2012), in contrast to pumping in the posterior terminal bulb in *C. elegans* (Fig. 10.1B, C). Pharyngeal pumping decreases both during the lethargus period before each moult and during the dauer stage of nematodes (Cassada & Russell, 1975). Normal adult pumping rates for *P. pacificus* are around 130 pumps min^{-1} (Kroetz *et al.*, 2012) and reduce notably during the dauer stage. In *C. elegans*, normal pumping rates of 150-

Table 10.1. *Dauer-specific behaviours in nematodes.*

Behaviour	Elicitor or sub-behaviour	Description	Species studied	References
A				
Chemosensory responses	Dissolved chemicals	Attraction to natural extracts from distinct hosts, or to synthetic dissolved chemicals, involves specific chemosensory receptors. In *C. elegans*, chemosensory attraction is mediated by ASE and amphid wing cell (AWC)	*Anguina agrostis* *Bursaphelenchus xylophilus* *Caenorhabditis elegans* *Globodera rostochiensis* *Globodera pallida* *Heterodera avenae* *Heterodera glycines* *Meloidogyne javanica* *Panagrellus redivivus* *Pristionchus pacificus* *Rotylenchulus reniformis*	Ward, 1973; Cassada & Russell, 1975; Albert & Riddle, 1983; Papademetriou & Bone, 1983; Balan, 1985; Riddle & Bird, 1985; Lung, 1993; Riga *et al.*, 1997; Hong *et al.*, 2005, 2008; Bargmann, 2006; Zhao *et al.*, 2007
	Volatiles	CO_2 attraction. In *C. elegans*, CO_2 response is mediated by amphid wing cells (AWA, AWB and AWC), and by ASE and BAG neurons at the tip of the nose	*Adoncholaimus thalassophygas* *Ancyclostoma caninum* *Caenorhabditis elegans* *Pristionchus pacificus* *Heterorhabditis bacteriophora* *Meloidogyne incognita* *Rotylenchulus reniformis* *Steinernema carpocapsae* *Steinernema glaseri*	Pline & Dusenbery, 1987; Riemann & Schrage, 1988; Granzer & Haas, 1991; Robinson, 1995; Roayaie *et al.*, 1998; Ashton *et al.*, 1999; Bargmann, 2006; Bretscher *et al.*, 2008; Hallem & Sternberg, 2008; Guillermin *et al.*, 2011; Hallem *et al.*, 2011

Table 10.1. *(Continued).*

	Thermo-sensitivity	Migration towards sources of warmth, within thermal tolerance for each species. In contrast to non-dauer stages, *C. elegans* dauers typically disperse from their optimum growth temperatures in behavioural assays rather than accumulate. Modulation of thermo-sensitivity in *C. elegans* occurs *via* amphid head neurons (AFD) and highly branched nociceptor neurons in cuticle (PVD and FLP)	*Ancyclostoma caninum* *Caenorhabditis briggsae* *Caenorhabditis elegans* *Meloidogyne incognita* *Strongyloides stercolaris*	Hedgecock & Russell, 1975; Albert & Riddle, 1983; Dusenbery, 1988; Granzer & Haas, 1991; Lopez *et al.*, 2000; Chatzigeorgiou *et al.*, 2010; Hallem *et al.*, 2011
B				
Pharyngeal pumping		Absence or reduction of pharyngeal pumping[*]. In *C. elegans*, exogenous serotonin has been shown to induce	*Caenorhabditis elegans* *Pristionchus pacificus*	Cassada & Russell, 1975; Avery & Horvitz, 1989; Gems *et al.*, 1998; Keane & Avery, 2003; Kroetz *et al.*, 2012

Table 10.1. *(Continued).*

		pharyngeal pumping in dauers but no specific neurons have been associated to this behaviour, suggesting that pharyngeal muscles may act autonomously		
C				
Locomotion	Crawling	Dauers maintain straight, coiled or kinked postures, and reduce locomotion noticeably**. Dauers retain the ability to respond to mechanical stimuli or dopamine signalling changes (Gaglia & Kenyon, 2009). Reduced locomotion of dauers may be mediated by similar neurons involved in lethargus, such as ALA (Van Buskirk & Sternberg, 2007) or RIS (Turek *et al.*, 2013)	*Caenorhabditis elegans* *Pristionchus pacificus*	Cassada & Russell, 1975; Gems *et al.*, 1998; Gaglia & Kenyon, 2009; Lee *et al.*, 2012

Table 10.1. *(Continued).*

	Initiation or start standing	Dauers and non-dauers lift their heads, but dauers exhibit this behaviour more frequently and it often terminates in tail standing or nictation (see next, behaviour). In *C. elegans*, it is related to cholinergic activation *via* IL2 neurons	*Caenorhabditis elegans* *Pristionchus pacificus*	Lee *et al.*, 2012
	Nictation	Complete tail standing. Dauers can either stay straight, wave, loop or curl, forming spirals. Aggregates of nictating worms can form dauer towers. In *C. elegans*, nictation is modulated by acetylcholine neurotransmission and IL2 neurons	*Caenorhabditis elegans* *Pristionchus pacificus* *Protorhabditis xylocola* *Rhabditis acarta* *Rhabditis buetschlii* *Rhabditis dolichura* *Rhabditis frugicola* *Rhabditis helversenorum* *Rhabditis inermis inermoides* *Rhabditis insectivora* *Rhabditis longispina* *Rhabditis papillosa* *Rhabditis pellioides*	Sudhaus, 1976; Kaya & Campbell, 2002; Brown *et al.*, 2011; Lee *et al.*, 2012

Table 10.1. *(Continued).*

			Rhabditis reciproca *Rhabditis stammeri* *Rhabditis typica* *Rhabditis viguieri* *Steinernema carpocapsae* *Steinernema ceratophorum* *Steinernema siamkayai* *Steinernema scapterisci*	
	Dauer tower	In an aggregate of nictating dauers a 'stable' mass can form centimetre-sized towers.	*Caenorhabditis elegans* *Pristionchus pacificus*	Félix & Duveau, 2012; Penkov *et al*., 2014

[*] Reduced pharyngeal pumping was also observed in *C. elegans* adults in *daf-2* mutant strains that resemble dauer behaviours or lethargus behaviours (*i.e.*, period before each moult) (Gems *et al*., 1998); however, no *daf-2* mutant strain is currently available for *P. pacificus*.

[**] *C. elegans daf-2* adults resemble locomotory behaviours of the dauer, *i.e.*, reduced locomotion and pharyngeal pumping.

250 pumps min^{-1} were reduced to 1 pump min^{-1} (Cassada & Russell, 1975). However, occasional contraction of the pharynx can be observed in dauers, and exogenous serotonin can induce pharyngeal pumping in *C. elegans* dauers of up to 75 pumps min^{-1} (Keane & Avery, 2003). Therefore, pharyngeal muscles and nerves remain functional in spite of feeding cessation of dauer juveniles.

Adult mouth dimorphism ratio in *P. pacificus* is affected by extrinsic and intrinsic environmental factors. Extrinsic signals such as starvation and dauer pheromone regulate mouth dimorphism ratios (Bento *et al.*, 2010); however, intrinsic signals such as reduced pharyngeal pumping and cessation of feeding may also play a role (Fig. 10.5) (see Ragsdale, Chapter 11, this volume). Indirect evidence of pharyngeal pumping stimulation suggests that this may indeed be the case, as starved juveniles in both *P. pacificus* and *C. elegans* show higher rates of pharyngeal pumping (Avery & Horvitz, 1990; Kroetz *et al.*, 2012), and starvation of early juveniles increased the proportion of eurystomatous mouth forms in adults of *P. pacificus* (Bento *et al.*, 2010). Therefore, behavioural alterations of feeding or pharyngeal pumping may serve as the intrinsic signals that regulate mouth dimorphism (Fig. 10.5). It would be interesting to test whether mutant strains with abnormal pharyngeal pumping, *e.g.*, serotonergic signalling mutants (Avery & Horvitz, 1990), also show disproportionate mouth dimorphism ratios in *P. pacificus* (Fig. 10.5C).

Active diapause behaviours

In contrast to dormancy states of other animals, the dauer juveniles in nematodes have the ability to respond to stimuli rapidly and even perform active behaviours regularly throughout the diapause stage. The non-feeding dauer juveniles activate several metabolic pathways that include lipid metabolism (β-oxidation of fatty acids) to guarantee enough energy processing for survival and locomotory responses for dispersal (Braeckman *et al.*, 2009). The dauer, as well as the infective stage of most parasitic nematodes, is relevant because it is during this stage that nematodes find their hosts. Non-dauer stages of *C. elegans* typically cluster near temperature optima during behavioural assays, whereas dauer juveniles were shown to disperse rather than accumulate (Hedgecock & Russell, 1975), suggesting that this developmental stage is more temperature-tolerant and may serve as an adaptation for dispersal. Host finding in nematodes can use two different strategies:

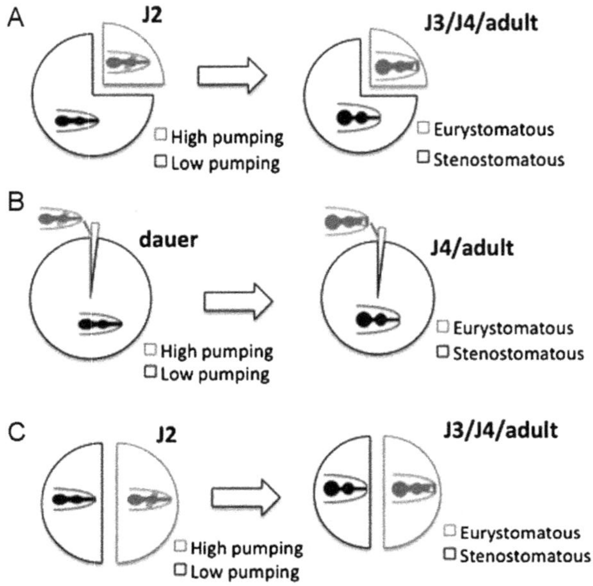

Fig. 10.5. *Hypothetical effects of early juvenile pharyngeal pumping on* Pristionchus pacificus *mouth dimorphism. Circle graphs represent proportion of worms that perform high/low pumping behaviours during juvenile stages (left), and proportion of worms that show eurystomatous or stenostomatous mouth forms (right). A: Direct development of adult worms under standard laboratory conditions bypass the dauer stage and show 75% stenostomatous mouth forms. Pumping rate proportions during early juvenile stages are hypothetical; B: Worms that undergo dauer development rarely show pharyngeal pumping behaviour as dauers, and develop nearly 100% stenostomatous mouth forms as adults; C: Increased high pumping behaviour in early juveniles, by starvation or genetic alterations, may show a corresponding increase in the proportion of eurystomatous mouth forms as adults.*

cruising or ambushing. Cruising dauer juveniles move actively and approach their hosts; ambushing dauers wait for the host to come into their vicinity before engaging in active attachment behaviours. Infective juveniles of some parasitic species show a specific ambushing behaviour, in which the animals stand on their tails in order to increase their chance of attachment to the potential host (Augustine, 1922; Payne, 1923).

Pristionchus pacificus dauer juveniles exhibit a 'stand-and-wave' behaviour reminiscent of ambushing behaviours first described for IJ of animal-parasitic nematodes. Dauer juveniles stand on their tails and remain erect, or begin to wave and loop in a behaviour referred to as 'nictation'.[3] Nictation is effective for host finding because it increases the surface area that may come into contact with passing hosts and reduces tension forces holding the worm to the substrate (Campbell & Gaugler, 1993; Crofton, 2009). In laboratory nematode cultures, the commonly used NGM agar plates prevent dauer juveniles nictating and waving because of the flat surface. When structure is added to the surface, either by adding sand grains or building artificial micro-dirt chips (Lee *et al.*, 2012), dauers begin exhibiting nictation behaviour (Table 10.1) (Brown *et al.*, 2011). In nature, studies of locomotion and behaviour in parasitic nematode species on natural substrates showed that peat or leaf litter are better substrates for nictation than sandy substrates (Hapca *et al.*, 2009; Neher, 2010). Nictation includes a range of lifting behaviours (Table 10.1). The 'initiation' phase of nictation (Lee *et al.*, 2012) occurs when nematodes raise their heads to 'start standing' (Ishibashi, 2002). Next, worms begin to nictate using four distinct motions (Table 10.1): straight standing, waving, looping and curling, *i.e.*, assuming a spiral form (Brown *et al.*, 2011). *Pristionchus pacificus* dauers may nictate in the straight standing posture for several minutes or even hours in a few cases, although waving, looping, or curling behaviours lasting only a few seconds may occur (Brown, unpubl. data). By contrast, the duration of nictation for *C. elegans* has been reported to last between 1-20 s approximately under laboratory conditions (Lee *et al.*, 2012). Greater nictation duration in *P. pacificus* may account for a stronger phoretic association to a scarab beetle host. During nictation, dauer juveniles may also attach to nearby obstacles and form 'body bridges' (Ishibashi & Kondo, 1990).

Environmental factors that induce dauer development in the early juvenile stages do not have an effect on downstream dauer-specific nictation behaviours. Studies on *C. elegans* showed that regardless of the mechanism of dauer induction in the early juvenile, *i.e.*, pheromone induction or starvation, the proportion of nictating dauers in a population remained the same (Lee *et al.*, 2012). Therefore, activation of distinct upstream dauer entry regulatory pathways leads to convergent and stereotypical morphological and behavioural phenotypes in the dauer juvenile.

[3] Nictation behaviour has been referred also as 'winken', 'tail standing' and 'body waving'. For a complete revision about the terminology of this behaviour see Kruitbos & Wilson (2010).

Some species of parasitic nematode infective juveniles also have the ability to 'leap' or 'jump' (Table 10.1) towards mechanical stimuli, such as air movement or host-associated volatile cues (Reed & Wallace, 1965; Ishibashi & Kondo, 1990; Campbell & Kaya, 1999a, b, 2000). A comparative study of leaping behaviour in several species of entomopathogenic steinernematids showed that it is far more prevalent than stable standing behaviours, and that jumpers generally tend to be more infective (Kaya & Campbell, 2002). Therefore, jumping may have evolved as a species-specific behavioural strategy for host infestation in nematodes that are highly dependent on a particular host to complete their life cycle; however, still standing in the absence of leaping also serves for ambushing in several species of steinernematids. More surprisingly, infective juveniles of cattle, sheep and horse lungworms (*i.e.*, *Dictyocaulus viviparus*) climb on sporangiophores of the coprophilous fungus *Pilobolus* and use sporangia discharge as a means of dispersal; lungworms thus disperse up to 3 m away from the original cattle dung to ensure new infections as cattle typically avoid grazing near their own faeces (Robinson, 1962; Biggane & Gormally, 1994). Leaping behaviour or sporangial dispersal have not yet been observed in *P. pacificus* or *C. elegans*; however, when dauer juveniles are found at high densities, adjacent worms that wave and coil come into contact, and often climb on each other to form towers that may reach 1 cm in height (Table 10.1). In summary, nictation and dauer tower formation represent host-finding behaviours in non-parasitic dauer stage nematodes, whereas leaping has only been reported for infective stages of parasitic nematodes.

At high population densities, dauer pheromones are known to promote other juveniles and dauers to enter and remain as dauers (Golden & Riddle, 1982; Hu, 2007). Therefore, it can be hypothesised that collective behaviours within populations are prone to respond to density effects, and may directly affect dauer-specific behaviours of individuals. To address this question we tested the proportion of nictation in populations of different densities (Fig. 10.6A). We placed increasing numbers of dauers (Fig. 10.6A) into arenas of constant size (*i.e.*, 6 cm NGM plate with grains of sand). The two highest densities (105 and 526 dauers cm^{-2}) of dauers resulted in higher proportions of nictation behaviour (5-8% of dauers in nictation) that increased through time (20 h post sand addition), in contrast to lowest densities (4 and 21 dauers cm^{-2}) that briefly decreased nictation rates to a minimum (5-30 h post sand addition fell to 0-1%). However, the highest nictation proportion

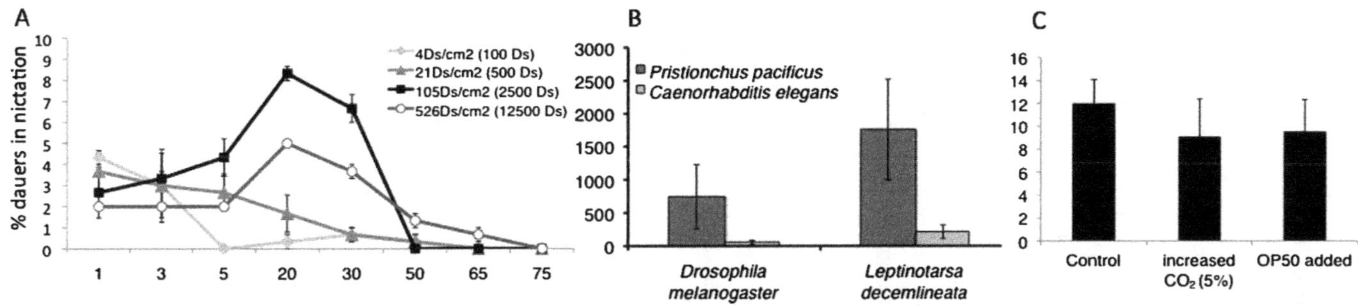

Fig. 10.6. *Environmental influence on nictation and host infestation of* Pristionchus pacificus. *A: Dauer density influence on the proportion of nictation behaviour; densities dauers (Ds cm^{-2}) are shown; in parenthesis the number of dauers added to the plate; B: Infestation assays on two hosts, the fruit fly,* Drosophila melanogaster *(wg, wingless strain), and the Colorado Potato Beetle (CPB),* Leptinotarsa decemlineata. *Bars show number of dauer present in the carcass of* P. pacificus *(dark grey) and* C. elegans *(light grey) for comparison (initial number,* n $=$ *50 000); C: Dauer nictation rates are not affected by* CO_2 *or the presence of bacterial food (OP50). All behavioural assays were performed in 6 cm diam. NGM plates with sand grains added to allow nictation of worms.*

occurred with approximately 100 dauers cm^{-2}, but not at 500 dauers cm^{-2} (Fig. 10.6A). The effect on behaviour of the presence of other nematode species that were in the same arena has been studied for two *Steinernema* species (Wang & Ishibashi, 1999). The ambusher species, *Steinernema carpocapsae*, is known to nictate at higher rates under the presence of the cruiser species *S. glaseri* (Wang & Ishibashi, 1999), suggesting that entomopathogenic IJ respond to the presence of other nematodes. Further experiments are needed to address the mechanisms of collective responses to nictation behaviour.

Sensing and neural system function in dauer behaviours

Nematodes are amenable to studies of behaviour and neuronal signalling that modulate behaviour. Dauer development in non-parasitic nematodes is phenotypically plastic and dispensable, in contrast to the obligate infective juvenile stage of parasitic nematodes. The independent evolution of parasitic life histories in nematodes suggests that dauers or infective juveniles must have evolved functional adaptations in sensorial and behavioural systems for species-specific phoretic interactions.

To test whether *P. pacificus* nictating dauers show host-specific attachment in the laboratory, we placed two host species on plates full of dauers. We used two hosts that differed dramatically in size and belonged to two different orders: *i*) the fruit fly *Drosophila melanogaster* (0.7 mg); and *ii*) the Colorado potato beetle *Leptinotarsa decemlineata* (190 mg). To increase contact of flies to the dauers and avoid flight-related escape, we used a flightless strain of *D. melanogaster* (wg). *Pristionchus pacificus* dauers ($n = 50\,000$) were left to interact with one host for a period of 6-8 h, and the number of dauers on each host were then counted (Fig. 10.6B); *C. elegans* dauers were also used for comparison. As expected, *P. pacificus* showed generally higher rates of infestation than *C. elegans*; however, we did not find the expected difference in *P. pacificus* infestation between *L. decemlineata* and the fruit fly. If there had been host specificity, we would have expected a much higher number of *P. pacificus* dauers on the beetle. Based solely on surface area exposure of each host, *i.e.*, 3.14 mm^2 for *D. melanogaster* and 314 mm^2 for *L. decemlineata*, we predicted at least 100-fold more infesting dauers on *L. decemlineata* than on the fruit fly. However, we only found two- to four-fold higher numbers of *P. pacificus* and *C. elegans* dauers, respectively (Fig. 10.6B). The unexpectedly high numbers of dauers that were recovered from the fruit fly may be due to a higher activity and crawling of the fly on the nictating arena, in contrast to *L. decemlineata* that repeatedly remained still for extended periods of time. Although nictation in the plates was not quantified, we observed very low proportions of nictation throughout the whole period of the assay. Therefore, our experimental assay is not conclusive about *P. pacificus* specificity; however, it shows that host activity plays an important role in nematode host infestation, and that nematodes are able to infest hosts in the absence of nictation.

As nematodes lack visual organs, they must sense their environment and potential hosts by chemotactic cues. CO_2 serves as an attractant for many parasitic nematode species (Klingler, 1965; Pline & Dusenbery, 1987; Riemann & Schrage, 1988; Gaugler *et al.*, 1991; Lewis *et al.*, 1993; Robinson, 1995; Haas, 2003); by contrast, non-dauer stages in *P. pacificus* and *C. elegans* do not respond to, and often avoid, CO_2 (Hallem & Sternberg, 2008; Hallem *et al.*, 2011). To test whether nictation behaviour was also negatively affected in dauer stage *P. pacificus* (Fig. 10.6C), we calculated the percentage of dauers in nictation in plates at atmospheric room levels of CO_2 (0.0035%) and at increased CO_2 levels in a cell culture chamber (5%). We found a slight decrease in nictation, supporting a general absence or slight avoidance response for all *P. pacificus* stages (Fig. 10.6C). An ancestral role for BAG neuron involvement in CO_2 response mechanisms in several species of parasitic and non-parasitic nematodes has been reported (Hallem *et al.*, 2011); however, the specific molecular differences of the neuronal circuitry or neuronal pathways involved in these opposite responses need further investigation. The *P. pacificus* genome has revealed a lower number of olfactory receptor genes compared with *C. elegans* (Hong & Sommer, 2006) that may reflect a more restricted array of responses to environmental stimuli. Other response behaviours to mechanical and chemical stimuli are reviewed in more detail by Hong (Chapter 12, this volume).

Locomotion can be affected by the presence of food. When *C. elegans* comes into contact with bacterial OP50 lawn, a slow-down response is activated by modulation of serotonin and dopamine signalling *via* amphid sensory neurons ASH, ADL, AWB and AWC+ASE neurons (Sawin *et al.*, 2000; Chao *et al.*, 2004; Ben Arous *et al.*, 2009). Slow down behaviour upon bacterial contact occurs either by mechanical or sensorial stimuli (Sawin *et al.*, 2000; Chao *et al.*, 2004). To test whether the presence of bacterial OP50 affected nictation behaviour in dauers in *P. pacificus*, we compared nictation rates in arenas with and without OP50 (Fig. 10.6C). We found no alteration, or a slight decrease, in the nictation rates when worms were exposed to food, suggesting that food stimulus does not immediately affect nictation behaviour (Fig. 10.6C). However, dauers probably sense food, as nictation rates decreased hours after adding OP50, most likely due to dauer exit of the juveniles (data not shown). How do *P. pacificus* dauers sense environmental signals? What neural pathways are involved? Which neurons mediate sensing and activation of dauer-specific behaviours?

The cholinergic system and the chemosensory inner labial (IL2) sensory neurons are main players in the regulation of nictation behaviour in *C. elegans* (Lee *et al.*, 2012). Loss of function by cholinergic disruption in these neurons decreased the rate of nictation initiation in dauers, *i.e.*, start standing, whereas targeted rescue experiments by optogenetic activation of acetylcholine in IL2 neurons resulted in higher rates of nictation (Lee *et al.*, 2012). What chemosensory signals actually trigger acetylcholine activation in IL2 neurons remains unknown. However, this study provides the intriguing possibility that IL2 neurons may have a mechanosensory function in dauers, in addition to the chemosensory functions of non-dauer stages. A change of function of IL2 neurons during development is also correlated with the reorganisation of dendritic ends in IL2 neurons during development. IL2 neurons are located in the head region of non-dauers with dendritic endings at the tips of the lips that are exposed externally and in direct contact to the environment through cuticular pores (Tabish *et al.*, 1995; Wolkow & Hall, 2012). These neurons are withdrawn within the lip cuticle during development of the dauer, but some cuticular pores remain open and exposed to the external environment (Albert & Riddle, 1983; Wolkow & Hall, 2012). How is the chemosensory or mechanosensory information processed within IL2 neurons during these two neuronal states? Are the neuropeptide signalling cascades involved in distinct neuronal functions maintained? Photoactivation of acetylcholine signalling in IL2 neurons in non-dauer stage worms did not result in nictation activation (Lee *et al.*, 2012). These results suggest that cholinergic activation alone in IL2 neurons is not sufficient to activate nictation behaviour in non-dauer stages. Therefore, additional neural pathways or distinct anatomical features of the dauer, *e.g.*, neuronal organisation, musculature or cuticle, may also be involved in facilitating nictation behaviour specifically in the dauer stage. Recent findings show that extracellular vesicles (ECV) that are shed and released by ciliated sensory neurons through cuticular pores are involved in worm communication, *i.e.*, induction of male mating behaviours (Sommer & Streit, 2011). If dauers also secrete ECV through open cuticular pores, then it may be hypothesised that nictation behaviour facilitates long-range dispersal of chemical communication signals. Other neurons with dual chemosensory and mechanosensory functions are: *i*) ASH neurons in *C. elegans* that are involved in the slow-down response of worms upon contact with bacterial food (Kaplan & Horvitz, 1993), also discussed in the previous paragraph; and *ii*) nociceptors in vertebrates involved in pain sensation that can transduce a variety of stimuli (Besson & Chaouch, 1987). However, the dual sensory response of IL2 neurons remains a particularly interesting case for further studies due to the temporal and developmental alteration of sensorial function of a single neuron, as well as the recent findings that IL2 neurons may directly function in animal communication.

Acetylcholine modulation of nictation behaviour in *P. pacificus* has not been studied; however, alternative neural pathways may also be regulating this behaviour. Previous forward genetic screens to identify nictation deficient mutant strains in *P. pacificus* (Brown *et al.*, 2011) and *C. elegans* (Brown, unpubl. data) showed that homozygous F2 mutant lines carrying the phenotype appeared fairly frequently, *i.e.*, 11 out of 1600 F2 screened for *P. pacificus* and five out of 1300 F2 screened for *C. elegans*. These relatively high mutation frequencies that result in nictation-deficient dauers suggest that a substantially large number of genes may be involved in regulating nictation behaviour either directly or indirectly. Therefore, additional neural pathways involved in the regulation of this behaviour in *P. pacificus* and *C. elegans* need to be examined further.

Transcriptome sequencing of dauer juveniles in *C. elegans* (Wang & Kim, 2003) and *P. pacificus* (Sinha *et al.*, 2012) also revealed several candidate genes important for the regulation of dauer-specific behaviours. In spite of the high divergence of the expression profiles of *P. pacificus* and *C. elegans* dauers, expression of members of the FMRF-like peptide family of neuropeptides was well represented and enriched in both species, including four likely to be involved in pharyngeal pumping (Sinha *et al.*, 2012). FMRFamide (Phe-Met-Arg-Phe-NH_2)-related peptides encoded by the *flp* genes are expressed in interneurons, motor neurons and amphid sensory neurons in the head of the worm, and are involved in many behaviours (Kim & Li, 2004). FLP neuropeptides are well-suited candidates to test their involvement in dauer sensing, locomotory behaviours such as nictation, direct regulation of reduced pharyngeal pumping and, therefore, indirect regulation of adult mouth dimorphism.

Ecological and evolutionary implications of dauer-specific behaviours

Host-finding strategies are similar across nematodes that present phoretic associations, but sensory profiles are highly specific. Many species within the genus *Pristionchus* have associations with scarab beetles, *e.g.*, *P. maupasi* is mainly associated with the European cockchafer *Melolontha*, *P. entomophagus* and *P. uniformis* are primarily found on the dung beetle *Geotrupes* or the Colorado potato beetle *L. decemlineata*, and *P. aerivorous* is associated with the lepidopteran *Helicoverpa zea*. Therefore, even species closely related to *P. pacificus* show differences in their chemoattraction profiles. Surprisingly, however, chemoattraction profiles varied significantly even at the population level, *i.e.*, across different strains of the same species (Hong & Sommer, 2006).

Specific associations between nematodes and insects are found in other genera of Diplogastridae, *e.g.*, *Micoletzkya chinaae* with bark beetles, or species in the genus *Parasitodiplogaster* that associate with specific fig wasps (Poinar & Herre, 1991; Hong & Sommer, 2006). Population- and species-specific chemorecognition profiles seem to be the norm across phoretic nematode species. Therefore, sensorial systems in nematodes show a diverse array of signal perception profiles that adapt rapidly to new environments. Distinct life histories in nematodes can be used to understand how diverse sensorial systems can evolve using only a handful of host-finding strategies and behaviours.

Across unrelated nematodes, molecular and morphological similarities suggest that entry and exit to the host-finding stage may be conserved in upstream effectors (Sinha *et al.*, 2012), *i.e.*, in core developmental genes of the dauer programme but not in downstream effectors. Comparisons of gene expression profiles of the host-infesting juveniles in three nematode species provide evidence for a low conservation of downstream effectors. Transcriptome studies were done for dauers of non-parasitic *P. pacificus* and *C. elegans* (Sinha *et al.*, 2012), and for the J2 of the plant-parasitic soybean cyst nematode, *Heterodera glycines* (Elling *et al.*, 2007). Therefore, developmental programmes may evolve hierarchically with upstream effectors acquiring very little change and downstream effectors diverging more rapidly. Do behavioural genetic networks also evolve in a hierarchical manner?

Host finding behaviours of parasitic and non-parasitic nematodes are essentially very similar. Little is known about genetic effectors of behaviour. Due to the facultative and highly environmentally influenced nature of behaviours, it is difficult to imagine that these may evolve in a hierarchical manner in the same way as developmental networks. Behavioural genes may therefore diverge more rapidly than developmental genes, perhaps using the same logic as downstream effectors of developmental regulatory networks. It is plausible that similar and convergent behaviours of parasitic and non-parasitic nematodes evolved independently from distinct genetic regulatory logics.

Acknowledgements

We thank current and former members of Sommer laboratory for discussion, and Carlos Winter (USP) for critical reading of this chapter. FDB would like to thank I. D'Anna for assistance with the nictation and infestation assays, and for dauer photographs. Metta Riebesell assisted with the *Pristionchus* cycle and dauer photographs. FDB obtained a DAAD Bilateral Exchange for Academics Fellowship and financial support from Universidad de los Andes to carry out some of this work in the Sommer laboratory. AO and FDB were granted postdoctoral research fellowships at the Max Planck Institute for Developmental Biology.

References

ALBERT, P.S. & RIDDLE, D.L. (1983). Developmental alterations in sensory neuroanatomy of the *Caenorhabditis elegans* dauer larva. *Journal of Comparative Neurology* 219, 461-481.

ALBERT, P.S. & RIDDLE, D.L. (1988). Mutants of *Caenorhabditis elegans* that form dauer-like larvae. *Developmental Biology* 126, 270-293.

ALBERT, P.S., BROWN, S.J. & RIDDLE, D.L. (1981). Sensory control of dauer larva formation in *Caenorhabditis elegans*. *Journal of Comparative Neurology* 198, 435-451.

ANTEBI, A., CULOTTI, J.G. & HEDGECOCK, E.M. (1998). *daf-12* regulates developmental age and the dauer alternative in *Caenorhabditis elegans*. *Development* 125, 1191-1205.

ANTEBI, A., YEH, W.H., TAIT, D., HEDGECOCK, E.M. & RIDDLE, D.L. (2000). *daf-12* encodes a nuclear receptor that regulates the dauer diapause and developmental age in *C. elegans*. *Genes & Development* 14, 1512-1527.

ASHTON, F.T., LI, J. & SCHAD, G.A. (1999). Chemo- and thermosensory neurons: structure and function in animal parasitic nematodes. *Veterinary Parasitology* 84, 297-316.

AUGUSTINE, D.L. (1922). Investigations on the control of hookworm disease. IX: on the position of the infective hookworm larvae in the soil. *American Journal of Epidemiology* 2, 172-176.

AVERY, L. & HORVITZ, H.R. (1989). Pharyngeal pumping continues after laser killing of the pharyngeal nervous system of *C. elegans*. *Neuron* 3, 473-485.

AVERY, L. & HORVITZ, H.R. (1990). Effects of starvation and neuroactive drugs on feeding in *Caenorhabditis elegans*. *Journal of Exerimental Zoology* 253, 263-270.

BAIRD, S.E., FITCH, D.H.A. & EMMONS, S.W. (1994). *Caenorhabditis vulgaris* sp. n. (Nematoda: Rhabditidae): a necromenic associate of pill bugs and snails. *Nematologica* 40, 1-11.

BALAN, J. (1985). Measuring minimal concentrations of attractants detected by the nematode *Panagrellus redivivus*. *Journal of Chemical Ecology* 11, 105-111.

BARGMANN, C.I. (2006). Chemosensation in *C. elegans*. In: The *C. elegans* Research Community (Ed.). *WormBook*. DOI:10.1895/wormbook.1.123.1.

BARGMANN, C.I. & HORVITZ, H.R. (1991). Control of larval development by chemosensory neurons in *Caenorhabditis elegans*. *Science* 251, 1243-1246.

BARRIERE, A. & FÉLIX, M.-A. (2005). High local genetic diversity and low outcrossing rate in *Caenorhabditis elegans* natural populations. *Current Biology* 15, 1176-1184.

BEN AROUS, J., LAFFONT, S. & CHATENAY, D. (2009). Molecular and sensory basis of a food related two-state behavior in *C. elegans*. *PLoS ONE* 4, e7584.

BENTO, G., OGAWA, A. & SOMMER, R.J. (2010). Co-option of the hormone-signalling module dafachronic acid-DAF-12 in nematode evolution. *Nature* 466, 494-497.

BESSON, J.M. & CHAOUCH, A. (1987). Peripheral and spinal mechanisms of nociception. *Physiological Reviews* 67, 67-186.

BIGGANE, R.P.J. & GORMALLY, M.J. (1994). The effect of dung beetle activity on the discharge of *Pilobolus* (Fungi, Mucorales) sporangia in cattle, sheep and horse faeces. *Entomophaga* 39, 95-98.

BOSE, N., OGAWA, A., VON REUSS, S.H., YIM, J.J., RAGSDALE, E.J., SOMMER, R.J. & SCHROEDER, F.C. (2012). Complex small-molecule architectures regulate phenotypic plasticity in a nematode. *Angewandte Chemie* 51, 12438-12443.

BRAECKMAN, B.P., HOUTHOOFD, K. & VANFLETEREN, J.R. (2009). Intermediary metabolism. In: The *C. elegans* Research Community (Ed.). *WormBook*. DOI:10.1895/wormbook.1.123.1.

BRETSCHER, A.J., BUSCH, K.E. & DE BONO, M. (2008). A carbon dioxide avoidance behavior is integrated with responses to ambient oxygen and food in *Caenorhabditis elegans*. *Proceedings of the National Academy of Sciences of the United States of America* 105, 8044-8049.

BROWN, F.D., D'ANNA, I. & SOMMER, R.J. (2011). Host-finding behaviour in the nematode *Pristionchus pacificus*. *Proceedings of the Royal Society B: Biological Sciences* 278, 3260-3269.

BUTCHER, R.A., FUJITA, M., SCHROEDER, F.C. & CLARDY, J. (2007). Small-molecule pheromones that control dauer development in *Caenorhabditis elegans*. *Nature Chemical Biology* 3, 420-422.

BUTCHER, R.A., RAGAINS, J.R., KIM, E. & CLARDY, J. (2008). A potent dauer pheromone component in *Caenorhabditis elegans* that acts synergistically with other components. *Proceedings of the National Academy of Sciences of the United States of America* 105, 14288-14292.

BUTCHER, R.A., RAGAINS, J.R., LI, W., RUVKUN, G., CLARDY, J. & MAK, H.Y. (2009). Biosynthesis of the *Caenorhabditis elegans* dauer pheromone. *Proceedings of the National Academy of Sciences of the United States of America* 106, 1875-1879.

CAMPBELL, J.F. & GAUGLER, R. (1993). Nictation behaviour and its ecological implications in the host search strategies of entomopathogenic nematodes (Heterorhabditidae and Steinernematidae). *Behaviour* 126, 155-169.

CAMPBELL, J.F. & KAYA, H.K. (1999a). Mechanism, kinematic performance, and fitness consequences of jumping behavior in entomopathogenic nematodes (*Steinernema* spp.). *Canadian Journal of Zoology* 77, 1947-1955.

CAMPBELL, J.F. & KAYA, H.K. (1999b). How and why a parasitic nematode jumps. *Nature* 397, 485-486.

CAMPBELL, J.F. & KAYA, H.K. (2000). Influence of insect associated cues on the jumping behavior of entomopathogenic nematodes (*Steinernema* spp.). *Behaviour* 137, 591-609.

CASSADA, R.C. & RUSSELL, R.L. (1975). The dauerlarva, a post-embryonic developmental variant of the nematode *Caenorhabditis elegans*. *Developmental Biology* 46, 326-342.

CHAO, M.Y., KOMATSU, H., FUKUTO, H.S., DIONNE, H.M. & HART, A.C. (2004). Feeding status and serotonin rapidly and reversibly modulate a *Caenorhabditis elegans* chemosensory circuit. *Proceedings of the National Academy of Sciences of the United States of America* 101, 15512-15517.

CHATZIGEORGIOU, M., YOO, S., WATSON, J.D., LEE, W.-H., SPENCER, W.C., KINDT, K.S., HWANG, S.W., MILLER, D.M., TREININ, M., DRISCOLL, M. ET AL. (2010). Specific roles for DEG/ENaC and TRP channels in touch and thermosensation in *C. elegans* nociceptors. *Nature Neuroscience* 13, 861-868.

CROFTON, H.D. (2009). The vertical migration of infective larvae of strongyloid nematodes. *Journal of Helminthology* 28, 35-52.

DALLEY, B.K. & GOLOMB, M. (1992). Gene expression in the *Caenorhabditis elegans* dauer larva: developmental regulation of Hsp90 and other genes. *Developmental Biology* 151, 80-90.

DIETERICH, C. & SOMMER, R.J. (2009). How to become a parasite – lessons from the genomes of nematodes. *Trends in Genetics* 25, 203-209.

DORMAN, J.B., ALBINDER, B., SHROYER, T. & KENYON, C. (1995). The *age-1* and *daf-2* genes function in a common pathway to control the lifespan of *Caenorhabditis elegans*. *Genetics* 141, 1399-1406.

DUMAS, K.J., GUO, C., WANG, X., BURKHART, K.B., ADAMS, E.J., ALAM, H. & HU, P.J. (2010). Functional divergence of dafachronic acid pathways in the control of *C. elegans* development and lifespan. *Developmental Biology* 340, 605-612.

DUSENBERY, D.B. (1988). Behavioral responses of *Meloidogyne incognita* to small temperature changes. *Journal of Nematology* 20, 351-355.

ELLING, A.A., MITREVA, M., RECKNOR, J., GAI, X., MARTIN, J., MAIER, T.R., MCDERMOTT, J.P., HEWEZI, T., BIRD, D. MCK., DAVIS, E.L. ET AL. (2007). Divergent evolution of arrested development in the dauer stage of *Caenorhabditis elegans* and the infective stage of *Heterodera glycines*. *Genome Biology* 8, R211.

ESTEVEZ, M., ATTISANO, L., WRANA, J.L., ALBERT, P.S., MASSAGUE, J. & RIDDLE, D.L. (1993). The *daf-4* gene encodes a bone morphogenetic protein-receptor controlling *C. elegans* dauer larva development. *Nature* 365, 644-649.

FÉLIX, M.-A. & DUVEAU, F. (2012). Population dynamics and habitat sharing of natural populations of *Caenorhabditis elegans* and *C. briggsae*. *BMC Biology* 10, 59.

FRIEDMAN, D.B. & JOHNSON, T.E. (1988). A mutation in the *age-1* gene in *Caenorhabditis elegans* lengthens life and reduces hermaphrodite fertility. *Genetics* 118, 75-86.

FUCHS, G. (1915). Die Naturgeschichte der Nematoden und eineger anderer Parasiten. 1. Des *Ips typographus*, 2. Des *Hylobius abietis*. *Zoologische Jahrbücher, Abteilung für Systematik, Geographie und Biologie der Tier* 38, 109-227.

GAGLIA, M.M. & KENYON, C. (2009). Stimulation of movement in a quiescent, hibernation-like form of *Caenorhabditis elegans* by dopamine signaling. *Journal of Neuroscience* 29, 7302-7314.

GAUGLER, R., CAMPBELL, J.F. & GUPTA, P. (1991). Characterization and basis of enhanced host-finding in a genetically improved strain of *Steinernema carpocapsae*. *Journal of Invertebrate Pathology* 57, 234-241.

GEMS, D., SUTTON, A.J., SUNDERMEYER, M.L., ALBERT, P.S., KING, K.V., EDGLEY, M.L., LARSEN, P.L. & RIDDLE, D.L. (1998). Two pleiotropic classes of *daf-2* mutation affect larval arrest, adult behavior, reproduction and longevity in *Caenorhabditis elegans*. *Genetics* 150, 129-155.

GEORGI, L.L., ALBERT, P.S. & RIDDLE, D.L. (1990). *daf-1*, a *C. elegans* gene controlling dauer larva development, encodes a novel receptor protein kinase. *Cell* 61, 635-645.

GERISCH, B. & ANTEBI, A. (2004). Hormonal signals produced by DAF-9/cytochrome P450 regulate *C. elegans* dauer diapause in response to environmental cues. *Development* 131, 1765-1776.

GERISCH, B., WEITZEL, C., KOBER-EISERMANN, C., ROTTIERS, V. & ANTEBI, A. (2001). A hormonal signaling pathway influencing *C. elegans* metabolism, reproductive development, and life span. *Developmental Cell* 1, 841-851.

GOLDEN, J.W. & RIDDLE, D.L. (1982). A pheromone influences larval development in the nematode *Caenorhabditis elegans*. *Science* 218, 578-580.

GOLDEN, J.W. & RIDDLE, D.L. (1984). The *Caenorhabditis elegans* dauer larva: developmental effects of pheromone, food, and temperature. *Developmental Biology* 102, 368-378.

GOTTLIEB, S. & RUVKUN, G. (1994). *daf-2*, *daf-16* and *daf-23*: genetically interacting genes-controlling dauer formation in *Caenorhabditis elegans*. *Genetics* 137, 107-120.

GRANT, W. & VINEY, M. (2011). The dauer phenomenon. In: Perry, R.N. & Wharton, D.A. (Eds). *Molecular and physiological basis of nematode survival*. Wallingford, UK, CAB International Publishing, pp. 99-125.

GRANZER, M. & HAAS, W. (1991). Host-finding and host recognition of infective *Ancylostoma caninum* larvae. *International Journal for Parasitology* 21, 429-440.

GUILLERMIN, M.L., CASTELLETTO, M.L. & HALLEM, E.A. (2011). Differentiation of carbon dioxide-sensing neurons in *Caenorhabditis elegans* requires the ETS-5 transcription factor. *Genetics* 189, 1327-1339.

HAAS, W. (2003). Parasitic worms: strategies of host finding, recognition and invasion. *Zoology* 106, 349-364.

HALLEM, E.A. & STERNBERG, P.W. (2008). Acute carbon dioxide avoidance in *Caenorhabditis elegans*. *Proceedings of the National Academy of Sciences of the United States of America* 105, 8038-8043.

HALLEM, E.A., DILLMAN, A.R., HONG, A.V., ZHANG, Y., YANO, J.M., DEMARCO, S.F. & STERNBERG, P.W. (2011). A sensory code for host seeking in parasitic nematodes. *Current Biology* 21, 377-383.

HAPCA, S., HERITAGE, S., KRUITBOS, L. & WILSON, M. (2009). Influence of substrate on the body-waving behaviour of nematodes. *Nematology* 11, 917-925.

HEDGECOCK, E.M. & RUSSELL, R.L. (1975). Normal and mutant thermotaxis in the nematode *Caenorhabditis elegans*. *Proceedings of the National Academy of Sciences of the United States of America* 72, 4061-4065.

HERRMANN, M., MAYER, W.E. & SOMMER, R.J. (2006). Nematodes of the genus *Pristionchus* are closely associated with scarab beetles and the Colorado potato beetle in Western Europe. *Zoology (Jena)* 109, 96-108.

HERTWECK, M., GOBEL, C. & BAUMEISTER, R. (2004). *C. elegans* SGK-1 is the critical component in the Akt/PKB kinase complex to control stress response and life span. *Developmental Cell* 6, 577-588.

HONG, R.L. & SOMMER, R.J. (2006). Chemoattraction in *Pristionchus* nematodes and implications for insect recognition. *Current Biology* 16, 2359-2365.

HONG, R.L., VILLWOCK, A. & SOMMER, R.J. (2005). Cultivation of the rhabditid *Poikilolaimus oxycercus* as a laboratory nematode for genetic analyses. *Journal of Experimental Zoology A: Comparative Experimental Biology* 303, 742-60.

HONG, R.L., SVATOS, A., HERRMANN, M. & SOMMER, R.J. (2008). Species-specific recognition of beetle cues by the nematode *Pristionchus maupasi*. *Evolution & Development* 10, 273-279.

HU, P.J. (2007). Dauer. In: The *C. elegans* Research Community (Ed.). *WormBook*. DOI:10.1895/wormbook.1.144.1.

ISHIBASHI, N. (2002). Behaviour of entomopathogenic nematodes. In: Lee, D.L. (Ed.). *The biology of nematodes*. London, UK, Taylor & Francis, pp. 511-520.

ISHIBASHI, N. & KONDO, E. (1990). Behaviour of infective juveniles. In: Gaugler, R. & Kaya, H. (Eds). *Entomopathogenic nematodes in biological control*. Boca Raton, FL, USA, CRC Press, pp. 139-150.

JEONG, P.Y., JUNG, M., YIM, Y.H., KIM, H., PARK, M., HONG, E., LEE, W., KIM, Y.H., KIM, K. & PAIK, Y.K. (2005). Chemical structure and biological activity of the *Caenorhabditis elegans* dauer-inducing pheromone. *Nature* 433, 541-545.

JEZYK, P.F. & FAIRBAIRN, D. (1967). Metabolism of ascarosides in ovaries of *Ascaris lumbricoides* (Nematoda). *Comparative Biochemistry and Physiology* 23, 707-719.

JIA, K., ALBERT, P.S. & RIDDLE, D.L. (2002). DAF-9, a cytochrome P450 regulating *C. elegans* larval development and adult longevity. *Development* 129, 221-231.

KAPLAN, J.M. & HORVITZ, H.R. (1993). A dual mechanosensory and chemosensory neuron in *Caenorhabditis elegans*. *Proceedings of the National Academy of Sciences of the United States of America* 90, 2227-2231.

KAPLAN, F., ALBORN, H.T., VON REUSS, S.H., AJREDINI, R., ALI, J.G., AKYAZI, F., STELINSKI, L.L., EDISON, A.S., SCHROEDER, F.C. & TEAL, P.E. (2012). Interspecific nematode signals regulate dispersal behavior. *PLoS ONE* 7, e38735.

KAYA, H. & CAMPBELL, J. (2002). Variation in entomopathogenic nematode (Steinernematidae and Heterorhabditidae) infective-stage jumping behaviour. *Nematology* 4, 471-482.

KEANE, J. & AVERY, L. (2003). Mechanosensory inputs influence *Caenorhabditis elegans* pharyngeal activity via ivermectin sensitivity genes. *Genetics* 164, 153-162.

KENYON, C. (2006). My adventures with genes from the fountain of youth. *Harvey Lectures* 100, 29-70.

KENYON, C.J. (2010). The genetics of ageing. *Nature* 464, 504-512.

KENYON, C.J., CHANG, J., GENSCH, E., RUDNER, A. & TABTIANG, R. (1993). A *C. elegans* mutant that lives twice as long as wild type. *Nature* 366, 461-464.

KIM, K. & LI, C. (2004). Expression and regulation of an FMRFamide-related neuropeptide gene family in *Caenorhabditis elegans*. *Journal of Comparative Neurology* 475, 540-550.

KIM, K., SATO, K., SHIBUYA, M., ZEIGER, D.M., BUTCHER, R.A., RAGAINS, J.R., CLARDY, J., TOUHARA, K. & SENGUPTA, P. (2009). Two chemoreceptors mediate developmental effects of dauer pheromone in *C. elegans*. *Science* 326, 994-998.

KIMURA, K.D., TISSENBAUM, H.A., LIU, Y.X. & RUVKUN, G. (1997). *daf-2*, an insulin receptor-like gene that regulates longevity and diapause in *Caenorhabditis elegans*. *Science* 277, 942-946.

KLINGLER, J. (1965). On the orientation of plant nematodes and of some other soil animals. *Nematologica* 11, 4-18.

KROETZ, S.M., SRINIVASAN, J., YAGHOOBIAN, J., STERNBERG, P.W. & HONG, R.L. (2012). The cGMP signaling pathway affects feeding behavior in the necromenic nematode *Pristionchus pacificus*. *PLoS ONE* 7, e34464.

LEE, D.L. (2002). Behaviour. In: Lee, D.L. (Ed.). *The biology of nematodes*. London, UK, Taylor & Francis, pp. 369-387.

LEE, H., CHOI, M., LEE, D., KIM, H., HWANG, H., KIM, H., PARK, S., PAIK, Y. & LEE, J. (2012). Nictation, a dispersal behavior of the nematode *Caenorhabditis elegans*, is regulated by IL2 neurons. *Nature Neuroscience* 15, 107-112.

LEE, R.Y., HENCH, J. & RUVKUN, G. (2001). Regulation of *C. elegans* DAF-16 and its human ortholog FKHRL1 by the *daf-2* insulin-like signaling pathway. *Current Biology* 11, 1950-1957.

LEWIS, E.E., GAUGLER, R. & HARRISON, R. (1993). Response of cruiser and ambusher entomopathogenic nematodes (Steinernematidae) to host volatile cues. *Canadian Journal of Zoology* 71, 765-769.

LI, J., BROWN, G., AILION, M., LEE, S. & THOMAS, J.H. (2004). NCR-1 and NCR-2, the *C. elegans* homologs of the human Niemann-Pick type C1 disease protein, function upstream of DAF-9 in the dauer formation pathways. *Development* 131, 5741-5752.

LI, W., KENNEDY, S.G. & RUVKUN, G. (2003). *daf-28* encodes a *C. elegans* insulin superfamily member that is regulated by environmental cues and acts in the DAF-2 signaling pathway. *Genes & Development* 17, 844-858.

LIN, K., HSIN, H., LIBINA, N. & KENYON, C. (2001). Regulation of the *Caenorhabditis elegans* longevity protein DAF-16 by insulin/IGF-1 and germline signaling. *Nature Genetics* 28, 139-145.

LOPEZ, P.M., BOSTON, R., ASHTON, F.T. & SCHAD, G.A. (2000). The neurons of class ALD mediate thermotaxis in the parasitic nematode, *Strongyloides stercoralis*. *International Journal for Parasitology* 30, 1115-1121.

LUNG, G. (1993). The role of phytosiderophores as an attractive substance of root exudates from several cereals for second stage juveniles of *Heterodera avenae*. *Mededelingen Faculteit Landbouwkundige en Toegepaste Biologische Wetenschappen Universiteit Gent* 58(2b), 729-735.

MACOSKO, E.Z., POKALA, N., FEINBERG, E.H., CHALASANI, S.H., BUTCHER, R.A., CLARDY, J. & BARGMANN, C.I. (2009). A hub-and-spoke circuit drives pheromone attraction and social behaviour in *C. elegans*. *Nature* 458, 1171-1175.

MAHANTI, P., BOSE, N., BETHKE, A., JUDKINS, J.C., WOLLAM, J., DUMAS, K.J., ZIMMERMAN, A.M., CAMPBELL, S.L., HU, P.J., ANTEBI, A. *ET AL*. (2014). Comparative metabolomics reveals endogenous ligands of DAF-12, a nuclear hormone receptor, regulating *C. elegans* development and lifespan. *Cell Metabolism* 19, 73-83.

MAK, H.Y. & RUVKUN, G. (2004). Intercellular signaling of reproductive development by the *C. elegans* DAF-9 cytochrome P450. *Development* 131, 1777-1786.

MARTIN, R., DABRITZ, F., ENTCHEV, E.V., KURZCHALIA, T.V. & KNOLKER, H.J. (2008). Stereoselective synthesis of the hormonally active (25S)-Delta7-dafachronic acid, (25S)-Delta4-dafachronic acid, (25S)-dafachronic acid, and (25S)-cholestenoic acid. *Organic and Biomolecular Chemistry* 6, 4293-4295.

MATYASH, V., ENTCHEV, E.V., MENDE, F., WILSCH-BRAUNINGER, M., THIELE, C., SCHMIDT, A.W., KNOLKER, H.J., WARD, S. & KURZCHALIA, T.V. (2004). Sterol-derived hormone(s) controls entry into diapause in *Caenorhabditis elegans* by consecutive activation of DAF-12 and DAF-16. *PLoS Biology* 2, e280.

MAUPAS, É. (1899). La mue et l'enkystement chez les nématodes. *Archives Zoologie Expérimentale et Générale* 7, 563-628.

MAYER, M.G. & SOMMER, R.J. (2011). Natural variation in *Pristionchus pacificus* dauer formation reveals cross-preference rather than self-preference of nematode dauer pheromones. *Proceedings of the Royal Society B: Biological Sciences* 278, 2784-2790.

MCGRATH, P.T., XU, Y., AILION, M., GARRISON, J.L., BUTCHER, R.A. & BARGMANN, C.I. (2011). Parallel evolution of domesticated *Caenorhabditis* species targets pheromone receptor genes. *Nature* 477, 321-325.

MORRIS, J.Z., TISSENBAUM, H.A. & RUVKUN, G. (1996). A phosphatidylinositol-3-OH kinase family member regulating longevity and diapause in *Caenorhabditis elegans*. *Nature* 382, 536-539.

MOTA, M.M. & VIEIRA, P.R. (Eds) (2008). *Pine wilt disease: a worldwide threat to forest ecosystems*. Berlin, Germany, Springer.

MOTOLA, D.L., CUMMINS, C.L., ROTTIERS, V., SHARMA, K.K., LI, T., LI, Y., SUINO-POWELL, K., XU, H.E., AUCHUS, R.J., ANTEBI, A. *ET AL.* (2006). Identification of ligands for DAF-12 that govern dauer formation and reproduction in *C. elegans*. *Cell* 124, 1209-1223.

NEHER, D.A. (2010). Ecology of plant and free-living nematodes in natural and agricultural soil. *Annual Review of Phytopathology* 48, 371-394.

NIJHOUT, H.F. (2003). Development and evolution of adaptive polyphenisms. *Evolution & Development* 5, 9-18.

OGAWA, A., STREIT, A., ANTEBI, A. & SOMMER, R.J. (2009). A conserved endocrine mechanism controls the formation of dauer and infective larvae in nematodes. *Current Biology* 19, 67-71.

OGAWA, A., BENTO, G., BARTELMES, G., DIETERICH, C. & SOMMER, R.J. (2011). *Pristionchus pacificus daf-16* is essential for dauer formation but dispensable for mouth form dimorphism. *Development* 138, 1281-1284.

OGG, S. & RUVKUN, G. (1998). The *C. elegans* PTEN homolog, DAF-18, acts in the insulin receptor-like metabolic signaling pathway. *Molecular Cell* 2, 887-893.

OGG, S., PARADIS, S., GOTTLIEB, S., PATTERSON, G.I., LEE, L., TISSENBAUM, H.A. & RUVKUN, G. (1997). The Fork head transcription factor

DAF-16 transduces insulin-like metabolic and longevity signals in *C. elegans*. *Nature* 389, 994-999.

OKUMURA, E., TANAKA, R. & YOSHIGA, T. (2013). Species-specific recognition of the carrier insect by dauer larvae of the nematode *Caenorhabditis japonica*. *Journal of Experimental Biology* 216, 568-572.

OSCHE, G. (1956). Die Präadaptation freilebender Nematoden an den Parasitismus. *Zoologische Anzeiger* 19, 391-396.

PAPADEMETRIOU, M.K. & BONE, L.W. (1983). Chemotaxis of larval soybean cyst nematode, *Heterodera glycines* Race 3, to root leachates and ions. *Journal of Chemical Ecology* 9, 387-396.

PARADIS, S. & RUVKUN, G. (1998). *Caenorhabditis elegans* Akt/PKB transduces insulin receptor-like signals from AGE-1 PI3 kinase to the DAF-16 transcription factor. *Genes & Development* 12, 2488-2498.

PARADIS, S., AILION, M., TOKER, A., THOMAS, J.H. & RUVKUN, G. (1999). A PDK1 homolog is necessary and sufficient to transduce AGE-1 PI3 kinase signals that regulate diapause in *Caenorhabditis elegans*. *Genes & Development* 13, 1438-1452.

PARK, D., O'DOHERTY, I., SOMVANSHI, R.K., BETHKE, A., SCHROEDER, F.C., KUMAR, U. & RIDDLE, D.L. (2012). Interaction of structure-specific and promiscuous G-protein-coupled receptors mediates small-molecule signaling in *Caenorhabditis elegans*. *Proceedings of the National Academy of Sciences of the United States of America* 109, 9917-9922.

PATEL, D.S., FANG, L.L., SVY, D.K., RUVKUN, G. & LI, W. (2008). Genetic identification of HSD-1, a conserved steroidogenic enzyme that directs larval development in *Caenorhabditis elegans*. *Development* 135, 2239-2249.

PATTERSON, G.I., KOWEEK, A., WONG, A., LIU, Y. & RUVKUN, G. (1997). The DAF-3 Smad protein antagonizes TGF-beta-related receptor signaling in the *Caenorhabditis elegans* dauer pathway. *Genes & Development* 11, 2679-2690.

PAYNE, F.K. (1923). Investigation on the control of hookworm disease. XXX. Studies on factors involved in migration of hookworm larvae in soil. *American Journal of Epidemiology* 3, 547-583.

PENKOV, S., AKIRA, O., ULRIKE, S., VYACHESLAV, Z., TATE, D., BOLAND, S., GRUNER, M., VORKEL, D., VERBAVATZ, J.-M., SOMMER, R.J. *ET AL*. (2014). A wax ester promotes collective host finding in the nematode *Pristionchus pacificus*. *Nature Chemical Biology* 10, 281-285.

PERRY, R.N. & MOENS, M. (2011). Survival of parasitic nematodes outside the host. In: Perry, R.N. & Wharton, D.A. (Eds). *Molecular and physiological basis of nematode survival*. Wallingford, UK, CAB International Publishing, pp. 1-27.

PETERSEN, C., DIRKSEN, P., PRAHL, S., STRATHMANN, E.A. & SCHULENBURG, H. (2014). The prevalence of *Caenorhabditis elegans* across 1.5 years in selected North German locations: the importance of substrate type, abiotic parameters, and *Caenorhabditis* competitors. *BMC Ecology* 14, 4.

PLINE, M. & DUSENBERY, D.B. (1987). Responses of plant-parasitic nematode *Meloidogyne incognita* to carbon dioxide determined by video camera-computer tracking. *Journal of Chemical Ecology* 13, 873-888.

POINAR JR, G.O. & HERRE, E.A. (1991). Speciation and adaptive radiation in the fig wasp nematode, *Parasitodiplagaster* (Diplogasteridae: Rhabditida) in Panama. *Revue de Nématologie* 14, 361-374.

REED, E.M. & WALLACE, H.R. (1965). Leaping locomotion by an insect-parasitic nematode. *Nature* 206, 210-211.

REN, P., LIM, C.S., JOHNSEN, R., ALBERT, P.S., PILGRIM, D. & RIDDLE, D.L. (1996). Control of *C. elegans* larval development by neuronal expression of a TGF-beta homolog. *Science* 274, 1389-1391.

RIDDLE, D.L. & ALBERT, P.S. (1997). Genetic and environmental regulation of dauer larva development. In: Riddle, D.L., Meyer, B.J. & Preiss, J.R. (Eds). C. elegans *II*. Cold Spring Harbor, NY, USA, Cold Spring Harbor Laboratory Press, pp. 739-768.

RIDDLE, D.L. & BIRD, A.F. (1985). Responses of the plant parasitic nematodes *Rotylenchulus reniformis*, *Anguina agrostis* and *Meloidogyne javanica* to chemical attractants. *Parasitology* 91, 185-195.

RIEMANN, F. & SCHRAGE, M. (1988). Carbon dioxide as an attractant for the free-living marine nematode *Adoncholaimus thalassophygas*. *Marine Biology* 98, 81-85.

RIGA, E., PERRY, R.N., BARRETT, J. & JOHNSTON, M.R.L. (1997). Electrophysiological responses of male potato cyst nematodes, *Globodera rostochiensis* and *G. pallida*, to some chemicals. *Journal of Chemical Ecology* 23, 417-428.

ROAYAIE, K., CRUMP, J.G., SAGASTI, A. & BARGMANN, C.I. (1998). The G alpha protein ODR-3 mediates olfactory and nociceptive function and controls cilium morphogenesis in *C. elegans* olfactory neurons. *Neuron* 20, 55-67.

ROBINSON, A.F. (1995). Optimal release rates for attracting *Meloidogyne incognita*, *Rotylenchulus reniformis*, and other nematodes to carbon dioxide in sand. *Journal of Nematology* 27, 42-50.

ROBINSON, J. (1962). *Pilobolus* spp. and the translation of the infective larvae of *Dictyocaulus viviparus* from faeces to pasture. *Nature* 353-354.

ROTTIERS, V., MOTOLA, D.L., GERISCH, B., CUMMINS, C.L., NISHIWAKI, K., MANGELSDORF, D.J. & ANTEBI, A. (2006). Hormonal control of *C. elegans* dauer formation and life span by a Rieske-like oxygenase. *Developmental Cell* 10, 473-482.

SAWIN, E.R., RANGANATHAN, R. & HORVITZ, H.R. (2000). *C. elegans* locomotory rate is modulated by the environment through a dopaminergic pathway and by experience through a serotonergic pathway. *Neuron* 26, 619-631.

SHARMA, K.K., WANG, Z., MOTOLA, D.L., CUMMINS, C.L., MANGELSDORF, D.J. & AUCHUS, R.J. (2009). Synthesis and activity of dafachronic acid ligands for the *C. elegans* DAF-12 nuclear hormone receptor. *Molecular Endocrinology* 23, 640-648.

SINHA, A., SOMMER, R.J. & DIETERICH, C. (2012). Divergent gene expression in the conserved dauer stage of the nematodes *Pristionchus pacificus* and *Caenorhabditis elegans*. *BMC Genomics* 13, 254.

SOMMER, R.J. & STREIT, A. (2011). Comparative genetics and genomics in nematodes: genome structure, development and life style. *Annual Review of Genetics* 45, 1-20.

SRINIVASAN, J., KAPLAN, F., AJREDINI, R., ZACHARIAH, C., ALBORN, H.T., TEAL, P.E., MALIK, R.U., EDISON, A.S., STERNBERG, P.W. & SCHROEDER, F.C. (2008). A blend of small molecules regulates both mating and development in *Caenorhabditis elegans*. *Nature* 454, 1115-1118.

STARICH, T.A., HERMAN, R.K., KARI, C.K., YEH, W.H., SCHACKWITZ, W.S., SCHUYLER, M.W., COLLET, J., THOMAS, J.H. & RIDDLE, D.L. (1995). Mutations affecting the chemosensory neurons of *Caenorhabditis elegans*. *Genetics* 139, 171-188.

SUDHAUS, W. (1976). Vergleichende Untersuchungen zur Phylogenie, Systematik, Okologie, Biologie und Ethologie der Rhabditidae (Nematoda). *Zoologica* 43, 1-229.

SUDHAUS, W. (2008). Evolution of insect parasitism in rhabditid and diplogastrid nematodes. In: Makarov, S.E. & Dimitrijević, R.N. (Eds). *Advances in arachnology and developmental biology. Monographs 12*. Belgrade, Serbia, Institute of Zoology, pp. 143-161.

TABISH, M., KIDWAI SIDDIQUI, Z., NISHIKAWA, K. & SIDDIQUI, S.S. (1995). Exclusive expression of *C. elegans osm-3* Kinesin gene in chemosnsory neurons open to the external environment. *Journal of Molecular Biology* 247, 377-389.

THOMAS, J.H., BIRNBY, D.A. & VOWELS, J.J. (1993). Evidence for parallel processing of sensory information controlling dauer formation in *Caenorhabditis elegans*. *Genetics* 134, 1105-1117.

TIAN, H., SCHLAGER, B., XIAO, H. & SOMMER, R.J. (2008). Wnt signaling induces vulva development in the nematode *Pristionchus pacificus*. *Current Biology* 18, 142-146.

TUREK, M., LEWANDROWSKI, I. & BRINGMANN, H. (2013). An AP2 transcription factor is required for a sleep-active neuron to induce sleep-like quiescence in *C. elegans*. *Current Biology* 23, 2215-2223.

Van Buskirk, C. & Sternberg, P.W. (2007). Epidermal growth factor signaling induces behavioral quiescence in *Caenorhabditis elegans*. *Nature Neuroscience* 10, 1300-1307.

Vowels, J.J. & Thomas, J.H. (1992). Genetic analysis of chemosensory control of dauer formation in *Caenorhabditis elegans*. *Genetics* 130, 105-123.

Wang, J. & Kim, S.K. (2003). Global analysis of dauer gene expression in *Caenorhabditis elegans*. *Development* 130, 1621-1634.

Wang, X.D. & Ishibashi, N. (1999). Infection of the entomopathogenic nematode, *Steinernema carpocapsae*, as affected by the presence of *Steinernema glaseri*. *Journal of Nematology* 31, 207-211.

Wang, Z., Zhou, X.E., Motola, D.L., Gao, X., Suino-Powell, K., Conneely, A., Ogata, C., Sharma, K.K., Auchus, R.J., Lok, J.B. *et al.* (2009). Identification of the nuclear receptor DAF-12 as a therapeutic target in parasitic nematodes. *Proceedings of the National Academy of Sciences of the United States of America* 106, 9138-9143.

Ward, S. (1973). Chemotaxis by the nematode *Caenorhabditis elegans*: identification of attractants and analysis of the response by use of mutants. *Proceedings of the National Academy of Sciences of the United States of America* 70, 817-821.

Weller, A.M., Mayer, W.E., Rae, R. & Sommer, R.J. (2010). Quantitative assessment of the nematode fauna present on *Geotrupes* dung beetles reveals species-rich communities with a heterogeneous distribution. *Journal of Parasitology* 96, 525-531.

West-Eberhard, M.J. (2003). *Developmental plasticity and evolution*. Oxford, UK, Oxford University Press.

Wolkow, C.A. & Hall, D.H. (2012). Dauer neuroanatomy. In: *WormAtlas*. DOI:10.3908/wormatlas.3.4.

Wolkow, C.A., Munoz, M.J., Riddle, D.L. & Ruvkun, G. (2002). Insulin receptor substrate and p55 orthologous adaptor proteins function in the *Caenorhabditis elegans daf-2*/insulin-like signaling pathway. *Journal of Biological Chemistry* 277, 49591-49597.

Yoshiyama-Yanagawa, T., Enya, S., Shimada-Niwa, Y., Yaguchi, S., Haramoto, Y., Matsuya, T., Shiomi, K., Sasakura, Y., Takahashi, S., Asashima, M. *et al.* (2011). The conserved Rieske oxygenase DAF-36/Neverland is a novel cholesterol-metabolizing enzyme. *Journal of Biological Chemistry* 286, 25756-25762.

Zhao, L.L., Wei, W., Kang, L. & Sun, J.H. (2007). Chemotaxis of the pinewood nematode, *Bursaphelenchus xylophilus*, to volatiles associated with host pine, *Pinus massoniana*, and its vector *Monochamus alternatus*. *Journal of Chemical Ecology* 33, 1207-1216.

Nematology Monographs & Perspectives, 2015, Vol. 11, 301-329

Chapter 11

Mouth dimorphism and the evolution of novelty and diversity

Erik J. RAGSDALE

Department of Biology, Indiana University, 915 E. 3rd Street, Bloomington, IN 47405, USA
ragsdale@indiana.edu

Introduction

The origin of complex traits and the generation of morphological diversity are two of the most enduring puzzles of evolutionary biology. A principal mission of evo-devo has been to solve them by determining how developmental pathways and their component factors are selectively used or re-used to produce different phenotypes. Yet the existence of developmental biology as a field and a concept shows us that the sum of parts of the genetic code is in itself insufficient to explain how complex traits are specified. The path from genotype to phenotype requires – and is susceptible to – interactions with the environment, which includes the intrinsic one set by the same genetic blueprint as well as external pressures from abiotic conditions or other organisms. Given this complication to how genes build traits, the search for mechanisms of divergence cannot be limited to genetic variation only. It must also consider developmental plasticity, or the variation of phenotype produced by a single genotype (West-Eberhard, 2003).

Pristionchus pacificus shows a particularly striking case of developmental plasticity, specifically a polyphenism, or a polymorphism due to an environmental response (Bento *et al.*, 2010). This nematode has two distinct feeding morphotypes that differ in the shape of the stoma (mouth) and complexity of their teeth. Moveable teeth characterise the family Diplogastridae, which includes *P. pacificus* and other nematode species with a mouth dimorphism (Fürst von Lieven & Sudhaus, 2000). Because teeth are a morphological novelty with respect to the simple mouth cavities of outgroups, developmental plasticity is coupled to a structural innovation in diplogastrids. Although plasticity manifested as continuous variation is ubiquitous in nature, the dimorphism of *P. pacificus* makes analyses of plasticity simpler by objectively categorising plasticity into binary states. Furthermore, the sophisticated genetic toolkit established for *P. pacificus* grants access to the specific genes involved, giving studies a level of detail difficult to achieve in non-model organisms. As the conditions of: *i*) exhibiting an obvious polyphenism; and *ii*) being amenable to detailed genetic analysis are both met in *P. pacificus*, this species presents an exciting opportunity to explore the origins of diversity and novelty through the study of developmental plasticity.

Morphology of dimorphic mouthparts

The first allusion to the diplogastrid mouth plasticity was made by Potts (1910), who drew the two forms of *P. maupasi* and remarked "how greatly the state of contraction of the mouth affects the buccal cavity." Hirschmann (1951) later determined the plasticity to be a discrete dimorphism, in a species that has since been called *P. lheritieri*. By propagating cultures from isofemale lines, she demonstrated that two forms, otherwise distinct enough to characterise separate species, belonged to the same breeding populations. She ascribed the terms "eurystomatous" (Eu) and "stenostomatous" (St) to the wide and narrow breadth, respectively, of the mouth cavities of the alternative forms, although it was also clear that the forms were accompanied by discrete differences in tooth morphology (Fig. 11.1).

The mouth dimorphism is expressed at the adult stage. The phenotype follows an irreversible decision during development and can be specified at least as late as the third-stage juvenile (J3) (Serobyan *et al.*, 2013). The resulting forms of all mouth-dimorphic diplogastrids show differences in stomatal width and the prominence of cuticular structures, most commonly a moveable dorsal tooth (Fig. 11.1A, B). Differences extend to a range of other structures, which include characters that diagnose species that are otherwise similar in their non-sexual morphology (Fig. 11.2). In *P. pacificus* and other *Pristionchus* species, the Eu form is distinguished by several additional structures, particularly in the stegostom (= pharyngeal region of stoma), that are of lower complexity or missing in the St form. First, the Eu form bears an additional,

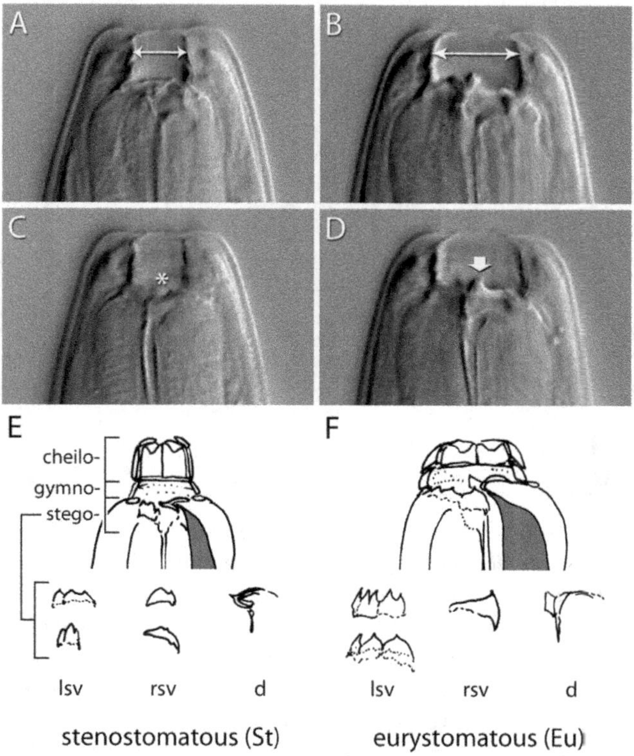

Fig. 11.1. *The mouth dimorphism of* Pristionchus. *A-D: Nomarski images of* Pristionchus pacificus. *A, C: A single stenostomatous (St) hermaphrodite in sagittal and right subventral planes, respectively; B, D: A single eurystomatous (Eu) hermaphrodite in the corresponding planes. Dorsal is left. The dimorphism is marked by a difference in the width of the stoma (arrows), in addition to the shape of the dorsal tooth (shown in A, B) and the absence (asterisk, C) or presence (arrow, D) of an opposing subventral tooth; E, F: Dimorphism in* P. quartusdecimus, *another species of the* pacificus *group of* Pristionchus *(see Ragsdale* et al., *Chapter 4, this volume). Stoma and anterior pharynx are drawn. Dorsal is right. In addition to having a tooth in the right subventral (rsv) sector of the stegostom (stego-), the Eu form (F) exhibits greater possible complexity in the left subventral (lsv) sector, as shown by stegostomatal structures and variants below whole drawings. Besides its differences in stegostomatal morphology, the mouth dimorphism extends to other regions of the stoma, including the gymnostom (gymno-) and cheilostom (cheilo-), indicating the coordination of several cell types in mouth form determination. Other abbreviation: d, dorsal. A-D: Modified from Ragsdale* et al. *(2013b); E, F: Modified from Kanzaki* et al. *(2013a).*

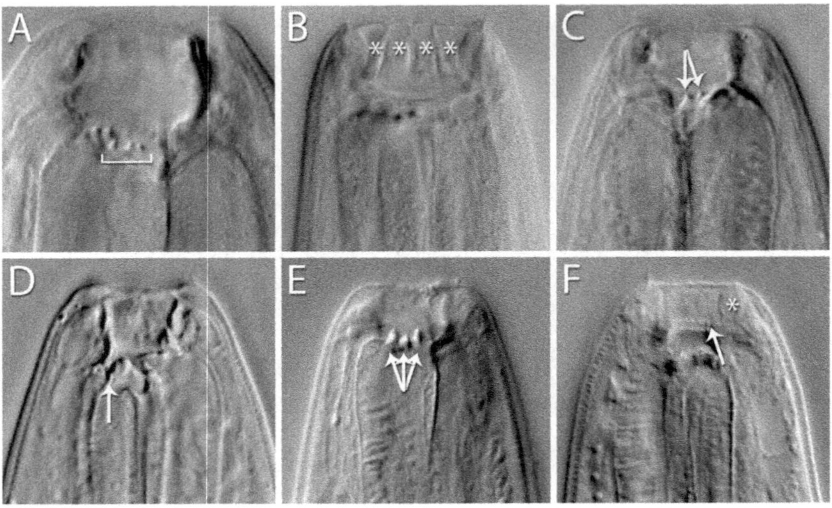

Fig. 11.2. *Complexity and diversity of mouthparts in* Pristionchus. *In spite of general morphological similarity among* Pristionchus *species, characters of the stoma, which are often dimorphic, vary across the genus. A: Adventitious plate (bracket) of denticles in Eu form of* P. hoplostomus *(Ragsdale* et al., *2013a); B: Radial division of cheilostom into 12 complete plates (asterisks indicate four lateral plates) in 'megastomatous' Eu form of* P. triformis *(Ragsdale* et al., *2013a); C: Right subventral ridge of denticles (arrows) that is, among* Pristionchus *species, unique to the Eu form of* P. fissidentatus *(Kanzaki* et al., *2012b); D: Adventitious denticles in the St form of* P. maxplancki *(Kanzaki* et al., *2013a); E: Three conical left subventral denticles of* P. bucculentus, *a putatively monomorphic species; F: Apomorphic morphology, including serrated gymnostom (arrow) and cheilostomatal bulges, of monomorphic species* P. elegans *(Kanzaki* et al., *2012b).*

opposing tooth in the right subventral sector, which in the St form is a low ridge armed at most with a minute denticle or cusp (Fig. 11.1C, D). The right subventral tooth of the Eu form assumes a distinct 'claw-like' shape (*i.e.*, with a sinusoidal anterior margin and arched posterior margin) and, like the dorsal tooth, is actuated by pharyngeal muscle. Second, a plate projecting from the left subventral wall of the stoma has more complex serration in the Eu form. In the St form of species in the *pacificus* group of *Pristionchus* (see Ragsdale *et al.*, Chapter 4, this volume), the number of peaks is regularly two or three. By contrast, the Eu form can have as many as eight, although the number of peaks is not consistent across individuals, nor are individual peaks necessarily homologous (Kanzaki *et al.*, 2012a, 2013a; Fig. 11.1E, F). In other species of *Pristionchus*, differences in the left subventral armature are even more pronounced. For example, the Eu forms of *P. fukushimae* and *P. hoplostomus* have a secondary plate projecting from the medial surface of the main plate, whereas any adventitious structures in the St forms are completely missing (Ragsdale *et al.*, 2013a; Fig. 11.2A). Among *Pristionchus* species, the complexity difference between forms is perhaps greatest in *P. fissidentatus*, in which the Eu form shows two novel structures, namely a ridge of denticles in the right subventral sector (Fig. 11.2C) and a large, separate denticle just left of the ventral midline (Kanzaki *et al.*, 2012b). Beyond discrete differences in the stegostom, alternative forms of *Pristionchus* species differ in the sclerotisation of the stomatal walls (*i.e.*, gymnostom, cheilostom) (Fig. 11.1A-D). Additionally, in *P. fukushimae*, *P. hoplostomus* and *P. triformis*, the plates of the cheilostom often have secondary divisions in the Eu form. In *P. triformis* in particular, these divisions are complete, enabling a 'megastomatous' Eu form with 12 plates to be characterised (Fig. 11.2B), in contrast to the St form, which almost always has six plates. Although the dimorphism is most clearly manifested in structures derived from the pharynx, differences extend throughout the rest of the stoma. By identifying specific stomatal tissues, it is possible to reveal precisely where different morphogenetic modules are executed during development.

The eutely, or cell constancy, of tissues in at least some nematodes (Sulston & Horvitz, 1977) enables developmental pathways to be localised to the level of single cells (*e.g.*, Sternberg, 1988; Eizinger & Sommer, 1997; see Rudel, Chapter 9, this volume). Consequently, the individual cells that secrete the cuticular structures of the stoma have been identified for taxa throughout the Rhabditida *sensu* De Ley & Blaxter (2002) (Wright & Thomson, 1981; White, 1988; De Ley *et al.*, 1995; Bumbarger *et al.*, 2006; Ragsdale *et al.*, 2008, 2011; Giblin-Davis *et al.*, 2010). Stomatal tissues comprise a highly conserved set of cells that have retained their relative positions and in most cases have changed only their length along the alimentary tract (Ragsdale & Baldwin, 2010). Although the complete cellular architecture is not yet published for *P. pacificus*, cells known in the diplogastrid *Acrostichus halicti* (= *Aduncospiculum halicti*) suggest similar conservation of stomatal tissue in Diplogastridae (Baldwin *et al.*, 1997). In particular, the cells that build the stegostom, the region of the stoma with complex dimorphic structures (Fig. 11.1), have been identified in *A. halicti*. In that species, the dorsal tooth is produced by cells homologous with the anterior two layers of muscle (pm1, pm2) in *Caenorhabditis elegans*. The tissue of the gymnostom, which secretes a ring of cuticle that lines the middle of the stoma, consists of another type of epithelium, the arcade syncytia. This epithelium is developmentally separate from the pharynx and connects to it only later during pharyngeal morphogenesis (Portereiko & Mango, 2001). Finally, the most anterior region of the stoma, the cheilostom, is derived from epidermis, although the precise identities of cheilostomatal cells in any diplogastrid have yet to be described.

Taken together, gross and fine-structural anatomy indicate that the mouth dimorphism results from developmental differences coordinated across several types of tissue. What qualitative differences in cellular architecture exist between mouth forms are still not clear, as none was reported for *A. halicti*. A complete anatomical reconstruction of stomatal tissues, including cell bodies and processes, may still reveal form-specific modules as differences in cellular connectivity. Because such reconstructions are feasible in nematodes, *P. pacificus* and other Diplogastridae are a promising system for articulated tests linking developmental plasticity to the diversity of form.

Evolutionary history of the dimorphism

Since Hirschmann's (1951) description of the dimorphism in *Pristionchus*, a similar dimorphism has been reported for species of other 'genera', including *Allodiplogaster* (Körner, 1954), *Micoletzkya* (Rühm, 1956), *Acrostichus* (Giblin & Kaya, 1984) and, more recently, several others (Kanzaki *et al.*, 2012c, 2013d; Susoy *et al.*, 2015). The presence of stomatal dimorphism in taxa with disparate morphologies immediately gives it a macroevolutionary context. Two scenarios can explain the taxonomic spread of the trait: either *i*) the dimorphism has been conserved across divergences deep enough to produce such distinct morphologies; or *ii*) it has evolved multiple times in parallel. In their meticulous study of stomatal morphology in Diplogastridae, Fürst von Lieven & Sudhaus (2000) favoured the latter scenario given the differences of the dimorphism among genera. They compared the clear distinction between one tooth (St) and two teeth (Eu), found in *Pristionchus* spp., with what they considered a simple difference of degree between the two forms in *A. halicti* and in *Allodiplogaster* spp., in which both forms had one tooth or two teeth, respectively. Furthermore, they had not observed the Eu form in males of *A. halicti*, in contrast to the Eu males of *Allodiplogaster* spp. This observation led them to hypothesise that the dimorphism is a unique phenomenon in each of these two genera. Given the regulatory machinery that must be in place to specify and execute alternative morphologies, the repeated recurrence of dimorphism would be a stunning exception to Dollo's law. On the other hand, the scenario in which the polyphenism appeared only once would require the long-term maintenance of an anciently acquired trait. Despite the colonisation of new or divergent ecological niches, conditional advantages for two distinct forms must have persisted in macroevolutionary time to avoid the loss of one form and the selective pressure to maintain a developmental switch.

The placement of Diplogastridae into a well-resolved phylogenetic infrastructure based on molecular characters, particularly a suite of rRNA and ribosomal protein gene sequences, has since enabled independent reconstructions of character histories (Mayer *et al.*, 2007, 2009; Ragsdale *et al.*, 2013a; Giblin-Davis & Kanzaki, Chapter 3, this volume). To infer the history of the dimorphism, Susoy *et al.* (2015) mapped the trait onto a phylogeny inferred for all known dimorphic genera and a broad representation of monomorphic Diplogastridae. In that study, the dimorphic condition was supported as the ancestral state for the family and thereafter lost at least ten times (Fig. 11.3). Thus, the developmental modules or switches for a dimorphism did not need to appear more than once. As a consequence, taxon-specific distinctions between forms, such as the number of teeth and denticles, must have accumulated separately in those taxa. Such differences in the dimorphism among various lineages are consistent with predictions about evolutionary modularity. Given sufficient separation of developmental pathways between the two forms, those forms should accumulate independent variability independently and thereby allow diversifying selection within a single species (West-Eberhard, 2003).

On a shorter evolutionary timescale, such as within *Pristionchus*, the persistence of the dimorphism as a single trait is more obvious. Among those *Pristionchus* species confirmed by genetic markers, most show two forms, even if in different ratios (Hirschmann, 1951; Sommer *et al.*, 1996; Sudhaus & Fürst von Lieven, 2003; Kanzaki *et al.*, 2012a, b, 2013a, b, 2014; Ragsdale *et al.*, 2013a). Moreover, each form has

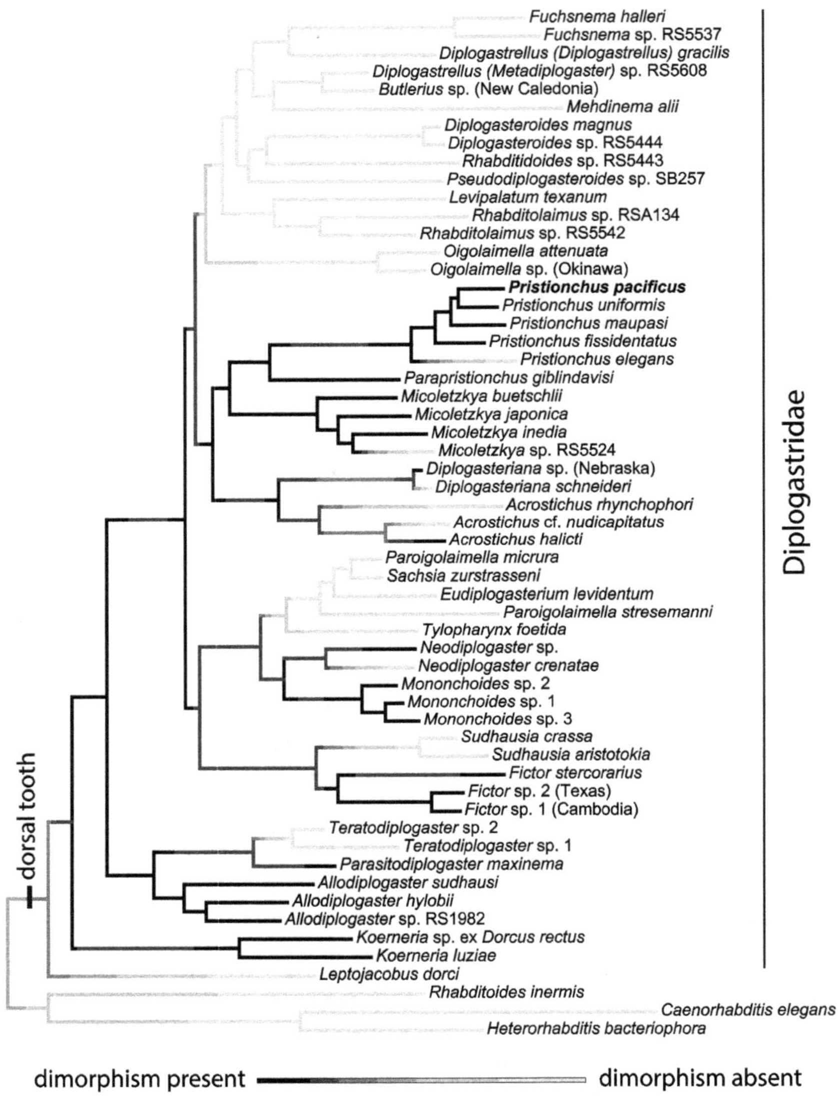

Fig. 11.3. *Evolutionary history of the dimorphism. History of character states was inferred by Bayesian stochastic character mapping on the posterior set of phylogenetic trees inferred from an alignment of SSU rRNA, LSU rRNA, and 11 ribosomal protein genes. The dimorphism evolved once and was independently lost at least ten times. According to the reconstructed history, the origin of moveable teeth (mapped here by simple parsimony) coincided with the appearance of the dimorphism. Tree modified from Susoy* et al. *(2015).*

greater similarity to the putatively homologous form in closely related species than to the alternative form of its own species (*e.g.*, Kanzaki *et al.*, 2012a, 2013a; Ragsdale *et al.*, 2013a). It is therefore unlikely that the stereotypic morphology for a given form evolved convergently within younger clades. Only among the most divergent taxa do similarities of a putatively homologous form break down. For example, a 'claw-like' tooth, which is only present in the Eu form of *Pristionchus* species, is present in both forms of the closest known outgroup to *Pristionchus*, *Parapristionchus giblindavisi*, although the differences between the two forms of that species are distinct (Kanzaki *et al.*, 2012c). Therefore, in lineages with ancient losses of the dimorphism (*e.g.*, in the clades defined by *Tylopharynx* and *Sachsia* or by *Oigolaimella* and *Fuchsnema*, Fig. 11.3), which form was originally fixed in those lineages cannot be inferred with any reliability.

Complete loss of the dimorphism is hypothesised for only two *Pristionchus* species, *P. elegans* and *P. bucculentus*, based on screens of hundreds of individuals (Kanzaki *et al.*, 2012b, 2013c; Fig. 11.2E, F). These two species are sister taxa in the set of all molecularly characterised *Pristionchus* species but, interestingly, they appear to have fixed alternative morphs (putative St in *P. elegans*, Eu in *P. bucculentus*) from a dimorphic common ancestor. Although the dimorphism has not been completely lost in other lineages of *Pristionchus*, it is apparent that the ratio of the two forms, measured as the frequency of one form in a population in a constant environment, varies among species. This ratio differs even among populations within a species, as reported for *P. pacificus* and its putative sister species, *P. exspectatus* (Ragsdale *et al.*, 2013b).

A phylogenetic framework and examination of hundreds or thousands of specimens per species have thus allowed a historical reconstruction of the dimorphism across Diplogastridae, among *Pristionchus* and within *P. pacificus*. As a result, this work has shown that the dimorphism was in the common ancestor of Diplogastridae and has since become associated with a diversity of associated morphologies and regulatory responses.

Ecological function and adaptive value

Diplogastrids are rapid colonisers of rich and ephemeral habitats (Sachs, 1950; Bongers, 1999), and *Pristionchus* species in particular are

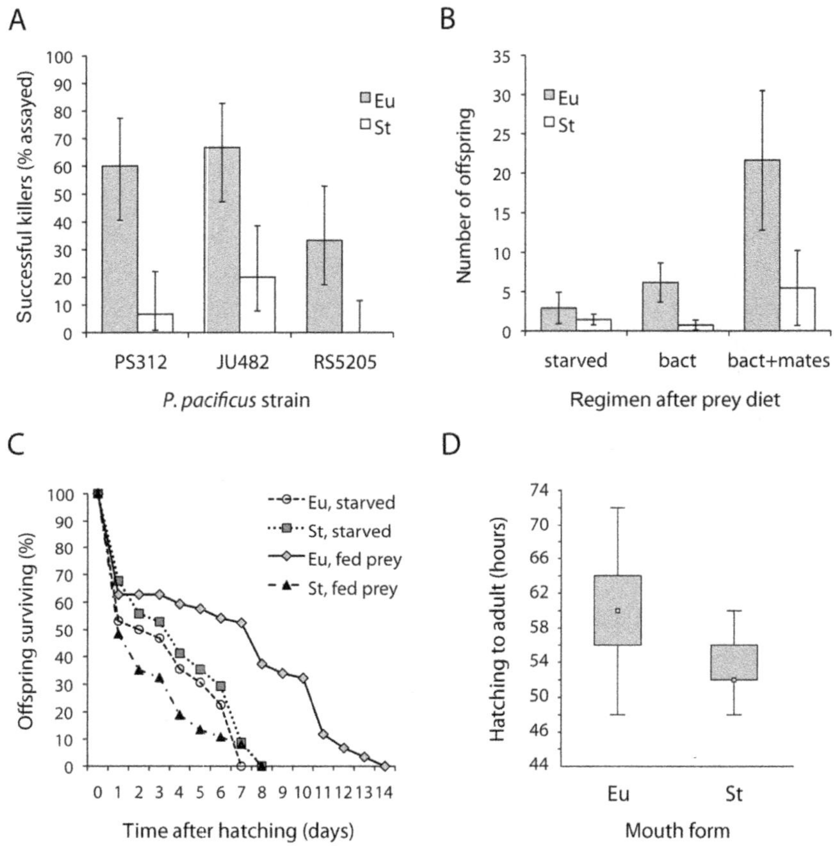

Fig. 11.4. *A putative fitness trade-off for a feeding polyphenism. A: Results of an assay where starved hermaphrodites of* Pristionchus pacificus *were placed into an arena with abundant prey (*Caenorhabditis elegans *L1) for a set time interval. Eu individuals were more successful predators, as indicated by the higher proportion of killers among individuals assayed for the Eu form of a given strain. Whiskers represent a 95% CI; B: Fitness of hermaphrodites, as measured by the number of offspring following an adult diet of only prey. When reintroduced to bacteria (bact) after this diet, the Eu form had more offspring than the St form, whether or not prey-fed hermaphrodites were also allowed to mate (bact+mates). Whiskers represent a 95% CI; C: Fitness of hermaphrodites, as measured by the survival of their starved, developmentally arrested offspring. Eu hermaphrodites showed an advantage over the St form, which fared no better than Eu or St hermaphrodites starved for their entire adulthood; D: Time required for development of hermaphrodites on a bacterial diet and at 20°C. St individuals reached maturity significantly more quickly,*

necromenic associates of insects, especially beetles (Herrmann *et al.*, 2006, 2007; see Ragsdale *et al.*, Chapter 4, this volume). *Pristionchus* nematodes are generalists in these habitats, feeding on bacteria, fungi and other nematodes. The ecological succession on a host insect cadaver often starts with an explosion of bacterial growth, followed by the proliferation of nematodes competing for those bacteria (Rae *et al.*, 2008; Weller *et al.*, 2010). On a broad potential diet, a dimorphism in stomatal morphology could allow specialisation on alternative food sources. For example, the presence of additional or larger teeth has implied an advantage of the Eu form for feeding on larger food items such as nematode prey (Fürst von Lieven & Sudhaus, 2000; Kiontke & Fitch, 2010).

Putative fitness advantages for each form of *P. pacificus* have been identified by empirical studies of dimorphism function. An assay to quantify predatory ability showed the Eu form to be a more successful predator than the St form (Serobyan *et al.*, 2014). In that assay, potential predators were starved and placed into an arena with a standardised density of prey, which consisted of *C. elegans* L1, and were then given a set interval of time to hunt and kill a prey item. Under these conditions, many more Eu than St individuals were successful killers in all *P. pacificus* strains tested (Fig. 11.4A). The difference was not attributed to hunting behaviour, as the two forms showed no differences in their rates of encountering or attacking prey. Instead, the time for successful predators to turn an attack into a kill was shorter in Eu individuals, suggesting a greater efficiency of mouthparts in that form. This functional advantage apparently translates to an adaptive benefit, as the Eu form showed a higher fitness than the St form on a prey diet (Serobyan *et al.*, 2014). Fitness advantages of the Eu over the St form were detected in: *i*) the fecundity of mothers when allowed to feed on bacteria following prey (Fig. 11.4B); *ii*) the longevity of mothers, which continued to be fecund when mated; and *iii*) the survival of

indicating an advantage for the St form under the condition of ample bacterial food. Box plots show the median (centre square, white), the lower and upper quartiles (bounds of grey box), and the range (whiskers) of developmental times. A-C: Modified from Serobyan et al. *(2014); D: Modified from Serobyan* et al. *(2013).*

their developmentally arrested offspring, presumably due to maternal provisioning (Fig. 11.4C).

Because the Eu form can access a broader diet than can the St form, a functional advantage for the St form is not obvious. The occurrence of conditions favouring both forms would nevertheless be required to maintain a dimorphism in evolution (Moran, 1992). So why has the St form persisted? To test for differences in fecundity on a bacterial diet, Eu and St hermaphrodites were allowed to feed freely on bacteria, but no form-specific differences in fecundity were detected (Serobyan *et al.*, 2014). However, the rate of development from hatching to maturity was faster for the St form when reared on a bacterial diet, indicating a shorter generation time for that form (Serobyan *et al.*, 2013; Fig. 11.4D). Moreover, the rate difference was apparently greatest during the final moult, exactly the period during which the dimorphic phenotype is executed, suggesting a higher cost in developmental time for the Eu form. The dimorphism therefore represents a putative fitness trade-off: whereas the Eu form can derive greater benefit from a diet of prey and limited bacteria, the St form can grow and thereby reproduce faster on an abundant bacterial food supply. The dimorphism may also confer contrasting benefits for other food sources such as fungi or unicellular eukaryotes, and its functional context will expand with more knowledge of *P. pacificus* ecology.

Environmental cues and conditional regulation

Whether a dimorphism is genetically or environmentally specified determines how morphological differences will be inherited and selected. The presence of both forms in inbred lines such as the reference ('California') strain of *P. pacificus* has implied that the dimorphism is not due to genetic variability in those strains. Furthermore, artificial selection of either form for ten generations was unable to change the ratio of forms under constant laboratory conditions, confirming that the dimorphism in this species is a indeed a polyphenism (Bento *et al.*, 2010). Because the mouth form decision is a product of the environment, the decision must reliably respond to external cues if fitness is to be optimised under different conditions.

Starvation pressure, or deprivation of a bacterial food source, was shown to increase the proportion of Eu individuals in a population (Bento *et al.*, 2010). Although a diet of *C. elegans* larvae alone is nutritionally challenging for *P. pacificus* (Serobyan *et al.*, 2014), the Eu advantage on a bacteria-poor, prey-rich diet makes this cue intuitive. Whether the response is triggered metabolically or additionally by olfactory cues has not been tested directly. The partial regulation of the mouth dimorphism by EGL-4 (Kroetz *et al.*, 2012), a cGMP-dependent protein kinase that regulates olfaction and satiety recognition in nematodes (L'Etoile *et al.*, 2002; Hong *et al.*, 2008; You *et al.*, 2008; Hong, Chapter 12, this volume), suggests that either type of signal transduction is likely. Because the mouth form decision is linked to dauer development (Bento *et al.*, 2010), which executes a non-constitutive dispersal stage also in response to starvation (see Brown & Ogawa, Chapter 10, this volume), the well-studied dauer pathway can predict candidate mechanisms to test (Sommer & Ogawa, 2011).

Crowding by conspecifics also promotes the Eu form. For example, the isolation of individual larvae from culture populations rendered them more likely to be St than Eu (Serobyan *et al.*, 2013). Crude pheromone extracted from dauer-conditioned medium increases the incidence of the Eu form, possibly as a response to competition for a diminishing food resource (Bento *et al.*, 2010). Small molecules that are active for mouth form decision have since been identified (Bose *et al.*, 2012; see Schroeder, Chapter 7, this volume). One compound promoting the Eu form was the ascaroside ascr#1, which was previously shown to induce dauer entry in *C. elegans* (Butcher *et al.*, 2007). This molecule did not, however, have dauer activity in *P. pacificus* (Bose *et al.*, 2012), revealing that an old molecule had been co-opted for a new, non-overlapping function. A novel dimer of ascr#1, the diascaroside dasc#1, was also identified in pheromone of *P. pacificus*. This molecule showed even higher activity than ascr#1 for promoting the Eu form, indicating that a new biosynthetic pathway is also involved in the regulation of a taxon-specific trait. Additionally, two paratosides, derivatives of ascarosides, were discovered to regulate mouth form development. One of these, npar#1, was highly active in dauer formation, consistent with the coupling of mouth form regulation to the dauer pathway (Bento *et al.*, 2010).

Other environmental signals affecting the *P. pacificus* dimorphism are unknown, although several have been tested. Conditions without an effect include temperature, pH, and even the DNA stain acridine orange, which was reported to alter the mouth form ratio in *P. lheritieri*

(Hirschmann, 1951). Of the cues that are known, none is sufficient completely to saturate a single form in a population. Even in the presence of ample bacteria, *P. pacificus* consistently produces both forms and most examined strains are biased toward the Eu form in laboratory culture (Ragsdale *et al.*, 2013b). Because conditional regulation by all known cues is incomplete, at least some degree of stochastic regulation of the dimorphism cannot be ruled out. In such a case, conditional regulation might be combined with bet-hedging (Philippi & Seger, 1989), particularly a strategy in which a threshold alternative state is tuned to the probability of encountering conditions where a given form is optimised (Moran, 1992). The matching of heritable thresholds to habitats would be consistent with evidence that *P. pacificus* genotypes correlate with environmental variables (McGaughran *et al.*, 2014). In the unpredictable or rapidly changing environments that *Pristionchus* nematodes inhabit, a partial bet-hedging strategy would allow the proliferation of at least some individuals if any suitable food source were present.

Exposure to pheromones or conspecifics has shown that nematodes can mediate a plastic response within their own lifetime. However, it is also possible that environmental information is transmitted across generations. In rapidly changing conditions, epigenetic inheritance could allow the immediate response of phenotypes without the timescale necessary for natural selection to take place. Experiments in a standardised environmental and genetic background showed that such inheritance is possible for the mouth dimorphism (Serobyan *et al.*, 2013). An apparently epigenetic effect was specifically found in *P. pacificus* males, which had a mouth form ratio biased according to maternal phenotype. Whereas males with Eu mothers were relatively highly St (~20% Eu in a culture population), those with St mothers were virtually fixed for the St form. How this bias occurs is unknown, but it is possible that gene expression is affected in offspring by maternally provided miRNAs (*e.g.*, Johnson & Spence, 2011; Rechavi *et al.*, 2011). The ability to study candidate mechanisms in sufficient detail in nematodes makes epigenetics an exciting topic of research on the mouth dimorphism.

Finally, one other determinant of mouth form development is sex. Although not a direct response to the external environment, mouth form bias by gender could provide regulation contingent on certain conditions, specifically those likely to be encountered by a given sex or conditions under which that sex would have higher fitness. In the reference strain of

P. pacificus, both sexes produce both forms, although hermaphrodites are Eu-biased and males are St-biased (Serobyan *et al.*, 2013). This phenomenon may be widespread in dimorphic Diplogastridae. In some species or strains no Eu males have been observed, for example *A. halicti* (Fürst von Lieven & Sudhaus, 2000), *Micoletzkya* spp. (Susoy *et al.*, 2013), and even in one clade of *Pristionchus* (Ragsdale *et al.*, 2013a). Functional advantages to a sexual dimorphism are possible. One hypothesis is that a disposal toward the St form would allow faster development of males, in which reaching maturity more rapidly and mating would be favoured over adult diet (Bonduriansky *et al.*, 2008). Sexual dimorphism might also promote niche partitioning between sexes (Shine, 1989). For example, isometric differences between them, namely the smaller stomata of males, might inhibit successful competition of Eu males with Eu females/hermaphrodites for Eu-optimised food resources. In the latter case, development of the Eu form would be an unprofitable investment for males. Rigorous investigation of the mouth plasticity in males, particularly in a gonochoristic *Pristionchus* species, will test implications for sexual selection of mouthparts.

Developmental regulation coupled to the dauer plasticity

The greatest utility of *P. pacificus* as a model for developmental plasticity is the tractability of genetic analysis in this organism. By putting analytical tools into practice, studies are already uncovering pathways for mouth form development. The first mechanistic discovery in mouth form determination was the co-option of a dauer-induction pathway (Bento *et al.*, 2010). In that study, dauer formation defective (*daf-d*) mutants of *P. pacificus* were unresponsive to pheromones and only weakly responsive to starvation, both of which conditions induced a higher Eu frequency in the wild-type strain. The mutants were loss-of-function alleles of *daf-12*, which encodes a nuclear hormone receptor that is a convergence point for other regulatory cascades, including transforming growth factor beta (TGF-β) and insulin/insulin-like signalling (Antebi *et al.*, 2000; Ogawa *et al.*, 2009; see Brown & Ogawa, Chapter 10, this volume). Consequently, treatment with a ligand of DAF-12, the steroid hormone analogue Δ7-dafachronic acid (DA), decreased the incidence of Eu individuals, in addition to inhibiting dauer entry. Together, these results showed that a pathway already mediating development of a dispersal stage was harnessed for a new function in mouth-dimorphic nematodes. This co-option is consistent with a shared response to similar environmental cues. Dauer-inducing stressors, namely a lack of bacterial food and crowding by potential competitors, are precisely those in which the promoted predatory form shows significant fitness advantages.

Nevertheless, cues to invest in better predatory equipment do not coincide completely with those inducing dauer entry. 'Crowding' is communicated differently in mouth form and dauer development, as individual molecules constituting whole pheromone show overlapping but unique activities for the two plastic responses (Bose *et al.*, 2012). A clear separation of developmental pathways was found in the analysis of *P. pacificus daf-16* (*forkhead box O*) mutants (Ogawa *et al.*, 2011; see Brown & Ogawa, Chapter 10, this volume). The transcription factor DAF-16/FOXO, the target of insulin signalling, is evolutionarily conserved as a regulator of dauer formation in *C. elegans* and *P. pacificus*. However, *P. pacificus daf-16* mutants showed a wild-type mouth dimorphism response to cues affecting dauer formation, indicating partial independence of the latter from mouth form development. Although cross talk may occur between DAF-12 and DAF-16 in *P. pacificus* dauer formation, as hinted by their interactions in *C. elegans* longevity regulation (Shen *et al.*, 2012), no such logic is predicted for mouth form induction. Either co-option of dauer formation signalling for the mouth form was originally incomplete or DAF-16 was lost from mouth form development after the co-option event.

Considering these mechanistic advances in the field, analyses of dauer mutants have been useful for applying a corpus of detailed knowledge in *C. elegans* to the *P. pacificus* mouth dimorphism. Naturally missing from these inferences is how mouth form regulation itself has specialised, as dauer genetics can only inform on dauer-relevant processes. Given the partial separation of regulatory machinery, the question arises as to whether any mouth form-specific factors exist and can be identified.

Regulation through a developmental switch

To determine whether mouth form development can be understood in terms of individual genes, forward genetics must specifically target the mouth-dimorphism phenotype. In such an approach, a screen for

Eu-form-defective (*eud*) mutants recovered lines with lesions in single genes or tightly linked genomic intervals (Ragsdale *et al.*, 2013b). One of the mutants was identified as *eud-1*, a homologue of a *C. elegans* gene (*sul-2*) that encodes an arylsulfatase of unknown function. Genetic and transgenic experiments then revealed EUD-1 to execute a master switch for the mouth form decision (Fig. 11.5A). Whereas mutants were completely defective for the Eu form (0% Eu in a clonal population), transgenic lines that over-expressed *eud-1* were fixed for that form (100% Eu). Moreover, the switch acts in a dosage-dependent manner, as expressivity of the Eu form was sensitive to doses of one *vs* two copies of *eud-1*. Specifically, the phenotype of heterozygous mutants (one copy) was highly St, but not Eud, whereas the wild-type strain (two copies) was highly but not all Eu. Even males, which carry only one copy of *eud-1* due to its location on the X chromosome, reflect this dosage effect: wild-type males are highly St (Serobyan *et al.*, 2013), although introducing more copies into males fully induced Eu formation. The EUD-1 switch is therefore sufficient to control sexual dimorphism of the mouth plasticity and thereby confer any adaptive benefits entailed by sexual differences.

A sensitive threshold for the EUD-1 switch predicts a similar sensitivity of other factors or elements regulating EUD-1. If this were the case, the switch could be adjusted according to specialised conditions encountered by a given strain. Consistent with this idea is the heritable variation of the mouth form ratio as observed across *P. pacificus* strains, some 80 of which were surveyed (Ragsdale *et al.*, 2013b). Indeed, lower levels of expression in highly St strains suggested that EUD-1 is regulated differently among populations, and transgenic experiments confirmed that EUD-1 executes the switch in divergent strains (Fig. 11.5B). The conserved activity of EUD-1 thus indicates the potential of the switch to effect change among them, possibly in response to selective pressures. Beyond intraspecific variation, phenotypic differences between species were identified and tested. Through an integration of transgenic and hybridisation experiments, functional analysis showed that EUD-1 controlled the dimorphism in *P. exspectatus* as well as *P. pacificus*. Regulatory differences between the two species were associated with both their X chromosomes and autosomal background, suggesting rapid evolution of a regulatory mechanism for an otherwise conserved developmental switch.

To explain the origin of this novel regulator, *eud-1* was found to result from lineage-specific gene duplications. This gene and two paralogues

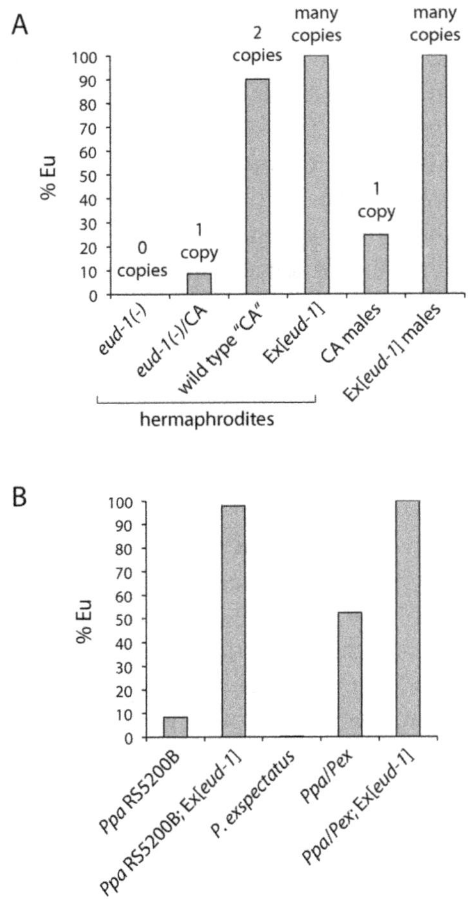

Fig. 11.5. *EUD-1 controls a dosage-dependent, evolutionarily conserved switch for the mouth form decision in* Pristionchus pacificus. *A: The developmental switch is sensitive to* eud-1 *copy number, as indicated by the phenotype of homozygous and heterozygous* eud-1 *mutants, the wild type reference strain ('California', CA), and transgenic nematodes over-expressing* eud-1 *by an extrachromosomal array (Ex[*eud-1*]). The dosage-dependence of EUD-1, encoded by an X-linked gene, is also reflected in the phenotypes of wild-type and transgenic males; B: The EUD-1 switch acts in other strains of* P. pacificus, *as shown by the phenotypes of wild-type and transgenic lines of the highly St strain RS5200B. Moreover, despite divergence of its regulation, the EUD-1 switch is present in other* Pristionchus *species: in contrast to the phenotype*

had arisen in a lineage including Diplogastridae, specifically since the divergence of *P. pacificus* from non-diplogastrid nematodes (Ragsdale *et al.*, 2013b). Duplications of an ancestral *sul-2* gene have thus made EUD-1 disposable for a specialised function in nematodes with stomatal dimorphism. One of the duplications, which resulted in two closely linked loci on the X chromosome, occurred in some ancestor within *Pristionchus* since the split of *P. pacificus* + *P. exspectatus* from the lineage of *P. elegans* (Fig. 11.3; see Ragsdale *et al.*, Chapter 4, this volume). However, the more ancient duplication between an autosomal copy (*sul-2.1*) and the X-linked copy (*sul-2.2*) was presumably the more critical event, as the presence of *eud-1* on the X chromosome is key to its dosage-dependent function. Functional tests of homologous loci in other species, including *C. elegans*, will enable a precise reconstruction of how the switch mechanism was co-opted from an ancestral sulfatase.

Epistasis tests have placed the EUD-1 switch into the developmental hierarchy of mouth form determination (Fig. 11.6). The switch acts downstream of, or in parallel to, other known regulators of the plasticity, including pheromone and hormone (Δ7-DA) signalling (Ragsdale *et al.*, 2013b). Given the similar developmental logic for dauer and mouth form regulation (Bento *et al.*, 2010), EUD-1 represents a terminal addition, or the modification of a developmental pathway downstream of other factors. Considering that evolution of known developmental pathways occurs mainly by tinkering with upstream components (Wilkins, 2002), this surprising result showed a way for a pre-existing pathway to be adapted to a novel function. Like the transcription factor DAF-16, a EUD-1 switch that operates downstream would in principle be regulated and evolve independently of other life-history traits.

A high-priority objective in the exploration of this uncharted and possibly new signalling pathway is to identify the ultimate target of EUD-1 function. Although the relevant substrate is unknown, product inhibition experiments support the enzymatic activity of EUD-1. Expression of EUD-1 in neurons suggests that the sulfatase might act in an unknown neuroendocrine signalling pathway. Possible mechanisms are still spec-

of interspecific hybrids of P. pacificus *and* P. exspectatus *(*Ppa/Pex*), hybrids over-expressing* eud-1 *(*Ppa/Pex *Ex[*eud-1*]) were saturated for the Eu form. The operation of EUD-1 in phenotypically divergent strains and species suggests the role of the switch in shaping patterns of micro- and macroevolution. A, B: Data from Ragsdale* et al. *(2013b).*

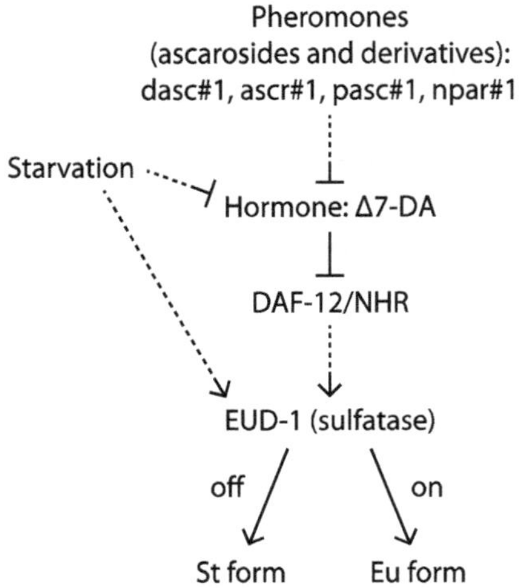

Fig. 11.6. *A regulatory model for the mouth dimorphism of* Pristionchus pacificus. *Environmental cues of starvation and crowding promote development of the Eu form through a steroid-hormone signalling module followed by a developmental switch executed by EUD-1.*

ulative, so how EUD-1 assumed a new function to regulate a new phenotype remains an outstanding problem for our understanding of the mouth dimorphism. Another problem is how exactly pheromones and hormone signalling are coupled to EUD-1. Further analysis of factors acting upstream of EUD-1 and a screen for suppressor mutants are needed to answer these questions.

The role of developmental plasticity in evolution

Studies of the mouth dimorphism have been largely limited to a single species. With a robust phylogeny to support a comparative approach, they can be expanded into a macroevolutionary context, allowing us to return to the question posed at the beginning of this chapter: what is the role, if any, of developmental plasticity in generating morphological complexity and diversity? Because developmental plasticity is the direct interaction between gene products and environmental inputs, it has been proposed as a facilitator of evolutionary novelty and rapid diversification (Brakefield *et al.*, 1998; Pigliucci, 2001; Schlichting, 2003; West-Eberhard, 2003; Suzuki & Nijhout, 2006; Moczek *et al.*, 2011). Developmental plasticity gives a genotype flexibility to explore fitness landscapes in response to different environmental conditions (Waddington, 1953). As a consequence, plasticity might accelerate the adaptive responses otherwise limited to the random accumulation of favourable genetic mutations. Tests of this hypothesis at a genetic level, which may be ultimately feasible in Diplogastridae, await mechanistic details for other species. However, a comparative analysis of the mouth dimorphism in general is already possible.

Objective phylogenetic tests showed the mouth plasticity to be associated with a burst of evolutionary diversification (Susoy *et al.*, 2015). First, the origin of the dimorphism coincided with the appearance of a suite of novelties. These novelties include an opposing tooth, bilateral asymmetry and possibly even the dorsal tooth itself (Fig. 11.3). A radiation of diverse forms then ensued (Fig. 11.7). Those lineages that retained the dimorphism had stomata that were significantly more complex, or host to more observable structures, than those that lost the dimorphism. However, when morphological differences are quantified by geometric morphometrics as differences in shape and size, a different pattern emerges. Evolutionary rates of change increased with the origin of the mouth dimorphism, but, following this 'pulse' of plasticity, lineages that subsequently lost the dimorphism were shown to evolve even faster than either dimorphic lineages or outgroup lineages with no dimorphism in their history. These results suggest a specific role for plasticity in tempo and mode of evolution. A polyphenism first facilitates the addition and maintenance of complex morphologies, but, after developmental character release or the elimination of pleiotropy for multiple phenotypes, evolutionary tempo can accelerate even further. In this model, plasticity provides the impetus to increase the degrees of freedom that can be selected upon to allow rapid diversification. Such a model was readily testable in a system with a discrete dimorphism. It is unclear whether similar principles also apply on a smaller scale to continuous plasticity, which is widespread in animals and plants, but if so, this model could reflect a general means for rapid evolutionary change.

Conclusions

Empirical research on the *P. pacificus* mouth dimorphism is still young but the tractability of the system is allowing rapid advances.

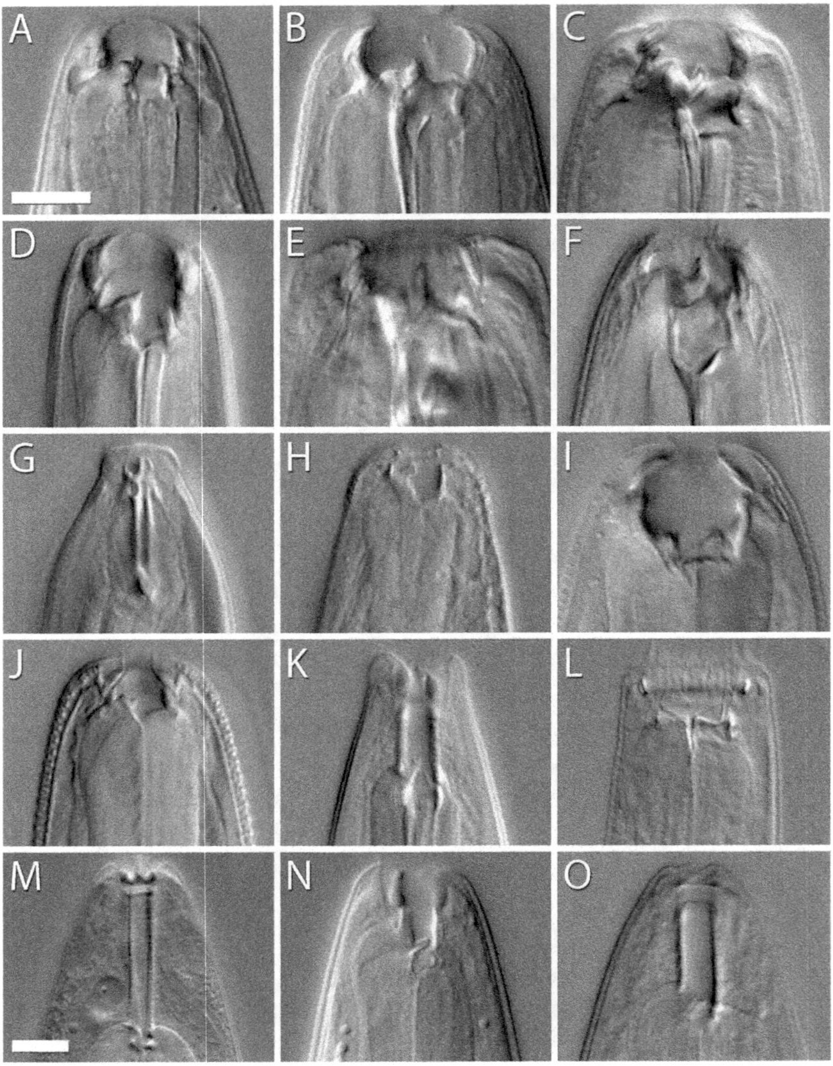

Fig. 11.7. *Correlation of mouth dimorphism and the diversity of stomatal form in Diplogastridae. A-F: Eu form of dimorphic species; G-O: Monomorphic species. Dimorphic species have significantly more complex mouthparts, as shown by the number of observable stomatal structures, than do monomorphic species. However, lineages that have secondarily lost the mouth dimorphism show a greater range of stomatal shape and size, suggesting that diversification*

Research on the dimorphism has answered questions at several levels of biological organisation, including morphology, feeding ecology, adaptive value, evolutionary history, and regulation by pheromones, hormones and a developmental switch. The strength of the system for understanding plasticity comes from the possibility of a multifaceted approach: *i*) multiple dimorphic species are available for a historical context and comparative analyses; *ii*) hundreds of *P. pacificus* strains allow inferences and tests of microevolution; *iii*) a short life-cycle has made direct assays of plastic responses and fitness advantages practical; and *iv*) advanced genetics tools permit functional tests that are still difficult in many animals with polyphenisms. It is particularly this latter feature, the promise of a genetic understanding, which opens an exciting frontier for ecological evo-devo: in the *P. pacificus* system we can hope to reveal mechanisms connecting species interactions to pheromones and neural pathways, hormones and developmental cascades, and epigenetics. Put into a comparative framework, these mechanistic details can give new life to the old riddles of evolutionary novelty and the origin of diversity.

Acknowledgements

I thank Vladislav Susoy, Matthias Herrmann and Natsumi Kanzaki for providing cultures or specimens of several diplogastrid species that are pictured in Figure 11.7. I also thank an external expert for review of the manuscript.

in Diplogastridae has proceeded by both gain and loss of developmental plasticity. Species pictured are in the tree in Figure 11.3 and were isolated as previously described (Susoy et al.*, 2015). All images except that in (M) are at same scale; scale bars = 10 μm. Dorsal is left. A:* Pristionchus pacificus*; B:* Parapristionchus giblindavisi*; C:* Micoletzkya *sp.; D:* Diplogasteriana *sp.; E:* Allodiplogaster sudhausi*; F:* Mononchoides *sp. RS5441; G:* Tylopharynx foetida*; H:* Eudiplogasterium levidentum*; I:* Paroigolaimella micrura*; J:* Sachsia zurstrasseni*; K:* Sudhausia aristotokia*; L:* Oigolaimella *sp.; M:* Rhabditolaimus *sp. RSA134; N:* Levipalatum texanum*; O:* Rhabditidoides *sp. RS5443.*

References

ANTEBI, A., YEH, W.-H., TAIT, D., HEGECOCK, E.M. & RIDDLE, D.L. (2000). *daf-12* encodes a nuclear receptor that regulates the dauer diapause and developmental age in *C. elegans*. *Genes & Development* 14, 1512-1527.

BALDWIN, J.G., GIBLIN-DAVIS, R.M., EDDLEMAN, C.D., WILLIAMS, D.S., VIDA, J.T. & THOMAS, W.K. (1997). The buccal capsule of *Aduncospiculum halicti* (Nemata: Diplogasterina): an ultrastructural and molecular phylogenetic study. *Canadian Journal of Zoology* 75, 407-423.

BENTO, G., OGAWA, A. & SOMMER, R.J. (2010). Co-option of the endocrine signaling module Dafachronic Acid-DAF-12 in nematode evolution. *Nature* 466, 494-497.

BONDURIANSKY, R., MAKLAKOV, A., ZAJITSCHEK, F. & BROOKS, R. (2008). Sexual selection, sexual conflict and the evolution of ageing and life span. *Functional Ecology* 22, 443-453.

BONGERS, T. (1999). The Maturity Index, the evolution of nematode life history traits, adaptive radiation, and cp-scaling. *Plant and Soil* 212, 13-22.

BOSE, N., OGAWA, A., VON REUSS, S.H., YIM, J.J., RAGSDALE, E.J., SOMMER, R.J. & SCHRODER, F.C. (2012). Complex small molecular architectures regulate phenotypic plasticity in a nematode. *Angewandte Chemie* 51, 12438-12443.

BRAKEFIELD, P.M., KESBEKE, F. & KOCH, P.B. (1998). The regulation of phenotypic plasticity of eyespots in the butterfly *Bicyclus anynana*. *American Naturalist* 152, 853-860.

BUMBARGER, D.J., CRUM, J., ELLISMAN, M.H. & BALDWIN, J.G. (2006). Three-dimensional reconstruction of the nose epidermal cells in the microbial feeding nematode, *Acrobeles complexus* (Nematoda: Rhabditida). *Journal of Morphology* 267, 1257-1272.

BUTCHER, R.A., FUJITA, M., SCHROEDER, F.C. & CLARDY, J. (2007). Small-molecule pheromones that control dauer development in *Caenorhabditis elegans*. *Nature Chemical Biology* 3, 420-422.

DE LEY, P. & BLAXTER, M.L. (2002). Systematic position and phylogeny. In: Lee, D.L. (Ed.). *The biology of nematodes*. London, UK, Taylor & Francis, pp. 1-30.

DE LEY, P., VAN DE VELDE, M.C., MOUNPORT, D., BAUJARD, P. & COOMANS, A. (1995). Ultrastructure of the stoma in Cephalobidae, Panagrolaimidae, and Rhabditidae, with a proposal for a revised stoma terminology in Rhabditida (Nematoda). *Nematologica* 41, 153-182.

EIZINGER, A. & SOMMER, R.J. (1997). The homeotic gene *lin-39* and the evolution of nematode epidermal cell fates. *Science* 278, 452-455.

FÜRST VON LIEVEN, A. & SUDHAUS, W. (2000). Comparative and functional morphology of the buccal cavity of Diplogastrina (Nematoda) and a first outline of the phylogeny of this taxon. *Journal of Zoological Systematics and Evolutionary Research* 38, 37-63.

GIBLIN, R.M. & KAYA, H.K. (1984). *Aduncospiculum halicti* n. gen., n. sp. (Diplogasterida: Diplogasteroididae), an associate of bees in the genus *Halictus* (Hymenoptera: Halictidae). *Revue de Nématologie* 7, 189-197.

GIBLIN-DAVIS, R.M., KANZAKI, N., DE LEY, P., WILLIAMS, D.S., SCHIERENBERG, E., RAGSDALE, E.J., ZENG, Y. & CENTER, B.J. (2010). Ultrastructure and life history of *Myolaimus byersi* n. sp. (Myolaimina: Myolaimidae); a phoretic associate of the crane fly *Limonia schwarzi* (Alexander) (Limoniidae) in Florida. *Nematology* 12, 519-542.

HERRMANN, M., MAYER, E.W. & SOMMER, R.J. (2006). Nematodes of the genus *Pristionchus* are closely associated with scarab beetles and the Colorado potato beetle in western Europe. *Zoology* 109, 96-108.

HERRMANN, M., MAYER, W., HONG, R., KIENLE, S., MINASAKI, R. & SOMMER, R.J. (2007). The nematode *Pristionchus pacificus* (Nematoda: Diplogastridae) is associated with the Oriental beetle *Exomala orientalis* (Coleoptera: Scarabaeidae) in Japan. *Zoological Science* 24, 883-889.

HIRSCHMANN, H. (1951). Über das Vorkommen zweier Mundhöhlentypen bei *Diplogaster lheritieri* Maupas und *Diplogaster biformis* n. sp. und die Entstehung dieser hermaphroditischen Art aus *Diplogaster lheritieri*. *Zoologische Jahrbücher, Abteilung für Systematik* 80, 132-170.

HONG, R.L., WITTE, H. & SOMMER, R.J. (2008). Natural variation in *P. pacificus* insect pheromone attraction involves the protein kinase EGL-4. *Proceedings of the National Academy of Sciences of the United States of America* 105, 7779-7784.

JOHNSON, C.L. & SPENCE, A.M. (2011). Epigenetic licensing of germline gene expression by maternal RNA in *C. elegans*. *Science* 333, 1311-1314.

KANZAKI, N., RAGSDALE, E.J., HERRMANN, M., MAYER, W.E., & SOMMER, R.J. (2012a). Description of three *Pristionchus* species (Nematoda: Diplogastridae) from Japan that form a cryptic species complex with the model organism *P. pacificus*. *Zoological Science* 29, 403-417.

KANZAKI, N., RAGSDALE, E., HERRMANN, M. & SOMMER, R.J. (2012b). Two new species of *Pristionchus* (Rhabditida: Diplogastridae) *P. fissidentatus* n. sp. from Nepal and La Réunion Island and *P. elegans* n. sp. from Japan. *Journal of Nematology* 44, 80-90.

KANZAKI, N., RAGSDALE, E., HERRMANN, M., MAYER, W.E., TANAKA, R. & SOMMER, R.J. (2012c). *Parapristionchus giblindavisi* n. gen., n. sp. (Rhabditida: Diplogastridae) isolated from stag beetles (Coleoptera: Lucanidae) in Japan. *Nematology* 14, 933-947.

KANZAKI, N., RAGSDALE, E., HERRMANN, M., RÖSELER, W. & SOMMER, R.J. (2013a). Two new species of *Pristionchus* (Nematoda: Diplogastridae) support the biogeographic importance of Japan for the evolution of the genus

Pristionchus and the model system *P. pacificus*. *Zoological Science* 30, 680-692.

KANZAKI, N., RAGSDALE, E., HERRMANN, M., SUSOY, V. & SOMMER, R.J. (2013b). Two androdioecious and one dioecious new species of *Pristionchus* (Nematoda: Diplogastridae): new reference points for the evolution of reproductive mode. *Journal of Nematology* 45, 172-194.

KANZAKI, N., RAGSDALE, E., HERRMANN, M., RÖSELER, W. & SOMMER, R.J. (2013c). *Pristionchus bucculentus* n. sp. (Rhabditida: Diplogastridae) isolated from a shining mushroom beetle (Coleoptera: Scaphidiidae) in Hokkaido, Japan. *Journal of Nematology* 45, 78-86.

KANZAKI, N., GIBLIN-DAVIS, R.M., YE, W., HERRE, E.A. & CENTER, B.J. (2013d). Description of *Parasitodiplogaster pharmaconema* n. sp. and redescription of *P. maxinema* from *Ficus maxima* Mill. (Moraceae). *Nematology* 15, 957-974.

KANZAKI, N., RAGSDALE, E., HERRMANN, M. & SOMMER, R.J. (2014). Two new and two recharacterized species resulting from a radiation of *Pristionchus* (Nematoda: Diplogastridae) in Europe. *Journal of Nematology* 46, 60-74.

KIONTKE, K. & FITCH, D.H.A. (2010). Phenotypic plasticity: different teeth for different feasts. *Current Biology* 20, R710-R712.

KÖRNER, H. (1954). Die Nematodenfauna des vergehenden Holzes und ihre Beziehungen zu den Insekten. *Zoologische Jahrbücher, Abteilung für Systematik* 82, 245-353.

KROETZ, S.M., SRINIVASAN, J., YAGHOOBIAN, J., STERNBERG, P.W. & HONG, R.L. (2012). The cGMP signaling pathway affects feeding behavior in the necromenic nematode *Pristionchus pacificus*. *PLoS ONE* 7, e34464.

L'ETOILE, N.D., COBURN, C.M., EASTHAM, J., KISTLER, A., GALLEGOS, G. & BARGMANN, C.I. (2002). The cyclic GMP-dependent protein kinase EGL-4 regulates olfactory adaptation in *C. elegans*. *Neuron* 36, 1079-1089.

MAYER, W., HERRMANN, M. & SOMMER, R.J. (2007). Phylogeny of the nematode genus *Pristionchus* and implications for biodiversity, biogeography and the evolution of hermaphroditism. *BMC Evolutionary Biology* 7, 104.

MAYER, W., HERRMANN, M. & SOMMER, R.J. (2009). Molecular phylogeny of beetle associated diplogastrid nematodes suggests host switching rather than nematode-beetle coevolution. *BMC Evolutionary Biology* 9, 212.

MCGAUGHRAN, A., MORGAN, K. & SOMMER, R.J. (2014). *Post hoc* association analyses in a hermaphroditic nematode support significant genotype-environmental interactions. *PLoS ONE* 9, e87317.

MOCZEK, A.P., SULTAN, S., FOSTER, S., LEDÓN-RETTIG, C., DWORKIN, I., NIJHOUT, F., ABOUHEIF, E. & PFENNIG, D.W. (2011). The role of developmental plasticity in evolutionary innovation. *Proceedings of the Royal Society B: Biological Sciences* 278, 2705-2713.

MORAN, N.A. (1992). The evolutionary maintenance of alternative phenotypes. *American Naturalist* 139, 971-989.

OGAWA, A., STREIT, A., ANTEBI, A. & SOMMER, R.J. (2009). A conserved endocrine mechanism controls the formation of dauer and infective larvae in nematodes. *Current Biology* 19, 67-71.

OGAWA, A., BENTO, G., BARTELMES, G., DIETERICH, C. & SOMMER, R.J. (2011). *Pristionchus pacificus daf-16* is essential for dauer formation but dispensable for mouth form dimorphism. *Development* 138, 1281-1284.

PHILIPPI, T. & SEGER, J. (1989). Hedging one's evolutionary bets, revisited. *Trends in Ecology & Evolution* 4, 41-44.

PIGLIUCCI, M. (2001). *Phenotypic plasticity: beyond nature and nurture*. Baltimore, MD, USA, Johns Hopkins University Press.

PORTEREIKO, M.F. & MANGO, S.E. (2001). Early morphogenesis of the *Caenorhabditis elegans* pharynx. *Developmental Biology* 233, 482-494.

POTTS, F.A. (1910). Notes on the free-living nematodes. *Quarterly Journal of Microscopical Science* 55, 433-485.

RAE, R., RIEBESELL, M., DINKELACKER, I., WANG, Q., HERRMANN, M., WELLER, A.M., DIETERICH, C. & SOMMER, R.J. (2008). Isolation of naturally associated bacteria of necromenic *Pristionchus* nematodes and fitness consequences. *Journal of Experimental Biology* 211, 1927-1936.

RAGSDALE, E.J. & BALDWIN, J.G. (2010). Reconciling phylogenetic incongruence to articulate homology and phenotypic evolution: a case study from Nematoda. *Proceedings of the Royal Society B: Biological Sciences* 277, 1299-1307.

RAGSDALE, E.J., CRUM, J., ELLISMAN, M.H. & BALDWIN, J.G. (2008). Three-dimensional reconstruction of the stomatostylet and anterior epidermis in the nematode *Aphelenchus avenae* (Nematoda: Aphelenchidae) with implications for the evolution of plant parasitism. *Journal of Morphology* 269, 1181-1196.

RAGSDALE, E.J., NGO, P.T., CRUM, J., ELLISMAN, M.H. & BALDWIN, J.G. (2011). Reconstruction of the pharyngeal corpus of *Aphelenchus avenae* (Nematoda: Tylenchomorpha), with implications for phylogenetic congruence. *Zoological Journal of the Linnean Society* 161, 1-30.

RAGSDALE, E.J., KANZAKI, N., ROESELER, W., HERRMANN, M. & SOMMER, R.J. (2013a). Three new species of *Pristionchus* (Nematoda: Diplogastridae) show morphological divergence through evolutionary intermediates of a novel feeding-structure polymorphism. *Zoological Journal of the Linnean Society* 168, 671-698.

RAGSDALE, E.J., MÜLLER, M.R., RÖDELSPERGER, C. & SOMMER, R.J. (2013b). A develomental switch coupled to the evolution of plasticity acts through a sulfatase. *Cell* 155, 922-933.

RECHAVI, O., MINEVICH, G. & HOBERT, O. (2011). Transgenerational inheritance of an acquired small RNA-based antiviral response in *C. elegans*. *Cell* 147, 1248-1256.

RÜHM, W. (1956). Die Nematoden der Ipiden. *Parasitologische Schriftenreihe (Jena)* 6, 1-437.

SACHS, H.G. (1950). Die Nematodenfauna der Rinderexkremente. *Zoologische Jahrbücher, Abteilung für Systematik* 79, 209-272.

SCHLICHTING, C.D. (2003). Origins of differentiation via phenotypic plasticity. *Evolution & Development* 5, 98-105.

SEROBYAN, V., RAGSDALE, E.J., MÜLLER, M.R. & SOMMER, R.J. (2013). Feeding plasticity in the nematode *Pristionchus pacificus* is influenced by gender and social context and is linked to developmental speed. *Evolution & Development* 15, 161-170.

SEROBYAN, V., RAGSDALE, E.J. & SOMMER, R.J. (2014). Adaptive value of a predatory mouth form in a dimorphic nematode. *Proceedings of the Royal Society B: Biological Sciences* 281, 20141334.

SHEN, Y.D., WOLLAM, J., MAGNER, D., KARALAY, O. & ANTEBI, A. (2012). A steroid receptor-microRNA switch regulates life span in response to signals from the gonad. *Science* 338, 1472-1476.

SHINE, R. (1989). Ecological causes for the evolution of sexual dimorphism: a review of the evidence. *Quarterly Review of Biology* 64, 419-461.

SOMMER, R.J. & OGAWA, A. (2011). Hormone signaling and phenotypic plasticity in nematode development and evolution. *Current Biology* 21, R758-R766.

SOMMER, R.J., CARTA, L.K., KIM, S.-Y. & STERNBERG, P.W. (1996). Morphological, genetic and molecular description of *Pristionchus pacificus* sp. n. (Nematoda, Diplogastridae). *Fundamental and Applied Nematology* 19, 511-521.

STERNBERG, P.W. (1988). Lateral inhibition during vulval induction in *Caenorhabditis elegans*. *Nature* 336, 551-554.

SUDHAUS, W. & FÜRST VON LIEVEN, A. (2003). A phylogenetic classification and catalogue of the Diplogastridae (Secernentea, Nematoda). *Journal of Nematode Morphology and Systematics* 6, 43-90.

SULSTON, J.E. & HORVITZ, H.R. (1977). Post-embryonic cell lineages of the nematode *Caenorhabditis elegans*. *Developmental Biology* 82, 41-55.

SUSOY, V., KANZAKI, N. & HERRMANN, M. (2013). Description of the bark beetle associated nematodes *Micoletzkya masseyi* n. sp. and *M. japonica* n. sp. (Nematoda: Diplogastridae). *Nematology* 15, 213-231.

SUSOY, V., RAGSDALE, E.J., KANZAKI, N. & SOMMER, R.J. (2015). Rapid diversification of a macroevolutionary pulse of developmental plasticity. *eLife*, in press.

SUZUKI, Y. & NIJHOUT, H.F. (2006). Evolution of a polyphenism by genetic accommodation. *Science* 311, 650-652.

WADDINGTON, C.H. (1953). Genetic assimilation of an acquired character. *Evolution* 7, 118-126.

WELLER, A., MAYER, W., RAE, R. & SOMMER, R.J. (2010). Quantitative assessment of the nematode fauna present on *Geotrupes* dung beetles reveals species-rich communities with a heterogenous distribution. *Journal of Parasitology* 96, 525-531.

WEST-EBERHARD, M.J. (2003). *Developmental plasticity and evolution*. Oxford, UK, Oxford University Press.

WHITE, J. (1988). The anatomy. In: Wood, W.B. (Ed.). *The nematode* Caenorhabditis elegans. Cold Spring Harbor, NY, USA, Cold Spring Harbor Laboratory Press, pp. 81-122.

WILKINS, A.S. (2002). *The evolution of developmental pathways*. Sunderland, MA, USA, Sinauer.

WRIGHT, K.A. & THOMSON, J.N. (1981). The buccal capsule of *Caenorhabditis elegans* (Nematoda: Rhabditoidea): an ultrastructural study. *Canadian Journal of Zoology* 59, 1952-1961.

YOU, Y.J., KIM, J., RAIZEN, D.M. & AVERY, L. (2008). Insulin, cGMP, and TGF-beta signals regulate food intake and quiescence in *C. elegans*: a model for satiety. *Cell Metabolism* 7, 249-257.

Nematology Monographs & Perspectives, 2015, Vol. 11, 331-352

Chapter 12

Pristionchus pacificus olfaction

Ray L. HONG

Biology Department, California State University,
Northridge, CA 91330, USA
ray.hong@csun.edu

Introduction

Olfaction allows animals to detect chemical molecules from the environment from a distance and is probably the most important sensory modality in nematodes. Unlike gustation, which detects water-soluble chemicals by direct contact, olfactory cues allow nematodes to sense minute amounts of chemicals associated with nutrients, danger and potential hosts before engaging in relevant chemotaxis. Because of their similar culturing requirements in the laboratory, it was presumed that much of the superficial anatomical resemblances between the model systems *Caenorhabditis elegans* and *Pristionchus pacificus* entailed similar odour preference profiles. It took almost a decade after the adoption of *P. pacificus* as a model system and the discovery of a species-specific association between *Pristionchus* species and beetles before a systematic survey of the olfactory preferences revealed not only strong differences in the types of molecules and direction of responses among *Pristionchus* species, but also diametrically opposed odour profiles between *P. pacificus* and *C. elegans* (Hong & Sommer, 2006). Ensuing genetic studies identified a highly conserved protein kinase involved in the natural variation for an insect sex pheromone, but many more genes and genetic mutants need to be identified and characterised before it is possible to estimate the level of conservation between the two nematode species. Below, I review recent advances and highlight gaps in our understanding of the molecular mechanisms of olfaction in these two nematode models.

Olfaction in *C. elegans* and *P. pacificus*

Investigations into the genetics of olfaction in *C. elegans* began in earnest when Bargmann and coworkers established the basic odour palate using the population chemotaxis assay (Bargmann *et al.*, 1993). Young adult worms were washed three times with water or M9 buffer to remove *Escherichia coli* food and placed onto large 10 cm diam. assay plates made of NGM (Nematode Growth Medium) agar without tryptone (Fig. 12.1A). The wild-type *C. elegans* N2 strain was tested for attraction to low molecular weight, commercially available compounds such as 2,3-butanedione (diacetyl), 2,3-pentanedione, benzaldehyde, pyrazine, isoamyl alcohol, 2-butanone and 3,4,5-trimethylthiazole (Fig. 12.1B). Because most chemicals were diluted in ethanol, 100% ethanol was used as the counter-attractant. After 1 h, the number of worms on the attractant and counter-attractant areas were scored. The result for each assay with 100-200 worms was then expressed as a chemotaxis

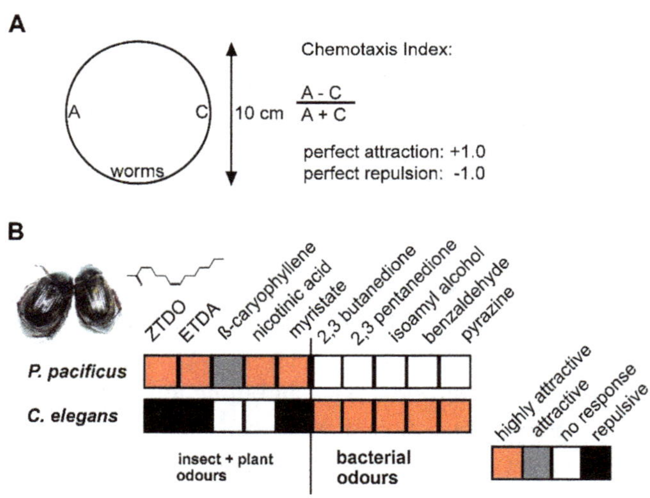

Fig. 12.1. Pristionchus *chemosensation. A: Chemotaxis assays are conducted on 10 cm NGM plates without tryptone. Washed worms are placed on a spot equidistant from the attractant and counter-attractant. Sodium azide-immobilised worms are scored after most of the worms have dispersed from the loading site; B: The Oriental beetle,* Exomala orientalis, *is a host for* P. pacificus *in Japan;* P. pacificus *is particularly attracted to insect pheromones (ZTDO, ETDA, methyl myristate) and plant volatiles (β-caryophyllene and nicotinic acid) whilst* C. elegans *is primarily attracted to bacterial catabolites.*

index (CI) that can vary from +1.0 (perfect attraction) to −1.0 (perfect repulsion), and is expressed as an average CI for a given compound and dilution after >10 replicates. Out of the 121 compounds tested on *C. elegans*, 2,3-butanedione emerged as one of the strongest attractants that worms responded to over a 1 million-fold dilution range. The other six compounds also showed CI of >0.6 over a thousand-fold dynamic range. As a necromenic nematode with species-specific associations with different species of beetles throughout the planet, one would expect *P. pacificus* to share some, but not all, of the same odour preferences as *C. elegans*. It still came as a surprise then, that only two out of the seven *C. elegans* strong attractants were weakly attractive to *P. pacificus* (CI > 0.6 for 2,3-butanedione and 2,3-pentanedione), and only at the highest undiluted concentration.

The lack of dynamic range and strong attractants prompted Hong and coworkers to test the *P. pacificus* odour palate using semiochemicals known to have communicative roles for inter-species and intra-species communication. Out of the 45 structurally diverse compounds tested, several plant- and insect-associated compounds were found to be weak to strong attractants (Hong & Sommer, 2006). As it was becoming increasingly evident that the wildtype strain California (PS312) and the mapping reference strain Washington (PS1843) behave very differently, characterisation of odour responses in *P. pacificus* was carried out routinely with both strains. β-caryophyllene, a sesquinterpene volatile compound released by many plants when attacked by herbivorous insects, was one of the most attractive compounds for both *P. pacificus* strains (CI ~ 0.7 at 50% dilution). While both California and Washington strains are attracted equally to β-caryophyllene, these two strains differ drastically in their responses to the insect-associated compounds. The most attractive known compound for *P. pacificus* is the sex pheromone for the leafworm moth, *Spodoptera litura*, *E*-11-tetradecenyl acetate (ETDA), but only for the Washington strain (CI > 0.8) over a ten-fold concentration range (Fig. 12.2) (Hong & Sommer, 2006). Another strong attractant is the sex pheromone for the corn earworm moth, *Helicoverpa zea*, *Z*-11-hexadecenal (ZHDA). However, the ecological significance of ETDA and ZHDA attraction in *P. pacificus* populations is unknown because the nematode load on lepidopterans has not been investigated. The prospect of *P. pacificus* being able to infect divergent insect hosts is bolstered by finding that another diplogastrid, *Chroniodiplogaster aerivora*, has been found to infect the juveniles of both scarab beetles

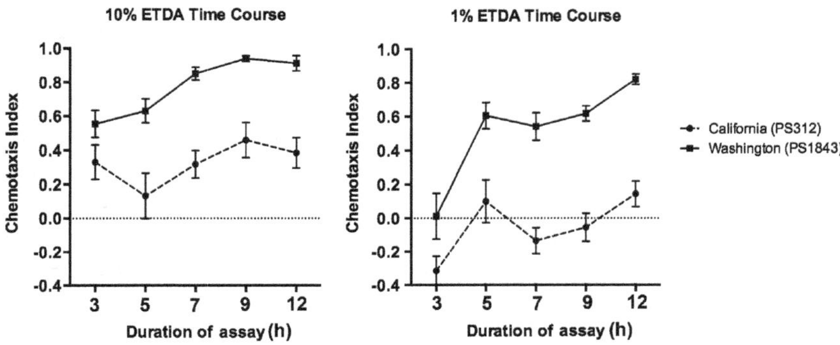

Fig. 12.2. *Time course of mean chemoattraction responses in two strains of* Pristionchus pacificus *to a moth pheromone,* E-*11-tetradecenyl acetate (ETDA). A sampling of five time points over a 24-h period shows clear attraction of the Washington strain (solid line) to ETDA compared to the insensitive reference strain California (dotted line). Error bars denote standard error of the means from six assay replicates.*

Phyllophaga spp. (Quebec) and *H. zea* (Arkansas) (Poprawski & Yule, 1990; Steinkraus *et al.*, 1993). In contrast to the orphaned pheromones ETDA and ZHDA that do not have a known insect host with *P. pacificus* strains, strong associations can be made between attraction to host pheromone and strain provenance of the pheromone in the Oriental beetle, *Exomala orientalis.* Following the discovery that *P. pacificus* populations are associated with the Oriental beetle, attraction to the beetle's sex pheromone Z-7-tetradecen-2-one (ZTDO) was found to correlate with strains isolated from Oriental beetles in Japan and northeastern USA (Fig. 12.1) (Hong *et al.*, 2008a). Hence, ZTDO is currently the most relevant host-derived compound for studying *P. pacificus* olfaction in the context of host ecology (Herrmann *et al.*, 2007).

The identification of the two insect pheromones with strong attraction, ETDA and ZTDO, enabled more in-depth descriptions of *P. pacificus* chemotaxis behaviour, particularly in comparison to *C. elegans* N2 chemotaxis. Several striking differences were found between *P. pacificus* and *C. elegans* toward their respective strongest attractants: *i*) *C. elegans* responds to 2,3-butanedione over a million-fold range, whereas *P. pacificus* reacts to ETDA only over a hundred-fold range; *ii*) whilst most *C. elegans* chemotaxis assays are complete after 90 min (*i.e.*, reaching their highest or lowest CI value), *P. pacificus* reach their highest CI value for the insect pheromones between 9-24 h. Part of this significantly slower chemotaxis is due to *P. pacificus* locomotion behaviour, such as slower forward velocity and higher reversal frequency (Kroetz *et al.*, 2012). However, slower locomotion off-food is not a sufficient explanation since *P. pacificus* reach their highest CI value between 3-4 h for all plant volatiles. These dichotomous odour responses between insect- and plant-derived compounds suggest different signalling pathways not observed in *C. elegans*; *iii*) *P. pacificus* does not seem to show odour adaptation after the 9 h duration in the time course analyses performed on ETDA and ZTDO; *iv*) whereas responses to most odours are similar between *C. elegans* N2 wild type and the mapping Hawaiian (CB4856) strain, the two *P. pacificus* strains showed a strong variation in response to the insect sex pheromones ETDA and ZTDO. The Washington allele is dominant for the attraction to ETDA, since F1 progeny between California and Washington are also attracted to ETDA. By contrast, the Washington allele is recessive for the attraction to the Oriental beetle pheromone ZTDO. Thus, the factors involved in the natural variation for insect pheromone preference are genetically distinct and may play an important role in the ability of *P. pacificus* populations to switch beetle hosts in diverse geographic locations (Hong *et al.*, 2008a); and *v*) a much smaller proportion of *P. pacificus* disperse during the chemotaxis assay (~60%) compared to *C. elegans* (95%), with also fewer worms ending up in the 1 cm diam. areas of the attractant and counter-attractant for scoring. Thus, several key differences in neurobiology contribute to the diametrically opposed olfactory profiles between *P. pacificus* and *C. elegans*, probably due to their distinct ecologies in insects and fruit composts, respectively.

Based on the response times for chemotaxis assays, *P. pacificus* seemed to have evolved two tempos of odour signalling, because only insect pheromone responses are found to be a major natural polymorphic trait sensitive to exogenous cGMP. One can consider the attraction to the plant volatile compounds to be fast (<3 h) and analogous to chemoattraction in *C. elegans* (1 h). This difference in chemotaxis response is due to *P. pacificus* requiring a longer time to reach odour sources because it is significantly slower in its forward velocity and reverses more often than *C. elegans* (Hong *et al.*, 2008a; Kroetz *et al.*, 2012). Equally important, *P. pacificus* responds to the sudden removal of food during chemotaxis assay by slowing down, whereas the same off-food condition speeds up *C. elegans* animals (Rivard *et al.*, 2010). By contrast, *P. pacificus* attraction to the molecularly larger insect pheromone compounds is slow (>9 h) and has no known corresponding chemotaxis behaviour in *C. elegans* since all worms disperse in less than 2 h. However, the expected lower odour volatility of the large ZTDO pheromone molecule is not the reason for slower chemotaxis response in *P. pacificus*, since *C. elegans* shows rapid avoidance behaviour to ZTDO within 1 h (Hong *et al.*, 2008a). Therefore, the slow, yet sustained, chemoattraction response to insect pheromones in *P. pacificus* may reflect a fundamentally different mechanism for odour signalling that requires the suppression of odour adaptation pathways through transcriptional regulation. In other words, insect pheromone sensing may require reduced ability to undergo odour adaptation.

Olfaction profiles reflect host preferences

The difference between *C. elegans* and *P. pacificus* extends beyond the species level. While *C. elegans* attractants such as 2,3-butanedione and isoamyl alcohol are attractive to the closest sister species *C. remanei* and *C. briggsae* (Hong & Sommer, 2006; Hong *et al.*, 2008a), *Pristionchus* species vary significantly in their responses to a panel of semiochemicals from their environment (Fig. 12.3). For example, while the plant volatile β-caryophyllene is attractive to all four *Pristionchus* species, another ubiquitous plant volatile, linalool, elicits both repulsive as well as attractive responses in the same genus. More strikingly, two *Pristionchus* species show strong preference for host-specific compounds. The unusual and caustic compound phenol is a known sex pheromone of the *Melolontha* cockchafer and is attractive only to *P. maupasi* (Ruther *et al.*, 2002), while Z-7-tetradecen-2-one (ZTDO) from the Oriental beetle is only attractive to *P. pacificus*. Such specificity is even more remarkable when one considers that only a subset of the 28 globally representative *P. pacificus* strains tested show attraction to ZTDO (Hong, 2008a and unpubl. data), suggesting that attraction to host pheromones can evolve quickly in global populations. It is thus possible, with just three odours such as linalool, phenol and ZTDO, to use chemotaxis responses to discriminate among *Pristionchus* species or even strains. Such exquisite olfaction profiles may stem from the need to find particular beetle hosts that have overlapping geographical ranges, as well as to discriminate between developmental stages of the host, such as

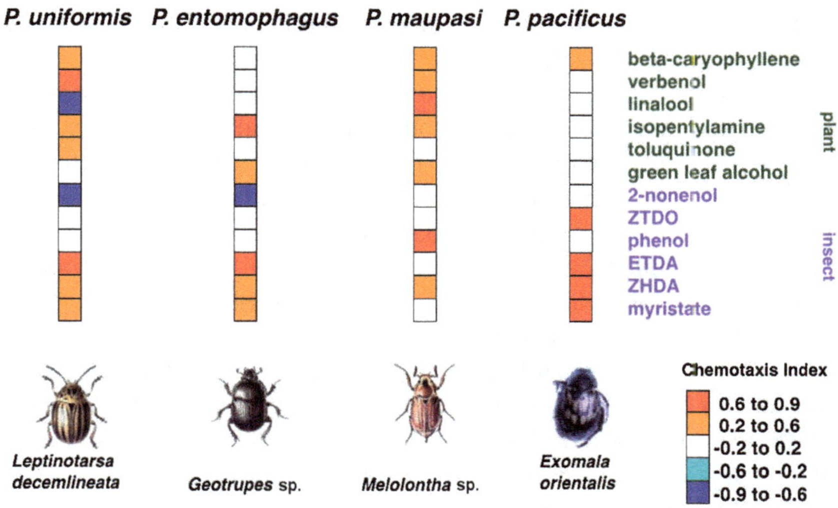

Fig. 12.3. *Distinct olfaction profiles among* Pristionchus *species. A heat map of chemoattraction or chemoavoidance responses of* Pristionchus *species toward a panel of representative plant and insect semiochemicals, including known beetle host sex pheromones, phenol from the May beetle,* Melolontha *sp. and Z-7-tetradecen-2-one (ZTDO) from the Oriental beetle,* Exomala orientalis.

ground-dwelling beetle grubs from flying adult beetles, the latter being better at disseminating *Pristionchus* nematodes to new locations.

Another approach used to identify ecologically important beetle host odours is to utilise whole body washes from beetles, rather than testing commercially available compounds. Nematodes, like most animals, encounter their environment through a bouquet of odours, some of which may be present in unique combinations depending on season, developmental state of the organisms, and density of both the nematode and host populations. Like all reductionist approaches, attempts to study complex, dynamic ecological systems from the wild in the laboratory require simplifying field conditions and generalising findings. With those caveats, efforts to identify complex host odours mediating the species-specific associations of *Pristionchus* nematodes to beetles are represented by two case studies in Germany and La Réunion Island.

The first study using complex odour mixtures was the identification of chemical cues involved in the association between *P. maupasi* and *Melolontha* cockchafers from Germany using gas chromatography and mass spectrometry (GC-MS), followed by chemotaxis assays in a reiterative fashion to obtain synergic compounds. GS-MS analysis of whole body cuticular washes of >250 *Melolontha* cockchafers in dichloromethane showed that the beetle's known sex pheromone, phenol, and the plant volatile compounds, linalool and green leaf alcohol, are weak attractants for *P. maupasi*. Phenol is attractive to *P. maupasi* over a thousand-fold dilution range, but is unattractive to *P. pacificus* and even repulsive to *P. entomophagus* and *P. uniformis* (Hong *et al.*, 2008b). A blend of diluted phenol with certain previously identified plant volatiles (Bargmann *et al.*, 1993; Reinecke *et al.*, 2002; Ruther & Hilker, 2003) can dramatically increase *P. maupasi* attraction. Similarly, the complex cuticular wash of recently eclosed female adults spiked with linalool is significantly more attractive than either the beetle or the plant compound alone. These data suggest that *P. maupasi* are more attracted to feeding, sexually receptive, adult beetles than to non-feeding juvenile beetles. This ability of *P. maupasi* to detect a mixture of host-relevant odours better than odours presented alone is likely to represent a common host-seeking strategy among *Pristionchus* species given the similarity of beetle habitats (phytophagous scarabs from temperate regions). However, so far, such synergetic attraction from two or more compounds has not been identified in *P. pacificus*. Such attraction to odour blends has also not been reported for *C. elegans*. Thus, identifying the genetic mechanisms important for *Pristionchus* nematodes to integrate multiple olfactory signals and execute concerted behavioural responses could be leveraged to understand a key aspect of nematode olfaction not represented in *C. elegans*.

A more extensive examination of organism-environment interactions between *P. pacificus* populations and several species of beetles from multiple sites on La Réunion Island was conducted to determine if the natural variation in *P. pacificus* response to whole beetle washes correlated with the beetle hosts from which the nematodes were isolated (McGaughran *et al.*, 2013; McGaughran & Morgan, Chapter 8, this volume). Both live beetles (*Hoplia retusa* and *Hoplochelus marginalis*) as well as whole beetle washes (*Adoretus* sp., *Oryctes borbonicus*, *Maladera affinis*, *H. retusa* and *H. marginalis*) were used as attractants in standard and multiple choice chemotaxis assays. This study also took into account the population structure, so that, of the 61 strains that were originally obtained from La Réunion beetles, only 21 *P. pacificus* strains from the five island beetle species belonging to the closely-related 'C' lineage were carefully chosen for testing (Morgan *et al.*,

2012). This island study confirmed that olfactory responses to organic compounds and host beetle washes are highly variable, even within this single population lineage, but these responses did not correspond to the beetle hosts from which the strains were isolated. Interestingly, the descendants of the founding *P. pacificus* strain found in the soil in Southern California 20 years ago still showed an attractive response to the body wash of *Hoplochelus* from La Réunion Island, even stronger than the two nematode strains from *Hoplochelus*. Nevertheless, cluster analysis show that nematode responses were more similar toward beetle host washes than toward any single organic compound (derived from beetles) among strains from the same host. As more potential beetle hosts are identified around the globe, using whole beetle washes is a straightforward first step toward determining the preferential olfactory responses of *Pristionchus* populations toward these mixtures, with the ultimate goal to match specific host odours with the genetic variants mediating these ubiquitous interactions.

Natural variation in the cGMP pathway

The whole beetle washes highlights a striking finding in *Pristionchus* olfaction: the rapid divergence of host beetle sex pheromone attraction in different *P. pacificus* populations. The reference strain from California, PS312, shows generally less attraction towards odours than the mapping strain from Washington, PS1843. This may be due to genetic drift in the laboratory in the absence of hosts, or to not yet having found the odours representing the potential host of the California strain (Oriental beetles are not found in California). The possibility that such divergent responses between two *P. pacificus* strains could represent an ecologically important natural variation in host preferences motivated the effort to identify a major-effect locus associated with host pheromone attraction. By mapping with recombinant inbred lines followed by locus confirmation using near isogenic lines, it was found that at least one locus involved in the natural variation between the California and the Washington isolates for attraction to the lepidopteran sex pheromone ETDA is a cGMP-dependent protein kinase, homologous to the *C. elegans* EGL-4 (Hong *et al.*, 2008a).

Surprisingly, the *Ppa*-EGL-4 protein sequences of both strains are 100% identical, suggesting that the difference in the alleles is probably due to changes in gene regulation. Quantitative PCR shows a distinctly higher *Ppa-egl-4* expression in the Washington isolate than the California isolate during the fourth-stage juvenile (J4) period, and introgression of the Washington *Ppa-egl-4* locus into the California background is sufficient to increase odour attraction. Furthermore, a brief exposure for 1 h to the cell-permeable 8-bromo-cGMP can strongly increase *Ppa-egl-4* transcription in the California strain, but such treatment has no effect on the already high *Ppa-egl-4* transcript level in the Washington strain. More importantly, the brief exposure to exogenous cGMP also dramatically increased not only California strain's attraction to ETDA, but also elicited positive responses in other ETDA insensitive *P. pacificus* strains from China and Madagascar. These China and Madagascar strains exhibited even less *Ppa-egl-4* expression than the California strain. This cGMP-dependent positive odour response is specific to cGMP, since using 8-bromo-cAMP had no effect. Immunostaining of the EGL-4 protein and the transcriptional reporter *Ppa-egl-4p::gfp* both show strongest expression in multiple head neurons similar to the expression pattern of *Cel-egl-4p::gfp* expression in *C. elegans* (Hong *et al.*, 2008a; Cinkornpumin & Hong, unpubl. data). However, because amphid neurons cannot be unambiguously assigned between *C. elegans* and *P. pacificus* neurons and neuron-specific transgenic markers, it is unclear whether or not *Ppa-egl-4* is expressed in the homologous AWC neurons. Taken together, the correlation between higher *Ppa-egl-4* expression and insect pheromone attraction in multiple *P. pacificus* strains, and the ability to upregulate *Ppa-egl-4* expression with a short, 1 h exposure to exogenous cGMP in low expressing strains such as California and China, strongly indicate the evolutionarily conserved role of a multi-functional regulator at both the micro- as well as the macro-evolutionary scale. The connection between olfaction and host preference may have originated with the need to coordinate food foraging and oviposition through a 'master' behavioural regulator such as EGL-4. Indeed, natural variations in the cGMP-dependent protein kinase expression level in the fruit fly *Drosophila melanogaster*, the western root worm *Diabrotica virgifera* and the honey bee *Apis mellifera* have all been found to be linked to foraging, suggesting that EGL-4 homologues have conserved function in regulating food-related behaviour through neural and physiological mechanisms across vast evolutionary trajectories (Osborne *et al.*, 1997; Ben-Shahar *et al.*, 2002; Garabagi *et al.*, 2008; Kaun & Sokolowski, 2009).

The natural variation in cGMP signalling prompts the next question: what is the evolutionary origin of insect pheromone reception in nematodes? The finding that a ZTDO-insensitive strain (California) can be rescued by exogenous cGMP treatment suggests the insensitive strain does not lack the proper ZTDO receptor, but rather is subdued in its ability to transduce the signal or to translate the odour stimulus into an attractive chemotaxis response. More intriguingly, a ZTDO receptor must also function in *C. elegans* because *C. elegans* shows strong avoidance behaviour towards ZTDO, such that ZTDO is also recognised by *C. elegans* but produces an avoidance response probably due to changes in its neuronal circuitry. Because general insect host cues, such as CO_2, elicit avoidance response in *C. elegans* non-dauers but attractive chemotaxis response in dauer juveniles (Hallem & Sternberg, 2008; Hallem *et al.*, 2011), the response of *C. elegans* dauers to ZTDO was also tested and found to be a repellent (Hong, unpubl. data). However, chemotaxis assays could not be performed on *P. pacificus* dauers because they fail to move on the assay plates. One primary reason for reduced dispersal of *P. pacificus* dauers may be due to the oily secretions known as nematoil (Penkov *et al.*, 2014; see Ogawa & Brown, Chapter 10, this volume). Therefore, a modified assay to accommodate the particular clumping behaviour of *P. pacificus* dauer juveniles will be necessary to determine if they respond to insect pheromones in the same way as non-dauers. The range of responses to the Oriental beetle sex pheromone between *P. pacificus* strains (neutral to attractive), as well as between nematode species such as *P. pacificus* and *C. elegans* (attractive to repulsive), strongly suggest that insect pheromone recognition preceded the divergence of *P. pacificus* and *C. elegans* and continues to be under strong selective pressure among different *P. pacificus* populations.

The identification of an EGL-4 homologue in *P. pacificus* provided an entry point for identifying upstream elements involved in odour recognition. In *C. elegans*, the loss-of-function *Cel-egl-4* alleles show larger body size, reduced fat storage, weaker odour adaptation, longer life-span and more roaming, as well as defects in dauer formation and egg-laying (Fujiwara *et al.*, 2002; L'Etoile *et al.*, 2002; Hirose *et al.*, 2003; Hong & Sommer, 2006). However, the pleiotropic nature of EGL-4 function makes it difficult to leverage this finding in a genetic screen for additional odour signalling components, *i.e.*, enhancers of a body size phenotype are unlikely to yield only additional components mediating EGL-4's role in olfaction. The only loss-of-function allele,

Ppa-egl-4(tu374), also overlaps with those phenotypes in *C. elegans* mutants but not necessarily in the same phenotypic direction. For example, while the null allele of *Ppa-egl-4* share with *Cel-egl-4* for reduced odour adaptation and increased roaming duration, the *Ppa-egl-4* mutant has a shorter body and increased fat storage similar to the gain-of-function alleles in *C. elegans* (Raizen *et al.*, 2006; Kroetz *et al.*, 2012). Thus, while the role of EGL-4 as a regulator is conserved, the direction of its functions can change more frequently. To look for regulators of cGMP-dependent ZTDO response, a behaviour-based genetic screen was carried out in the California background following exogenous cGMP treatment for mutants that are Oriental beetle pheromone insensitive (*obi*). One outcome of the screen was *Ppa-obi-1(tu404)*, a mutant that would not chemotaxis towards the beetle pheromone ZTDO after the 1 h cGMP treatment (Cinkornpumin *et al.*, 2014). Double mutant analysis of chemotaxis behaviour in *Ppa-egl-4(tu374)*, *Ppa-obi-1(tu404)* worms suggests that *Ppa*-OBI-1 acts upstream of *Ppa*-EGL-4 in its olfactory function. Rather than encoding for components of the G-protein signalling pathway such as G-alpha proteins or guanylate cyclases, as one would expect based on past *C. elegans* research, *Ppa-obi-1* was found to encode for a protein with lipid-binding motifs not previously connected with any aspects of olfaction (Cinkornpumin *et al.*, 2014).

ZTDO as a volatile attractant and developmental regulator

A possible function for *Ppa*-OBI-1 is indirect modulation of pheromone signals through unknown aspects of the cGMP pathway. Defects in cGMP-dependent ZTDO attraction and locomotion in *Ppa-obi-1(tu404)* mutants strongly implicate *Ppa*-OBI-1 as feeding into the cGMP signalling pathway. Because the intracellular cGMP level cannot be measured directly and instantaneously in the neurons, studies in *C. elegans* utilised both genetic and drug approaches to dissect the link between cGMP and EGL-4 activity. Specifically, simultaneous knockdown of three phosphodiesterases or treatments of worms with phosphodiesterase inhibitor 3-isobutyl-1-methylxanthine (IBMX) can increase cytoplasmic cGMP level and thereby block EGL-4 nuclear translocation (O'Halloran *et al.*, 2012). Through unknown mechanisms, interference with phosphodiesterase activity results in higher intracellular cGMP level, lower nuclear EGL-4 and increased odour adaptation. In *P. pacificus*, it is not known if *Ppa*-EGL-4 can also translocate to the nucleus following prolonged odour exposure, but up-regulation of *Ppa-egl-4* following exogenous cGMP treatment is consistent with the notion that the time needed for insect pheromone chemosensation in *P. pacificus* may involve transcriptional regulation of certain target genes.

Another model for *Ppa*-OBI-1 function is the direct mediation of pheromone binding. GFP expression studies of the *Ppa-obi-1* promoter indicate expression in diverse cell-type lineages, including several cells with known excretory function in *C. elegans*. *Ppa-obi-1* promoter is activated in the amphid sheath cells, seam and hypodermal cells, vulval cells and cells in the excretory duct system (Cinkornpumin *et al.*, 2014). In *Drosophila*, the extracellular odourant binding protein LUSH is found in the sensillar lymph that is analogous to the lumen formed by the amphid sheath cells (Kim *et al.*, 1998; Kim & Smith, 2001; Ruther *et al.*, 2002; Xu *et al.*, 2005; Gomez-Diaz *et al.*, 2013). LUSH recognises the aggregation and mating pheromone 11-*cis* vaccenyl acetate (VA) and transfers VA to the G-protein coupled receptors (GPCR) Or67d, along with the CD36-like SNMP on the neuronal membrane (Benton *et al.*, 2007; Jin *et al.*, 2008; Martin *et al.*, 2011). In lepidopterans such as the wild silk moth, *Antheraea polyphemus*, and the Cotton leafworm, *Spodoptera littoralis*, pheromone-degrading and odourant-degrading enzymes are esterases specifically expressed in the sensillar lymph of male antennae (Vogt & Riddiford, 1981; Maibeche-Coisne *et al.*, 2004; Ishida & Leal, 2005). The termination of the signal by removing odours that have already triggered a response is an important process for preventing odour adaptation and increasing odour sensitivity (Fig. 12.4). Although it is not known if *Ppa*-OBI-1 has catalytic activity, it is possible that *Ppa*-OBI-1 acts as a solubilising factor for the lipid pheromones in the excretion-filled amphid sheath glia.

Quite unexpectedly, the *Ppa-obi-1* mutant also revealed another ecologically relevant function of the Oriental beetle pheromone as a developmental regulator that coevolved with necromeny. Exposure of J4 stage *Ppa-obi-1* mutants to volatile ZTDO as a suspended drop on the assay plate resulted in paralysed J4, and even minute level of the ZTDO pheromone (0.001%) in the agar can also prevent both *Ppa-obi-1* mutant and wild-type dauers from resuming reproductive development in the presence of food (Cinkornpumin *et al.*, 2014).

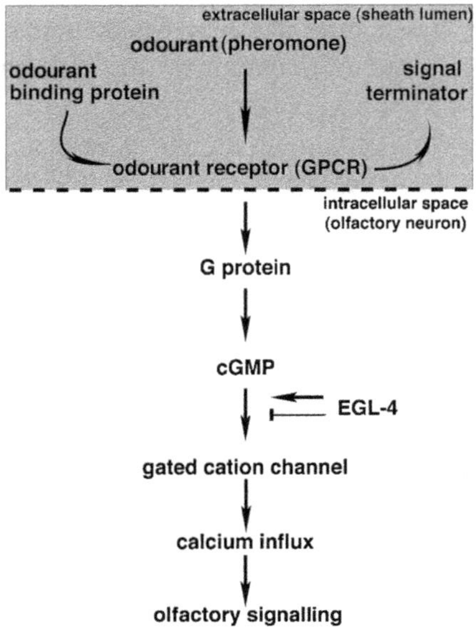

Fig. 12.4. *Overview of olfactory signalling in nematodes. A model for G-protein/cGMP mediated olfactory signalling in* C. elegans, *whereby the cGMP-dependent protein kinase EGL-4 has both positive and negative regulatory roles in odour sensing. Less known in nematodes are upstream regulators of receptor binding, such as odourant binding proteins and signal terminators. These regulators may be secreted into the sheath cell lumen/chamber to help concentrate odour molecules onto G-protein coupled receptors and subsequently disperse odour molecules to terminate olfactory stimuli.*

Based on the expression profile of *Ppa-obi-1p::gfp*, peak *Ppa-obi-1* expression coincides with hypersensitivity to the negative effects of the Oriental beetle pheromone, suggesting a protective role of *Ppa*-OBI-1 in dauers and J4. Feeding recombinant *E. coli* expressing *Ppa*-OBI-1 protein or soaking of the *Ppa*-OBI-1 protein to *Ppa-obi-1* mutants can ameliorate its hypersensitivity to Oriental beetle pheromone-induced paralysis (Cinkornpumin *et al.*, 2014). This ability to compensate the *Ppa-obi-1* mutation with exogenous *Ppa*-OBI-1 strongly suggests that *Ppa*-OBI-1 activity is required in cell types exposed to the environment, such as the *Ppa-obi-1*-expressing amphid sheath and excretory cells, whose pores remain open even as dauers. One interpretation of this tantalising finding is that *Ppa*-OBI-1 may act as a protector of the paralysing effects of volatile ZTDO exposure, perhaps by reducing over-excitation in cognate neurons. The exact mechanism for Oriental beetle pheromone-induced paralysis, as well as its likely link with the signalling pathway for chemoattraction, should be a high priority for future studies.

The Oriental beetle pheromone also acts through *Ppa-obi-1*-independent pathways during early *P. pacificus* developmental stages. Both wildtype and *Ppa-obi-1* J2 succumb to ZTDO-induced embryonic arrest and paralysis (Cinkornpumin *et al.*, 2014). ZTDO-exposed embryos arrest in various stages of development, though not when exposure is limited to eggs *in utero*. The susceptibility to ZTDO in wildtype embryos and J2 suggests a possible mechanism to suppress over-infestation of the beetle by limiting the number of viable embryos *per se*, or by favouring the population distribution towards older juveniles (wild-type J4 are resistant to ZTDO-induced paralysis). As would be expected of ecologically important traits, there is also evidence for natural variations in ZTDO-sensitivity, such as those found for ZTDO chemoattraction among various geographical isolates. Furthermore, these negative effects of ZTDO do not affect *C. elegans* and appear to be *P. pacificus*-specific, though more species have not yet been surveyed. The ecological implications of this apparent dual role of the host pheromone, both as an attractant and as a stage-specific developmental suppressor, compel us to rethink the mechanisms of the nematode-beetle interaction. It is also not obvious if ZTDO-induced embryonic arrest and dauer exit inhibition ultimately benefits the beetle hosts or *P. pacificus*. The beetle pheromone may represent a more widespread attenuated antagonism between necromenic nematodes and their insect hosts.

The importance of the sheath glia

Given the species-specific odour preferences for insect pheromones and plant volatiles in *Pristionchus* nematodes, as well as the strain-specific *P. pacificus* olfactory differences between ecologically distinct populations, one wonders if transmembrane odour receptors may be responsible for the rapid evolution of olfaction profiles? Analysis of the *P. pacificus* genome revealed a significantly lower number of predicted seven-transmembrane G-protein coupled receptors than the genome of *C. elegans* (Dieterich *et al.*, 2008), so the expansion of potential odourant receptors is not likely to account for the diversification of olfactory preferences in *P. pacificus* and relatives. Another possible source for changes in odour preference may occur in the organ environment in which the odourants interact with the neurons, such as in the lumen of the amphid sheath cells. The most-studied chemosensory neuron type in *C. elegans* is the AWC neuron, which is one of 12 chemosensory neuronal types arranged bilaterally around the worm's head. The two AWC neurons express the cGMP-dependent kinase, EGL-4, and mediate olfaction sensing of attractants, such as isoamyl alcohol and pentanedione. The dendritic endings of three pairs of winged neurons, together with a pair of neurons mediating thermotaxis, are fully embedded in the amphid sheath cells (Ward, 1973). Genetic ablation of the *C. elegans* amphid sheath glia results in developmental and chemotaxis defects of amphid neurons, whereby the characteristic neuronal functions involved in behaviour generation depend on the presence of sheath glia (Bacaj *et al.*, 2008). Unlike *C. elegans*, which has three pairs of winged chemosensory neurons (AWA, AWB, AWC), transmission electron microscopy (TEM) revealed that *P. pacificus* lacks amphid neurons with wing-shaped endings but shares the same number of amphid neurons with ciliated dendritic endings inside a single large amphid chamber (Riebesell, Hong & Sommer, unpubl. data). Sagittal electron micrograph sections indicate several electron dense regions in the amphid sheath cells that drain into the dendritic ends of amphid neurons (Fig. 12.5). Therefore, it remains to be determined whether or not the reduced ciliated surface area of homologous chemosensory AWA and AWC neurons in *P. pacificus* reflects possible different functional roles between the neurons and the amphid glia in chemosensation.

The importance of glial cells in *C. elegans* olfaction is under studied and only a handful of genetic mutants in amphid sheath glia function have been characterised. One of the most important factors for amphid development in *C. elegans* is DAF-6, which is a relative of the *Drosophila* patched-related protein and important for lumen morphogenesis (Perkins *et al.*, 1986; Perens & Shaham, 2005; Oikonomou *et al.*, 2011). Amphid sheath cells in *C. elegans* are also enriched in transcripts predicted to encode transmembrane and secreted proteins that could interact with odour molecules and sensory neurons (Osborne *et al.*, 1997; Bacaj *et al.*, 2008). Other glia-enriched proteins include the DEG/ENaC channel ACD-1 (Acid sensitive Channel Degenerin-like-1) and DEG-1, both of which are required for proper acid avoidance behaviour (Wang *et al.*, 2008). Conserved expression of the *daf-6* orthologue in the amphid sheath cells

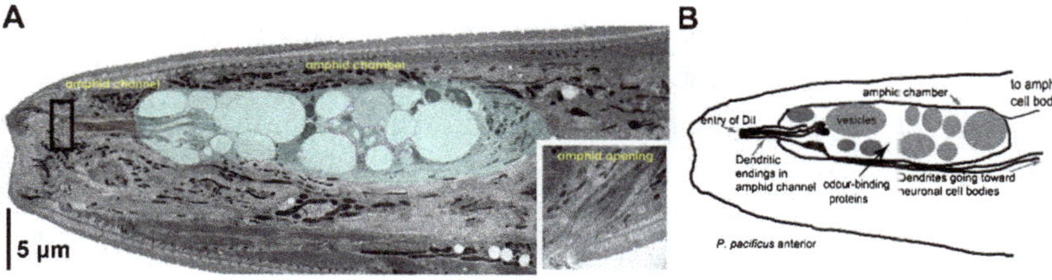

Fig. 12.5. Pristionchus pacificus *amphid sheath probably mediates olfaction. A: Transmission electron micrographs showing cilia of the amphid neurons in the amphid channel (shaded in brown) and the amphid chamber (shaded in blue) containing electron dense vesicles. The amphid finger neuron is outlined in purple. Inset shows possible chamber secretions flowing through the amphid opening (black outline) (Riebesell, Hong & Sommer, unpubl. data); B: Schematic depiction of the amphid chambers occupying a significant volume of the head region. Odours likely enter the amphid channels through the two amphid openings and may require secreted odour-binding proteins for proper odour receptor recognition on amphid neurons.*

would help to determine if the amphid sheath cell lumen in *C. elegans* is homologous to the amphid sheath chamber observed in TEM If so, these cells would represent a significant morphological difference between the lumen enveloping the cilia of winged amphid neuron in *C. elegans* and the amphid chamber occupying most of the head region in *P. pacificus*. Furthermore, if the electron-dense bodies observed in the electron micrographs of *P. pacificus* sheath cells represent secreted proteins that confer proper neuronal function (Riebesell, Hong & Sommer, unpubl. data), then future studies should aim to identify factors specifically expressed in the amphid sheath cell chamber together with genetic mutations that affect characteristic functions of the amphid neurons.

Open questions and challenges

Unlike the forward genetics approach used to identify odourant mutants in *C. elegans*, studies into *P. pacificus* olfaction began with the search for ecologically relevant attractive compounds before proceeding to take advantage of natural variants to identify the key regulator EGL-4 in insect pheromone sensing. The forward genetic screen for upstream factors mediating host pheromone attraction is leveraged by the identification of ecologically relevant host pheromones. This path of inquiry aspires to identify the biotic factors, genes and expression patterns that promote the *P. pacificus* necromenic lifestyle. Given the insect origins of the most attractive compounds and the ability to respond synergistically to multiple odourants, the rapidly evolving odour preferences found in *Pristionchus* species and in *P. pacificus* populations represent a possible paradigm for understanding how nematodes adapt to new hosts through olfactory rewiring. Yet, even evolutionarily distant nematodes such as *P. pacificus* and *C. elegans* are likely to be constrained in the number of chemosensory neurons, so differences in synaptic connectivity patterns and the luminal environments surrounding the sensory neurons probably play crucial roles in dictating the tempo and spectrum of olfaction profiles (Bumbarger *et al.*, 2013). Multi-facetted regulators of developmental and behavioural pathways such as EGL-4 and OBI-1 may be both a pivoting component as well as a system level constraint in behavioural diversity. Future studies in *P. pacificus* olfaction will probably yield both conserved factors in the GPCR signalling pathway, such as transmembrane receptors and G proteins, as well as factors previously not associated with nematode olfaction, such as those that interact with the lipid-binding protein OBI-1. Moreover, further explorations into the developmental role of host-derived chemicals on necromenic nematodes may reveal aspects of their ecology that are difficult to observe directly in nature.

References

BACAJ, T., TEVLIN, M., LU, Y. & SHAHAM, S. (2008). Glia are essential for sensory organ function in *C. elegans*. *Science* 322, 744-747.

BARGMANN, C.I., HARTWIEG, E. & HORVITZ, H.R. (1993). Odorant-selective genes and neurons mediate olfaction in *C. elegans*. *Cell* 74, 515-527.

BEN-SHAHAR, Y., ROBICHON, A., SOKOLOWSKI, M.B. & ROBINSON, G.E. (2002). Influence of gene action across different time scales on behavior. *Science* 296, 741-744.

BENTON, R., VANNICE, K.S. & VOSSHALL, L.B. (2007). An essential role for a CD36-related receptor in pheromone detection in *Drosophila*. *Nature* 450, 289-293.

BUMBARGER, D.J., RIEBESELL, M., RODELSPERGER, C. & SOMMER, R.J. (2013). System-wide rewiring underlies behavioral differences in predatory and bacterial-feeding nematodes. *Cell* 152, 109-119.

CINKORNPUMIN, J.K., WISIDAGAMA, D.R., RAPOPORT, V., GO, J.L., DIETERICH, C., WANG, X., SOMMER, R.J. & HONG, R.L. (2014). A host beetle pheromone regulates development and behavior in the nematode *Pristionchus pacificus*. *eLife*. DOI:10.7554/eLife.0322.

DIETERICH, C., CLIFTON, S.W., SCHUSTER, L.N., CHINWALLA, A., DELEHAUNTY, K., DINKELACKER, I., FULTON, L., FULTON, R., GODFREY, J., MINX, P. *ET AL.* (2008). The *Pristionchus pacificus* genome provides a unique perspective on nematode lifestyle and parasitism. *Nature Genetics* 40, 1193-1198.

FUJIWARA, M., SENGUPTA, P. & MCINTIRE, S.L. (2002). Regulation of body size and behavioral state of *C. elegans* by sensory perception and the EGL-4 cGMP-dependent protein kinase. *Neuron* 36, 1091-1102.

GARABAGI, F., FRENCH, B.W., SCHAAFSMA, A.W. & PAULS, K.P. (2008). Increased expression of a cGMP-dependent protein kinase in rotation-adapted western corn rootworm (*Diabrotica virgifera virgifera* L.). *Insect Biochemistry and Molecular Biology* 38, 697-704.

GOMEZ-DIAZ, C., REINA, J.H., CAMBILLAU, C. & BENTON, R. (2013). Ligands for pheromone-sensing neurons are not conformationally activated odorant binding proteins. *PLoS Biology* 11, e1001546.

HALLEM, E.A. & STERNBERG, P.W. (2008). Acute carbon dioxide avoidance in *Caenorhabditis elegans*. *Proceedings of the National Academy of Sciences of the United States of America* 105, 8038-8043.

HALLEM, E.A., DILLMAN, A.R., HONG, A.V., ZHANG, Y., YANO, J.M., DEMARCO, S.F. & STERNBERG, P.W. (2011). A sensory code for host seeking in parasitic nematodes. *Current Biology* 21, 377-383.

HERRMANN, M., MAYER, W., HONG, R.L., KIENLE, S., MINASAKI, R. & SOMMER, R.J. (2007). The nematode *Pristionchus pacificus* (Nematoda: Diplogastridae) is associated with the Oriental beetle *Exomala orientalis* (Coleoptera: Scarabaeidae) in Japan. *Zoological Science* 24, 883-889.

HIROSE, T., NAKANO, Y., NAGAMATSU, Y., MISUMI, T., OHTA, H. & OHSHIMA, Y. (2003). Cyclic GMP-dependent protein kinase EGL-4 controls body size and lifespan in *C. elegans*. *Development* 130, 1089-1099.

HONG, R.L. & SOMMER, R.J. (2006). Chemoattraction in *Pristionchus* nematodes and implications for insect recognition. *Current Biology* 16, 2359-2365.

HONG, R.L., WITTE, H. & SOMMER, R.J. (2008a). Natural variation in *Pristionchus pacificus* insect pheromone attraction involves the protein kinase EGL-4. *Proceedings of the National Academy of Sciences of the United States of America* 105, 7779-7784.

HONG, R.L., SVATOS, A., HERRMANN, M. & SOMMER, R.J. (2008b). Species-specific recognition of beetle cues by the nematode *Pristionchus maupasi*. *Evolution & Development* 10, 273-279.

ISHIDA, Y. & LEAL, W.S. (2005). Rapid inactivation of a moth pheromone. *Proceedings of the National Academy of Sciences of the United States of America* 102, 14075-14079.

JIN, X., HA, T.S. & SMITH, D.P. (2008). SNMP is a signaling component required for pheromone sensitivity in *Drosophila. Proceedings of the National Academy of Sciences of the United States of America* 105, 10996-11001.

KAUN, K.R. & SOKOLOWSKI, M.B. (2009). cGMP-dependent protein kinase: linking foraging to energy homeostasis. *Genome* 52, 1-7.

KIM, M.S. & SMITH, D.P. (2001). The invertebrate odorant-binding protein LUSH is required for normal olfactory behavior in *Drosophila. Chemical Senses* 26, 195-199.

KIM, M.S., REPP, A. & SMITH, D.P. (1998). LUSH odorant-binding protein mediates chemosensory responses to alcohols in *Drosophila melanogaster. Genetics* 150, 711-721.

KROETZ, S.M., SRINIVASAN, J., YAGHOOBIAN, J., STERNBERG, P.W. & HONG, R.L. (2012). The cGMP signaling pathway affects feeding behavior in the necromenic nematode *Pristionchus pacificus. PLoS ONE* 7, e34464.

L'ETOILE, N.D., COBURN, C.M., EASTHAM, J., KISTLER, A., GALLEGOS, G. & BARGMANN, C.I. (2002). The cyclic GMP-dependent protein kinase EGL-4 regulates olfactory adaptation in *C. elegans. Neuron* 36, 1079-1089.

MAIBECHE-COISNE, M., NIKONOV, A.A., ISHIDA, Y., JACQUIN-JOLY, E. & LEAL, W.S. (2004). Pheromone anosmia in a scarab beetle induced by *in vivo* inhibition of a pheromone-degrading enzyme. *Proceedings of the National Academy of Sciences of the United States of America* 101, 11459-11464.

MARTIN, C., CHEVROT, M., POIRIER, H., PASSILLY-DEGRACE, P., NIOT, I. & BESNARD, P. (2011). CD36 as a lipid sensor. *Physiology & Behavior* 105, 36-42.

MCGAUGHRAN, A., MORGAN, K. & SOMMER, R.J. (2013). Natural variation in chemosensation: lessons from an island nematode. *Ecology and Evolution* 3, 5209-5224.

MORGAN, K., MCGAUGHRAN, A., VILLATE, L., HERRMANN, M., WITTE, H., BARTELMES, G., ROCHAT, J. & SOMMER, R.J. (2012). Multi locus analysis of *Pristionchus pacificus* on La Réunion Island reveals an evolutionary history shaped by multiple introductions, constrained dispersal events and rare out-crossing. *Molecular Ecology* 21, 250-266.

O'HALLORAN, D.M., HAMILTON, O.S., LEE, J.I., GALLEGOS, M. & L'ETOILE, N.D. (2012). Changes in cGMP levels affect the localization of EGL-4 in AWC in *Caenorhabditis elegans. PLoS ONE* 7, e31614.

OIKONOMOU, G., PERENS, E.A., LU, Y., WATANABE, S., JORGENSEN, E.M. & SHAHAM, S. (2011). Opposing activities of LIT-1/NLK and

DAF-6/patched-related direct sensory compartment morphogenesis in *C. elegans. PLoS Biology* 9, e1001121.

OSBORNE, K.A., ROBICHON, A., BURGESS, E., BUTLAND, S., SHAW, R.A., COULTHARD, A., PEREIRA, H.S., GREENSPAN, R.J. & SOKOLOWSKI, M.B. (1997). Natural behavior polymorphism due to a cGMP-dependent protein kinase of *Drosophila. Science* 277, 834-836.

PENKOV, S., OGAWA, A., SCHMIDT, U., TATE, D., ZAGORIY, V., BOLAND, S., GRUNER, M., VORKEL, D., VERBAVATZ, J.-M., SOMMER, R.J. *ET AL.* (2014). A wax ester promotes collective host finding in the nematode *Pristionchus pacificus. Nature Chemical Biology* 10, 281-285.

PERENS, E.A. & SHAHAM, S. (2005). *C. elegans daf-6* encodes a patched-related protein required for lumen formation. *Developmental Cell* 8, 893-906.

PERKINS, L.A., HEDGECOCK, E.M., THOMSON, J.N. & CULOTTI, J.G. (1986). Mutant sensory cilia in the nematode *Caenorhabditis elegans. Developmental Biology* 117, 456-487.

POPRAWSKI, T.J. & YULE, W.N. (1990). A new small iridescent virus from grubs of *Phyllophaga anxia* (Leconte) (Col., Scarabaeidae). *Journal of Applied Entomology* 110, 63-67.

RAIZEN, D.M., CULLISON, K.M., PACK, A.I. & SUNDARAM, M.V. (2006). A novel gain-of-function mutant of the cyclic GMP-dependent protein kinase *egl-4* affects multiple physiological processes in *Caenorhabditis elegans. Genetics* 173, 177-187.

REINECKE, A., RUTHER, J. & HILKER, M. (2002). The scent of food and defence: green leaf volatiles and toluquinone as sex attractant mediate mate finding in the European cockchafer *Melolontha melolontha. Ecology Letters* 5, 257-263.

RIVARD, L., SRINIVASAN, J., STONE, A., OCHOA, S., STERNBERG, P.W. & LOER, C.M. (2010). A comparison of experience-dependent locomotory behaviors and biogenic amine neurons in nematode relatives of *Caenorhabditis elegans. BMC Neuroscience* 11, 22.

RUTHER, J. & HILKER, M. (2003). Attraction of forest cockchafer *Melolontha hippocastani* to (Z)-3-hexen-1-ol and 1,4-benzoquinone: application aspects. *Entomologia Experimentalis et Applicata* 107, 141-147.

RUTHER, J., REINECKE, A., TOLASCH, T. & HILKER, M. (2002). Phenol – another cockchafer attractant shared by *Melolontha hippocastani* Fabr. and *M. melolontha* L. *Journal of Biosciences* 57, 910-913.

STEINKRAUS, D.C., BOYS, G.O., KRING, T.J. & RUBERSON, J.R. (1993). Pathogenicity of the facultative parasite, *Chroniodiplogaster aerivora* (Cobb) (Rhabditida, Diplogasteridae) to corn earworm (*Helicoverpa zea* (Boddie)) (Lepidoptera, Noctuidae). *Journal of Invertebrate Pathology* 61, 308-312.

VOGT, R.G. & RIDDIFORD, L.M. (1981). Pheromone binding and inactivation by moth antennae. *Nature* 293, 161-163.

WANG, Y., APICELLA, A., LEE, S.-K.K., EZCURRA, M., SLONE, R.D., GOLDMIT, M., SCHAFER, W.R., SHAHAM, S., DRISCOLL, M. & BIANCHI, L. (2008). A glial DEG/ENaC channel functions with neuronal channel DEG-1 to mediate specific sensory functions in *C. elegans*. *EMBO Journal* 27, 2388-2399.

WARD, S. (1973). Chemotaxis by the nematode *Caenorhabditis elegans*: identification of attractants and analysis of the response by use of mutants. *Proceedings of the National Academy of Sciences of the United States of America* 70, 817-821.

XU, P., ATKINSON, R., JONES, D.N.M. & SMITH, D.P. (2005). *Drosophila* OBP LUSH is required for activity of pheromone-sensitive neurons. *Neuron* 45, 193-200.

Nematology Monographs & Perspectives, 2015, Vol. 11, 353-383

Chapter 13

Anatomy and connectivity in the pharyngeal nervous system

Dan BUMBARGER [1] and Metta RIEBESELL [2]

[1] *Allen Institute for Brain Science, Seattle, WA 98103, USA*
danb@alleninstitute.org
[2] *Department for Evolutionary Biology, Max-Planck Institute for Developmental Biology Tübingen, 72073 Tübingen, Germany*
metta.riebesell@tuebingen.mpg.de

Introduction

The nematode pharynx is a pumping organ that transports food from the outside environment to the intestine. It is one of the most prominent and functionally important organ systems in a nematode, and variation in the behaviour and anatomy of the pharynx between species correlates with the tremendous phylogenetic and ecological diversity within the phylum. There are anatomical specialisations for processing various food types, ranging from bacterial grinders in nematodes such as the model organism *Caenorhabditis elegans* to the hypodermic needle-like stomatostylets of many plant-parasitic nematodes. Along with these anatomical specialisations, nematodes vary greatly in details of the behaviour and functioning of the pharynx during feeding. Detailed information pertaining to neural circuits within the pharyngeal nervous system are lacking, greatly limiting the applicability of the comparative method to questions related to the form and function of the nervous system in *C. elegans* and other nematodes.

The cellular level anatomy of the pharynx is best described for *C. elegans*. In this species, as well as most others that have been examined, the pharynx is composed of eight sets of muscle cells, named pm1-pm8 from anterior to posterior (Fig. 13.1). With the exception of the most posterior muscle cell, pm8, all sets of muscle cells exhibit triradiate symmetry, with one dorsal and two ventrosublateral segments.

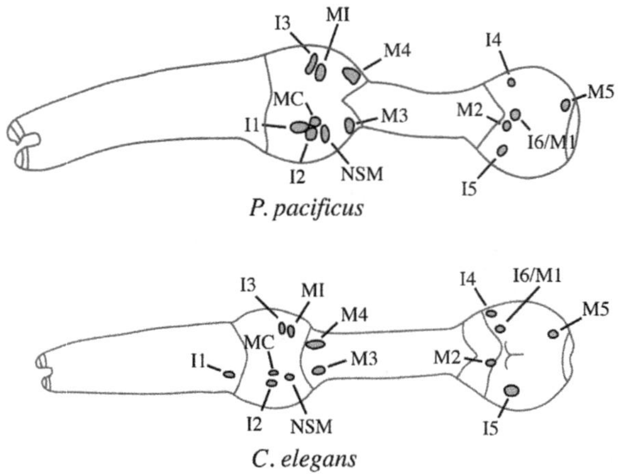

Fig. 13.1. *Positions of neuron nuclei in* Pristionchus pacificus *and* Caenorhabditis elegans. *Grey lines outline approximate boundaries of the pharynx and major muscle groups. Black shapes filled with grey indicate the location, size and shape of neuron cell classes. Cell classes with more than one member (I1, I2, M2, M3, MC and NSM) are bilaterally symmetrical. In both species I6 is found on the left hand side of the pharynx and M1 on the right.*

Between these muscle segments are triradiate ventral and superolateral marginal cells. The nematode pharynx is divided into four, functionally specialised, regions, with the first three sets of muscle cells (pm1-pm3) in *C. elegans* making up the procorpus, the next set (pm4) forming the metacorpus, the fifth set of cells (pm5) forming the isthmus and the final three sets of muscle cells (pm6-pm8) forming the terminal bulb. The first two sets of cells in the corpus (pm1, pm2) form part of the mouth opening, or buccal cavity, the rest of the buccal cavity being formed by the tip of pm3 and anterior epidermal cells that are not part of the pharynx. Five gland cells have their nuclei in the terminal bulb of *C. elegans*, two of which open in the terminal bulb, two in the metacorpus and one in the buccal cavity.

A wiring diagram of synaptic connectivity has long been available for the pharyngeal nervous system of *C. elegans* (Albertson & Thomson, 1976). This was, in fact, the first large-scale description of synaptic connectivity based on ultrastructural data and was completed 10 years before the full-animal wiring diagram (White *et al.*, 1986). Recently, a pharyngeal nervous system wiring diagram was completed for a second species, the diplogastrid nematode *Pristionchus pacificus* (Bumbarger *et al.*, 2013). In both species this nervous system consists of 20 neurons, a nerve ring neuropil in the metacorpus of the pharynx that facilitates communication between neurons. Communication between the pharyngeal and somatic nervous systems appears to be limited to a single pair of gap junctions towards the anterior of the pharynx, as well as through volume transmission *via* biogenic amines such as serotonin.

This chapter will examine the anatomical details of the pharyngeal nervous system of *P. pacificus* within a comparative context. Due to the limited information available, the discussion will focus on comparisons with the model organism *C. elegans*. Further anatomical details about non-neuronal cells of the pharynx in *P. pacificus* will be described elsewhere. For the purpose of this chapter, it is sufficient to mention that *P. pacificus* has the same number and approximately the same arrangement of cell nuclei as *C. elegans*, with the exception that *P. pacificus* lacks two gland cell nuclei present in *C. elegans*.

Overview of the *P. pacificus* nervous system

Although they are not closely related, the general structure of the pharyngeal nervous system is remarkably conserved between *C. elegans* and *P. pacificus* (Bumbarger *et al.*, 2013). The relative positions of neuron nuclei are nearly the same (Fig. 13.1), and similarity in the details of neurite position and arborisation within the pharynx makes statements of homology for individual neurons reasonably unambiguous (Chiang *et al.*, 2006; Bumbarger *et al.*, 2013).

A nerve ring commissure embedded in the metacorpus at the junction between the muscle cells pm4 and pm5 encircles the pharyngeal lumen and serves as the primary point of communication between neurons. In both species, three nerves extend anteriorly from the nerve ring into the procorpus in the dorsal and subventral sectors, and two nerves extend posteriorly into the isthmus in the ventrosublateral sectors (Fig. 13.2). The neurons I6 and M1 enter the posterior nerve ring individually, the former dorsally and the latter dorsally and slightly to the right.

The entry/exit locations of nerves into the nerve ring in *P. pacificus* differ slightly from *C. elegans*, most probably as a result of differences in the anatomy of the interface between pharynx muscles pm4 and pm5. Muscle pm5 forms a larger portion of the median bulb in *C. elegans* than it does in *P. pacificus*. In *C. elegans*, the posteriorly directed

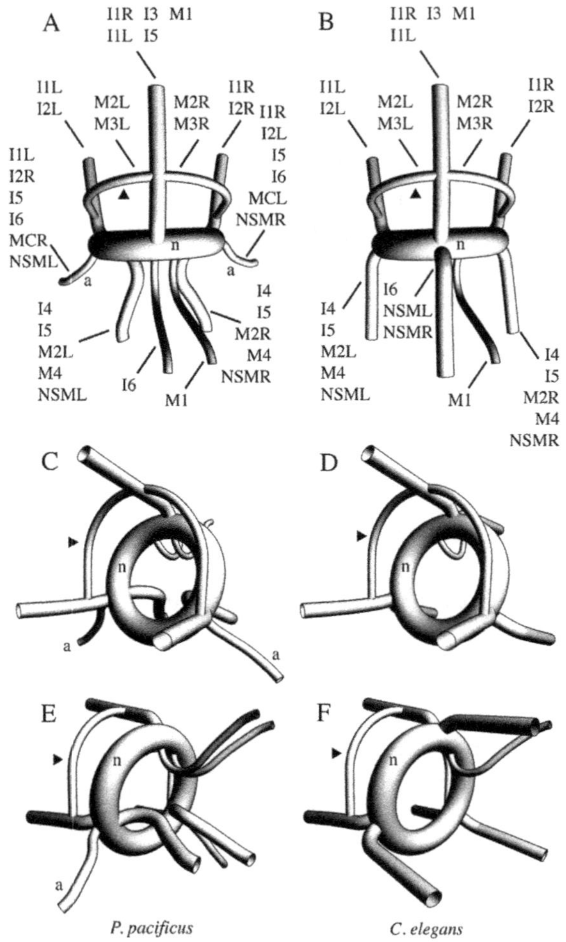

Fig. 13.2. *Comparison of pharyngeal nerve ring structure between* Pristionchus pacificus *and* Caenorhabditis elegans. *A, C, E: The pharyngeal nerve ring of* P. pacificus*; B, D, F: The pharyngeal nerve ring of* C. elegans. *A, B: Dorsal perspective with anterior at the top and posterior at the bottom; C, D: View from anterior and left of the nerve ring: E, F: View from posterior and left of the nerve ring. Capital letters with lines indicate neurons found in each nerve at that location. Black arrowheads indicate the location of the anterior metacorpus commissure. Important differences include the presence of accessory nerves in* P. pacificus, *and medial (*P. pacificus*) vs peripheral (*C. elegans*) connection points for the dorsal and ventrosublateral posterior nerves. Abbreviations: n = pharyngeal nerve ring; a = pharyngeal nerve ring accessory nerve.*

ventrosublateral nerves exit the nerve ring and pass anterior to the M3 cell body as they extend peripherally through the median bulb, then turn to extend posteriorly along the outside of the isthmus. In the ventrosublateral sectors of *P. pacificus*, there are short accessory nerves that extend peripherally anterior to the M3 cell bodies before extending posteriorly (Fig. 13.2), just as the main ventrosublateral nerves do in *C. elegans*, but do not enter the isthmus. These accessory nerves in *P. pacificus* consistently contain processes from I2, I6 and neurosecretory motor (NSM) neurons, and can variably also contain processes from I1, I5 and MC. The neurites originating from the cell bodies of M3 neurons enter the nerve ring at the same location as the accessory nerves.

The relative positions of neurons within the nerve ring have previously been examined only in transverse sections through the animal. This orientation is not ideal, as interpretation will differ according to where in the nerve the section is observed. In order to examine neuron positions in a more relevant way, we constructed 3D models of the nerve ring for both species and used these to examine virtual ventral-radial and subdorsal-radial sections through the nerve ring (Fig. 13.3). These virtual slices reveal some amount of conservation in the placement of neurons within the nerve ring. For example, M4 is conserved in having a medial position in both the ventral and subdorsal regions of the nerve ring. MI is anterior in the ventral nerve ring of both species.

Due to evolutionary changes in neurite morphology, the neurites present in each region of the nerve ring differ between species. In *P. pacificus*, the neurons I4, M4 and MI cross the ventral midline of the nerve ring. In *C. elegans*, two additional neuron classes consistently cross ventrally (NSM and I5). Two others (I2 and I1) vary between individuals in whether they cross or not. NSM in *P. pacificus* is highly unusual in that it is the only pharynx neuron that does not send a process oriented circumferentially within the nerve ring or the anterior metacorpus commissure. It passes through the nerve ring with a longitudinal orientation and does not form synapses at this location; nor does it cross the ventral midline as it does in *C. elegans*. Processes from the neurons I3, I4 and I6 are consistently found in the subdorsal nerve ring of *P. pacificus* but not in *C. elegans*.

A second, smaller, commissure composed of the neuron classes M2 and M3, here termed the anterior metacorpus commissure, bridges each ventrosublateral sector by encircling the pharynx dorsally just anterior to

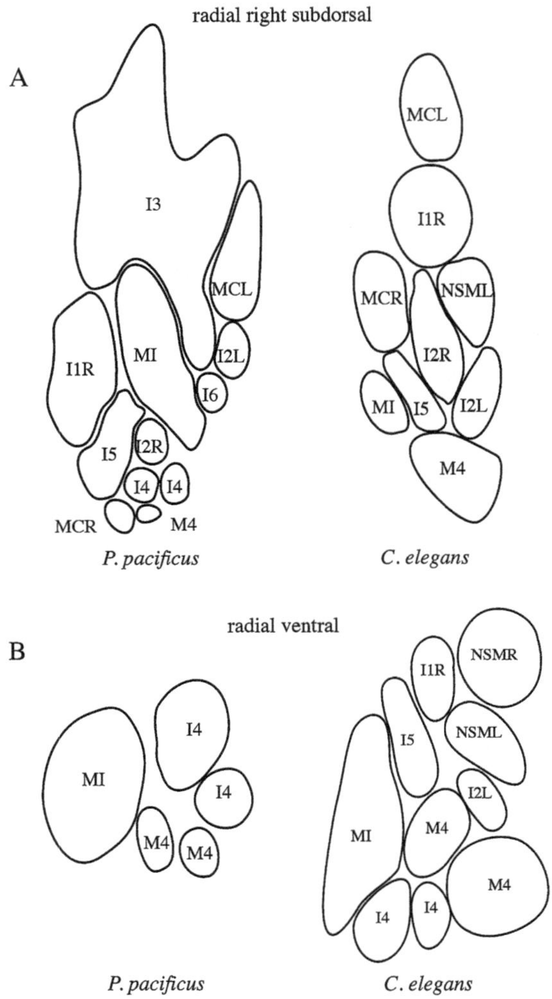

Fig. 13.3. *Virtual radial sections through the pharyngeal nerve rings of* Pristionchus pacificus *and* Caenorhabditis elegans. *A: Virtual radial section through the right subdorsal nerve ring; B: Virtual radial section through the ventral nerve ring. In both panels, slices are oriented with anterior to the left, peripheral above, and medial below.*

the nerve ring. In both species this commissure passes through the pm4 muscle cell. In *P. pacificus* it travels between the mc1 and mc2 marginal epithelial cells, whereas our observations indicate that in *C. elegans* it travels through the mc1 cell and does not come in contact with mc2.

A terminal bulb commissure that bridges the ventrosublateral and dorsal regions of the pharynx has been described in *C. elegans* as consisting of processes from I4, I5 and M5. In *P. pacificus*, this commissure is formed primarily by I4 and I5 and appears to be in a similar location. The single process of M5 in *P. pacificus* travels only briefly through the commissure before extending anteriorly with a neurite from I6 into the isthmus. The exact location of the commissure is not described for *C. elegans* but in *P. pacificus* it is located in between the pm5 and pm6 muscle cells and passes through the g1 dorsal gland cell.

The two species appear to have nervous systems constructed of the same component parts with similar anatomy, yet there are significant differences in pharynx function and behaviour that raise questions about structure-function relationships in nervous systems. The regions of the pharynx that are coupled in their activity differ between major groups of nematodes. The metacorpus and terminal bulb are coupled in *C. elegans*, whereas the isthmus and terminal bulb are coupled in *P. pacificus* (Chiang *et al.*, 2006). The terminal bulb of *P. pacificus* and other diplogastrid nematodes has undergone an evolutionary shift from exhibiting pumping activity to being a peristaltic region (Chiang *et al.*, 2006). Perhaps most interestingly, the anterior region of the pharynx in *P. pacificus* has evolved more complex behaviour in order to accommodate multimodal feeding. While feeding on bacteria, the anterior pharynx of *P. pacificus* appears to function very similarly to *C. elegans* and other nematodes, but when predatory on other nematodes a specialised tooth in the mouth opening is actuated and the pumping rate decreases substantially.

Sensory input in the pharynx

Little is known, even for *C. elegans*, about sensory input within the pharyngeal nervous system. No ciliated neurons have been reported within the pharynx of either *C. elegans* or *P. pacificus*. Neurons are designated as putative mechanosensory or proprioceptive neurons when a branch terminates in a subcuticular ending with an attachment to the cuticle lining the pharynx lumen. In *C. elegans*, the neuron classes M3, I1, I2, I3, I5 and I6 contain such subcuticular endings. With the exception of I5, the same neurons have subcuticular endings in *P. pacificus*, indicating the likelihood that sensory inputs into the pharynx are largely conserved.

The lack of a subcuticular ending in the I5 neuron of *P. pacificus* does not completely rule out a sensory function. I5 in this species forms a ring around the pharynx lumen at the junction between pm4 and pm5, travelling through the terminal bulb commissure on the dorsal side. Here I5 is unusual in that it displays numerous varicosities not correlated with the presence of synapse locations. These varicosities are not present in the equivalent region of I5 in *C. elegans* and could be indicative of specialised function.

M3 was erroneously reported to have two subcuticular endings in *P. pacificus* (Bumbarger *et al.*, 2013). There is a posteriorly directed process but it lacks the darkly staining junctions reported for other putative sensory endings. In *C. elegans*, the M3 putative sensory ending contacts the cuticle of the pharyngeal lumen just posterior to the nerve ring. In *P. pacificus*, however, the sensory ending is associated instead with the cuticle of the subventral gland cell ducts rather than the pharyngeal lumen and forms junctions with the pm4 and pm5 muscle cells.

I1, I2, I3 all have subcuticular endings in similar locations in both species. I1 and I3 attach to the cuticle in between the pm1 and pm2 muscle cells. I2 attaches adjacent to pm1, just anterior to I1; in *P. pacificus*, but not *C. elegans*, it differs in morphology from other pharynx subcuticular endings in that it penetrates into the cuticle, resulting in less of a barrier between it and the outside environment (Fig. 13.4). All

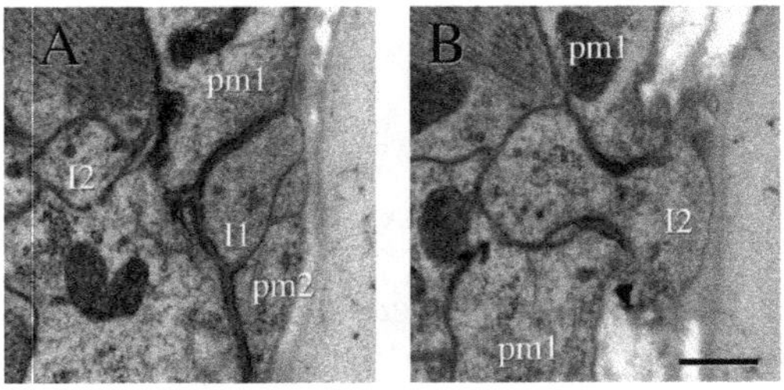

Fig. 13.4. *Subcuticular endings for the neuron cell classes I1 and I2 in* Pristionchus pacificus. *A: I1 projects between muscle cells pm1 and pm2 and terminates adjacent to the cuticle; B: I2 projects through a small pocket formed by the folding on the anterior side of pm1 and terminates within the cuticle, rather than adjacent as for other subcuticular endings in the pharynx. This terminus is immediately anterior to that of I1. (Scale bar = 0.5 μm.)*

three of these neuron classes in *P. pacificus* were observed to contain an accumulation of vesicles close to the ending, although no presynaptic densities are nearby.

In *C. elegans*, the subcuticular ending of I6 is found at the distal end of a neurite that originates at the posterior neuron cell body and attaches to the cuticle between muscle cells pm5 and pm6. In *P. pacificus* the equivalent neurite has a more complex morphology, described in more detail in the individual neuron descriptions below. The cuticle attachment point of I6 in *P. pacificus* is between pm6 and pm7, a more posterior attachment point than in *C. elegans*. In addition, at this subcuticular ending the neurite in *P. pacificus* forms a branch that projects to the posterior end of the terminal bulb. Here it wraps around the pharyngeal lumen where the pharynx and pharyngeal-intestinal valve meet. At the same location, the distal tips of NSM also wrap around the pharynx. The location expands as food is transported into the intestine, raising the possibility that I6 and/or NSM may have a sensory function at this location that is evolutionarily derived in *P. pacificus*.

General observations on connectivity

Maps of synaptic connectivity in the pharynx of *P. pacificus* resulted in the first comparison between species of complete anatomically determined maps of wiring for a large functional unit of any nervous system (Bumbarger *et al.*, 2013). The *P. pacificus* data were compared to the first study of its kind from *C. elegans* (Alberton & Thomson, 1976). A graph theoretical treatment of these data in the nematode comparative connectomics study (Bumbarger *et al.*, 2013) was restricted to analysis of cell classes rather than individual cells, as the available *C. elegans* data neglected to identify individual neurons as postsynaptic partners in their diagrams, instead only listing the cell class. Furthermore, an ongoing re-analysis of the original transmission electron micrographs for *C. elegans* (Cook *et al.*, 2014; www.wormwiring.org) identified serious flaws that make it regrettably difficult to put much weight on comparisons with *C. elegans* at present. Specifically, they identify numerous connections missed in the original analysis, particularly those to muscle cells. This re-analysis is likely to show that *C. elegans* exhibits a higher level of conservation in connectivity with respect to *P. pacificus* and will certainly impact our view of how these circuits function.

In light of this forthcoming re-analysis, it is best to focus on patterns that are based on conservation rather than divergence in the connectivity matrix between *P. pacificus* and *C. elegans*. Several patterns can be observed where the synapse classes present are very similar, but the locations of synapses are likely to be indicative of differences in developmental signalling pathways and circuit function. For example, in *P. pacificus* the muscle cell pm3 receives synapses along its entire length, whilst in *C. elegans* the homologous muscle cell receives synapses only close to the mouth opening (Cook *et al.*, 2014; www.wormwiring.org). Similarly, the neuron M5 is presynaptic to the muscle cell pm5 along its entire length in *C. elegans*, but the synapses are restricted to the anterior portion of pm5 in *P. pacificus*.

How much does synaptic connectivity between homologous neurons change over evolutionary time? Comparative physiology in systems such as the crustacean stomatogastric ganglion or swimming circuits in leaches has demonstrated that the formation or removal of connections is not necessary for major evolutionary modifications in animal behaviour (Katz & Harris-Warrick, 1999; Newcomb & Katz, 2009; Baltzley *et al.*, 2010; Sakurai *et al.*, 2011). Instead, explanations for behavioural differences between species derive primarily from changes in the physiological properties of synapses (*e.g.*, a change in neurotransmitter) rather than a change in a connectivity matrix. However, comparative circuit descriptions in the visual systems of flies, using serial transmission electron microscopy, found that connectivity between homologous neurons was surprisingly different between distantly related species. A previous study had identified the evolutionary addition of specific photoreceptor synapses with amacrine cells that the authors hypothesised were associated with increased time resolution in visual processing (Shaw & Meinertzhagen, 1986). Comparison between *P. pacificus* and *C. elegans* has suggested a great deal of evolutionary malleability in synaptic wiring (Bumbarger *et al.*, 2013) but forthcoming higher quality *C. elegans* data may change our interpretation.

Potential connectivity

Nervous system connectivity networks are sparsely connected, meaning that only a small portion of possible connections in a connectivity matrix are realised. Spatial proximity is a clear prerequisite to the formation of synapses not involving volume transmission. Potential connectivity can be defined as the subset of possible connections that meet this proximity requirement and has been explored in the context of improving understanding of the potential for synaptic plasticity (Stepanyants *et al.*, 2002).

In order to examine differences in potential connectivity between *P. pacificus* and *C. elegans*, we re-examined the anatomical data for one individual from each species and represented it as adjacency matrices that indicate whether or not neuron cell classes in the pharynx nerve ring come close enough to one another potentially to form a synapse (Table 13.1). Rather than representing observed synaptic connectivity, this table represents the potential connectivity based on proximity of neurites.

The nervous system of *P. pacificus* is more highly connected in terms of both actual and potential connectivity. Thirteen cell classes enter the nerve ring in both species and there are 78 possible adjacency relationships representing the total possible synaptic connectivity. In *P.*

Table 13.1. *Potential connectivity in the pharyngeal nerve rings of* Caenorhabditis elegans *and* Pristionchus pacificus. *A potential connection is observed if two neuron classes are adjacent to each other in at least one region of the nerve ring. White cells indicate where no connection was observed. Dark grey indicates potential connections observed in both species. Cells with a P or C indicate potential connections observed only in* P. pacificus *or* C. elegans, *respectively.*

	I1	I2	I3	I4	I5	I6	M1	M2	M3	M4	MC	MI	NSM
I1	–			P		P	P						
I2		–										P	
I3			–	P	P		P			C	P	P	C
I4	P		P	–		P					P		
I5			P		–	P			P				
I6	P			P	P	–	P	P	P		P	P	
M1	P		P			P	–		P		P		
M2						P		–					
M3					P	P	P		–				
M4			C							–	P		
MC			P	P		P	P			P	–		
MI		P	P			P						–	P
NSM			C									P	–

pacificus, 92.3% of these potential connections are realised, whereas in *C. elegans* they are realised in only 67.0%. In *C. elegans*, no neuron class has potential connectivity with every other neuron class. In *P. pacificus*, I2, I4, I5, I6, M3, MCL and MI come in close proximity to all other neuron classes in the pharyngeal nerve ring. All potential connections in *C. elegans* are present in *P. pacificus*, with the exception of potential I3-M4 and I3-NSM connections.

It is difficult to know the functional implications of potential connectivity comparisons, but the patterns evident between the two species at the very least demonstrate that potential connectivity should be examined together with actual connectivity in future comparative work. With larger networks or with comparisons between more taxa you can, in principle, ask whether potential connectivity places a constraint on the evolutionary formation and removal of connections. The relationship between potential connectivity and synaptic plasticity could also be investigated.

Phylogenetic comparison

Modern treatments of nematode phylogeny divide nematodes into two classes, the Chromadorea and the Enoplea (De Ley & Blaxter, 2002). Most of our knowledge of nematode nervous systems comes from the order Rhabditida, which is within the Chromadorea and includes both *P. pacificus* and *C. elegans*. Neuron number and cell body position in the pharynx is highly conserved within the Rhabditida (Zhang & Baldwin, 2000; Chiang *et al.*, 2006; Ragsdale *et al.*, 2011; Bumbarger *et al.*, 2013). Studies outside of this group are unfortunately not complete enough to evaluate homology for most individual neurons. At least 18 neurons are known to be present in the pharynx of *Ascaris suum*, a non-Rhabditida vertebrate parasite that is also placed in the Chromadorea (Cowden *et al.*, 1993). Cell bodies were identified with antibody staining and it is thought likely that the remaining two neurons may also be present in *A. suum* (Antony Stretton, pers. comm.). No reliable counts of pharyngeal neuron nuclei have been published for nematodes belonging to Enoplea.

While the fundamental pharynx nerve ring structure is conserved between *P. pacificus* and *C. elegans*, this is not the case for all nematodes. *Leptonemella juliae* (Chromadorea, Desmodorida) has six, rather than three, major nerves interacting anterior and posterior to the central nerve ring (Hoschitz *et al.*, 2001) and appears to have additional ring commissures near the mouth opening. Though more comparative work is needed to examine these additional rings, they may be homologous with the anterior metacorpus commissure and the anterior terminus of the neuron classes I1, I2, I3 and M1 in *C. elegans*. Less complete studies for a nematode belonging to Enoplea, *Longidorus leptocephalus* and *Xiphinema diversicaudatum* (Robertson, 1975, 1979) found three major nerves on each side of the nerve ring, indicating that the configuration of the nerve ring in Rhabditida is likely to be the ancestral pattern.

The presence of connections between somatic and pharyngeal nervous systems close to the mouth opening is conserved in all species examined. In *C. elegans*, *P. pacificus* and *Aphelenchus avenae* there are two bilaterally symmetric neurons (RIP) from the somatic nervous system connecting with the pharynx, suggesting that this may be conserved within the Rhabditida. In *P. pacificus*, RIP connects only to the I1 pharyngeal neurons. In *A. avenae*, it connects to both I1 and I2 (Ragsdale *et al.*, 2011). In *C. elegans*, it connects with I1 and variably to I2. There are, however, at least six connections between the somatic and pharyngeal nervous systems of *Leptonemella juliae* (Hoschitz *et al.*, 2001) and *Longidorus leptocephalus* (Robertson, 1979). This represents another indicator that ancestral lineages of nematodes may have had more complex nervous systems than those commonly studied.

The anatomy of pharyngeal sensory structures has not been widely surveyed in nematodes. The neuron class I1, which has subcuticular endings in both *P. pacificus* and *C. elegans*, lacks a subcuticular ending in *A. avenae* (Ragsdale *et al.*, 2011). *Aphelenchus avenae* has an otherwise highly derived pharyngeal anatomy, so it is not likely that it is representative of an ancestral species. The apparent loss of this subcuticular ending in *A. avenae*, as well as the loss of a subcuticular ending in the I5 neuron of *P. pacificus*, indicates that changes in the function of sensory neurons may not be uncommon in the phylum.

In *L. juliae*, three sensory structures located in the anterior pharynx have no obvious homologues to structures found in *P. pacificus* and other members of the Rhabditida. Surprisingly, each has a glial cell and associated multiple sensory neurons. Although ciliated pharyngeal nerve endings have not been observed in the Rhabditida, an undetermined number of ciliated neurons are present in *L. juliae*. The authors speculate that the cell bodies for these sensory structures may be found outside of the pharynx, which would also represent a significant difference from more commonly studied nematodes. In several members of the Enoplea, a number of authors have identified complex sensory structures in the anterior pharynx. The term 'endolid' was used to describe sensory structures observed with light microscopy in Dorylaimida (Siddiqi, 1970). Additionally, electron microscopy studies identified complex putatively chemosensory structures in the anterior pharynx of *X. diversicaudatum* and *L. leptocephalus* (Robertson, 1975). Ciliated neurons were observed in the endolids of *X. diversicaudatum* but not in *L. leptocephalus*. The greater complexity of sensory inputs into the pharynx of the Enoplea *vs* the Chromadorea is consistent with a greater complexity of sensory input into the somatic nervous system. Members of the Enoplea can have more sensory neurons in anterior sensilla and their bodies are typically covered in sensory setae.

The presence of a bilaterally symmetrical pair of serotonergic NSM appears to be a broadly conserved feature in nematodes (Rivard *et al.*, 2010). Antibodies against serotonin work well, making comparative observations on the anatomy of this class of neurons relatively simple. Common features of NSM neurons are the presence of a ventrosublateral process travelling on the same side of the animal as the cell body, and a process that crosses the animal ventrally, wraps around the nerve ring and then extends dorsally. Both processes contain numerous serotonin-containing synapses directed outwardly into the body cavity close to the somatic nerve ring. In most species, these processes are directed posteriorly from the cell body. In *A. suum*, the processes are directed anteriorly instead (Johnson *et al.*, 1996). NSM in *Haemonchus contortus*, another vertebrate parasite more closely related to *C. elegans*, has processes extending both anteriorly and posteriorly to the cell body (Rao *et al.*, 2010). These modifications appear to be in part tied to the somatic nerve ring, being anterior to the NSM cell bodies in large nematodes. In *P. pacificus*, NSM is highly unusual in that it does not interact significantly with the pharyngeal nerve ring and lacks dorsal processes. Furthermore, its ventrosublateral processes extend to the posterior end of the pharynx where they wrap around at the junction between the pharynx and pharyngeal-intestinal valve. Taken together, this seems to indicate that the differences in the morphology in NSM correlate with function.

Individual neuron descriptions

I1

The I1 cell class in *P. pacificus* is largely conserved in anatomy with I1 in *C. elegans* (Fig. 13.5). They are a bilaterally symmetrical pair of bipolar neurons with cell bodies located anterior and medial to those of I2 in the ventrosublateral nerves. They have the most anterior neuron cell bodies in the pharynx. An anteriorly directed process extends to a point slightly posterior to where pm2 attaches to the cuticle lining the mouth opening, where it divides into two short branches. One of these extends anteriorly to the junction between pm1 and pm2, where it forms a subcuticular ending (Figs 13.4, 13.5). The other branch extends ventrosublaterally to the periphery of the pharynx, where it forms a gap junction with the somatic interneuron RIP. This pair of gap

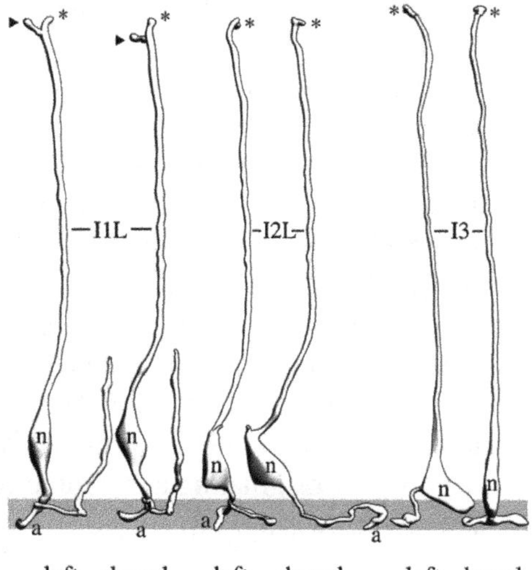

Fig. 13.5. *3D representation of pharynx neuron cell classes I1, I2 and I3. Each neuron is represented by renderings from both the left and dorsal perspectives. The cell class name is indicated with capital letters in between the left and dorsal views. Neurite thickness is exaggerated slightly for clarity and to resemble more closely how they might appear with fluorescence microscopy. A grey bar indicates the approximate boundaries of the pharyngeal nerve ring. * = subcuticular ending. Abbreviations: a = nerve ring accessory nerve; n = nucleus; black arrowhead = gap junction with RIP neuron in the somatic nervous system.*

junctions represents the only direct connection between the pharyngeal and somatic nervous systems. For most of its length, this process of I1 is placed at the most medial position of the anterior ventrosublateral nerve. This anterior process receives input from a large number of synapses originating from M1, as well as from a smaller number of synapses originating from I2. In *C. elegans*, there are no synapses in the region of the nerve where these occur in *P. pacificus*.

The posterior process exits the cell body and enters the nerve ring. A short branch extends with variable length into the adjacent accessory nerve of the nerve ring. The main branch of the neurite projects dorsally, wrapping around the nerve ring to meet the dorsal nerve. There, it turns anteriorly and extends into pm3, terminating adjacent to the nucleus of the epithelial cell e1D. This posterior process both gives and receives synapses, with many of the outgoing synapses being directed at M3, the I1 neuron from the opposite side, and the dorsal gland cell. In *C. elegans* this process does not terminate as far anterior and synaptic output is primarily toward M2, M3 and MC.

In *P. pacificus*, presynaptic densities are not restricted to the neurites. Next to the nucleus, they form connections to the posterior regions of the pm1 muscle cells. These synapses occur where the I1 cell body is narrowing to form the anterior neurite and form a connection to a short process extending posteriorly from the ventrosublateral pm1 cell bodies. The synaptic vesicles close to the presynaptic densities are typically few in number and poorly contrasted. These motor synapses are unusual in that they are located very far from the contractile region in pm1. While no such synapses have been identified in published annotations of *C. elegans* data (Albertson & Thomson, 1976), a re-analysis has identified synapses directed towards pm1 originating at a similar location along I1 (Cook *et al.*, 2014; www.wormwiring.org).

I2

The I2 neurons (Fig. 13.5) are a bilaterally symmetrical and bipolar pair of neurons with cell bodies located in the anterior ventrosublateral nerves posterior to I1, anterior to NSM, and peripheral to MC. The neurite extending anteriorly from the cell body extends through the nerve to terminate in a subcuticular ending just anterior to the subcuticular ending of I1 and adjacent to pm1 (Figs 13.4, 13.5). This terminus differs from that of I1 in that, rather than terminating adjacent to the cuticle, it enters and expands into the cuticle, leaving the wall of the cuticle very thin at its terminus. There are a number of tiny cellular processes extending into the tooth-like denticles lining the mouth opening. Although it is not possible with available micrographs to trace these fine processes, it is likely that they extend from I2. Along this anterior neurite, I2 receives extensive input from M1, and has a few synapses directed primarily to the pm3 and pm4 muscle cells, I1 and M1. Similar synaptic input and output in the anterior pharynx has recently been identified in the pharynx of *C. elegans* (Cook *et al.*, 2014; www.wormwiring.org).

The neurite exiting from the posterior side of the cell body enters and travels through the pharyngeal nerve ring. It crosses the midline dorsally and continues around the nerve ring to terminate at the distal tip of the accessory nerve on the side opposite from where it entered the nerve ring. This neurite receives very little synaptic input and most of the synaptic output from this neurite is directed towards the pm4 muscle.

I3

I3 is a bipolar neuron with a cell body located in the anterior dorsal nerve posterior to MI and anterior to M4 (Fig. 13.5). The anteriorly directed neurite is in contact with the dorsal gland almost its entire length, and extends to the mouth opening where it forms a subcuticular ending between the pm1 and pm2 muscle cells and at the base of the dorsal tooth. Along this process I3 receives extensive input from M1.

A short, posteriorly directed process extends from the cell body, originating just anterior to the anterior metacorpus commissure. It extends into the nerve ring, where it forms two bilaterally symmetrical branches that extend a short distance into either side of the nerve ring to terminate subdorsally. This posteriorly directed process, though short, gives and receives a number of synapses, with many of the outgoing synapses being directed towards the pm4 muscle cell.

I4

The cell body for the neuron I4 (Fig. 13.6) is located in the terminal bulb on the dorsal margin of the pharynx just to the right of the midline and at the transition between the pm5 and pm6 muscle cells. Two bilaterally symmetrical neurites exit the cell body and extend through the terminal bulb commissure to the posterior ventrosublateral nerves. Here, they travel through the isthmus to the nerve ring. As in *C. elegans*, this

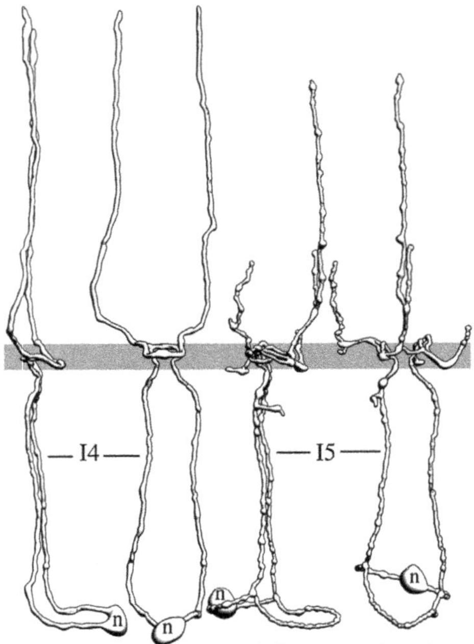

Fig. 13.6. *3D representation of pharynx neuron cell classes I4 and I5. Caption details and abbreviations as in Figure 13.5.*

region of the neurite in *P. pacificus* receives synaptic input from NSM. Upon entering the pharyngeal nerve ring, each neurite crosses ventrally and wraps around the nerve ring to the anterior ventrosublateral nerve on the opposite side. In the nerve ring, I4 does not receive synaptic input, and synaptic output is dominated by dyadic synapses to I5 and MI. These neurites continue through the corpus along the ventrosublateral nerves to terminate posterior to the buccal cavity. Synaptic input or output do not appear to be functions of the extensions through the corpus.

I5

The cell body of I5 (Fig. 13.6) is located on the ventral side of the terminal bulb close to the transition between pm5 and pm6. Two neurites originate at the cell body, one on each side. They are unusual in that they extend to the ventrosublateral nerves and continue through the terminal bulb commissure, where they fuse to form a closed loop. Where the loop passes each ventrosublateral nerve, an additional neurite branch is formed that extends through the isthmus to the nerve ring. These neurites receive extensive input from NSM along their length and extensive input from M2 in the anterior isthmus. Upon reaching the nerve ring, they wrap around the same side to the anterior dorsal nerve where they fuse into a single process. Unlike in *C. elegans*, I5 neurites in *P. pacificus* do not fuse on the ventral side of the nerve ring. Within the nerve ring, I5 receives significant synaptic input from I4 and MI, while producing synapses that are directed primarily toward the pm4 pharynx muscle cell.

Three neurites branch from I5 in the pharyngeal nerve ring. Two processes that originate from the subdorsal sector have a highly variable morphology. In one individual where they were fully reconstructed, these neurites branched again, with one extending into the nerve ring accessory nerves and the other extending into the anterior region of pm4 on the peripheral margin of the anterior ventrosublateral nerve. In another individual, these subdorsal branches extend to the peripheral margin of the terminal bulb without migrating to the ventrosublateral side. In both individuals, the subdorsal branches direct numerous synapses towards the pm4 muscle cell. The third process originates in the anterior dorsal nerve and extends through most of the corpus, forming an additional short branch that may be a remnant of the fusion of neurites in the nerve ring. This anterior dorsal branch produces numerous synapses directed towards pm4 in the metacorpus and towards the e1D epithelial cell in the anterior region of the pharynx. It receives extensive input from I1.

Outside the pharyngeal nerve ring, neurites of I5 contain numerous varicosities having the appearance of pearls on a string. In other neurons, similar varicosities are typically associated with a synaptic density, but in I5 they occur in regions of the neuron where no synapses occur.

I6

The cell body of I6 (Fig. 13.7) is located on the left subdorsal side of the terminal bulb, slightly posterior and dorsal to that of the left M2 neuron and at the transition between the pm5 and pm6 pharynx muscle cells. Two neurites originate at the cell body.

The neurite originating on the posterior side of the I6 cell body projects to the posterior margin of the dorsal pm6 pharynx muscle cell and forms a subcuticular ending on the right subdorsal sector of the cuticle lining the pharyngeal lumen. This putative sensory ending forms junctions with the muscle cell pm6 and the marginal epithelial

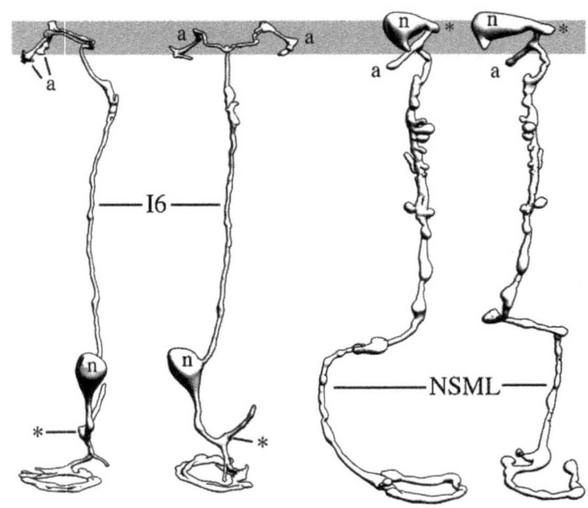

Fig. 13.7. *3D representation of pharynx neuron cell classes I6 and NSML. * = subcuticular ending. Caption details and abbreviation as in Figure 13.5.*

cell mc3DR. In *C. elegans*, the position of this junction differs slightly, instead being located between the pm5 and pm6 muscle cells. Close to the junction, the neurite divides to form anterior and posterior branches. The anterior branch projects to the cell body of M1 where it forms a putative gap junction. This branch and connection to M1 was erroneously omitted from the drawings in the supplemental material of Bumbarger and co-workers (Bumbarger *et al.*, 2013). The branch directed posteriorly from the subcuticular ending projects to the cell body for M5, where it divides into two or more branches with variable morphology that wrap around the posterior pharynx next to the junction with the pharyngeal-intestinal valve. As the neurites wrap around the pharynx, they receive synaptic input from NSM. In one individual, synapses directed at pm8 and a single synapse directed outside of the pharynx were observed.

The neurite originating on the anterior side of the I6 cell body projects through the isthmus in the dorsal nerve with M5 and adjacent to the dorsal gland cell. In most of the isthmus, this dorsal nerve is located midway between the periphery and centre of the pharynx. Close to the somatic nerve ring, both neurons in the dorsal nerve move to a peripheral position. As it nears the metacorpus, the I6 neuron moves again to a more central position, where it projects into the pharyngeal nerve ring on the dorsal side. Here it divides into two symmetrical branches that travel to the ventrosublateral sector of the pharyngeal nerve ring on either side. They then turn posteriorly into the accessory nerves, where they project to the distal end and terminate. These branches within the nerve ring are absent in *C. elegans*. This anterior neurite of I6 receives only a small amount of synaptic input and apparently has a primarily axonal function. It forms multiple synapses directed towards pm5 in the anterior isthmus, towards I1 in the pharyngeal nerve ring, and towards M3 in the accessory nerves.

M1

The cell body of M1 (Fig. 13.8) is located on the left subdorsal side of the terminal bulb, slightly posterior and dorsal to that of the left M2 neuron and at the transition between the pm5 and pm6 pharynx muscle cells. It occupies the equivalent position of I6 on the other side of the body.

A single neurite projects from the anterior side of the cell body and travels through the isthmus on the right subdorsal side of the pharynx, in between the pharynx muscle cell pm5 and the marginal epithelial cell mc2. Like the other dorsal neurons, as it nears the somatic nerve ring it moves to the periphery of the pharynx and moves back to a more central location as it approaches the metacorpus. M1 enters the pharyngeal nerve ring on the midway between the centre and right subdorsal region of the nerve ring. When it reaches the anterior side of the nerve ring, it splits into two branches. A short branch extends to the subdorsal corner of the nerve ring, while a longer branch projects anteriorly into the dorsal nerve. It passes medially to the anterior metacorpus commissure and on the anterior side an additional short branch forms and projects posteriorly on the peripheral side of the commissure to terminate near the pharyngeal nerve ring. This short, posteriorly-projecting branch is not found in *C. elegans*. The longer branch continues through the corpus to the transition between pm1 and pm2, where it divides into two symmetrical branches. These branches wrap around the pharyngeal lumen along either side of the animal between pm1 and pm2 until they reach the ventrosublateral nerves, where they turn to project posteriorly. In *C. elegans*, these terminal neurites of M1 project only a short distance, but in *P. pacificus* they project through the corpus to terminate at variable locations within the anterior metacorpus.

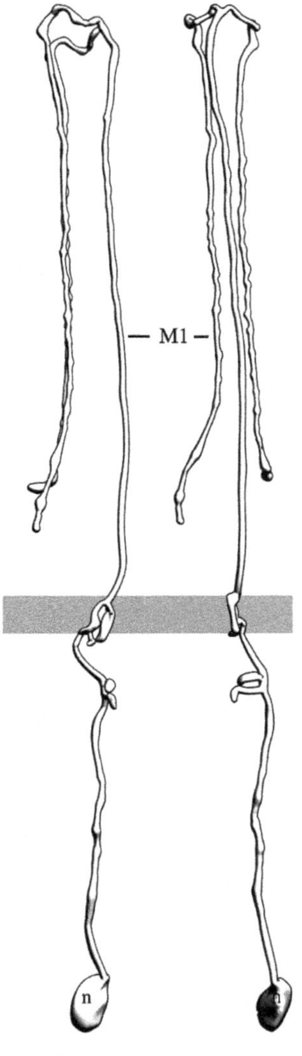

Fig. 13.8. *3D representation of pharynx neuron cell class M1. Caption details and abbreviation as in Figure 13.5.*

In both *C. elegans* (Cook *et al.*, 2014; www.wormwiring.org) and *P. pacificus*, M1 receives synaptic input from I1, I2 and MI in the pharyngeal nerve ring, and most of the output is directed towards pm1, pm2, pm3, I1, I2, I3 and e3d in the corpus. The motor output in the corpus of *C. elegans* is restricted to the anterior region of the pharynx close to the buccal cavity, whereas in *P. pacificus* M1 forms synapses with pm3 along its entire length.

M2

M2 (Fig. 13.9) is a bilaterally symmetrical class of neurons with cell bodies lateral and slightly dorsal. They are just anterior to and ventral to those of I6 on the right side and M1 on the left side. A single neurite projects from the anterior side of the cell body into the ventrosublateral nerve on the same side of the body. From there it projects through the entire isthmus. In the anterior isthmus, the neurite exhibits a complex morphology with the appearance of a mesh network that extends between the ventral midline and the subdorsal region of the pharynx periphery. There, it makes numerous synapses directed towards the dorsal mc2 marginal epithelial cells and the pm5 pharynx muscle cell. Within the isthmus, it receives a small amount of synapses from

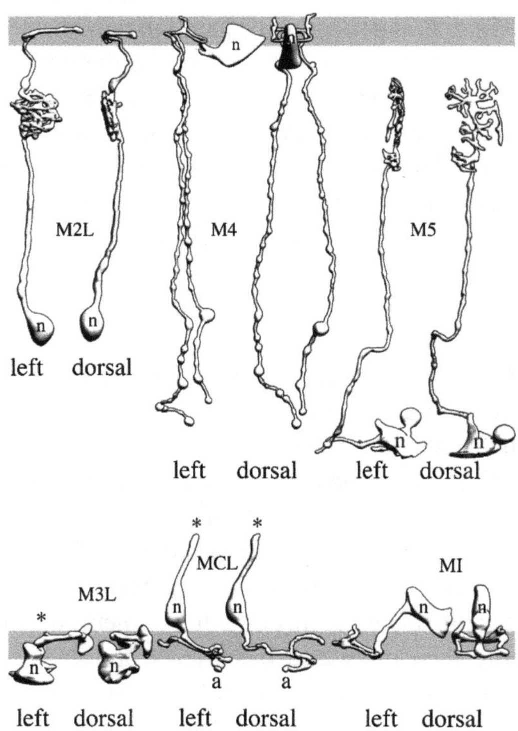

Fig. 13.9. *3D representation of pharynx neuron cell classes M2L, M3, M4, M5, MC forward and MI. Caption details and abbreviation as in Figure 13.5.*

NSM at variable locations along the neurite. As it nears the pharyngeal nerve ring, M2 resumes a more typical neurite morphology. It projects anteriorly through the medial side of the pharyngeal nerve ring. In front of the nerve ring, M2 contains between 30 and 50 microtubules that stain unusually dark in our preparations as compared to those in other neurons, giving them a distinct appearance. Upon reaching the anterior metacorpus commissure, each M2 projects along the same side of the body to the dorsal side, where it forms a junction with its partner from the opposite side. It has very little connectivity in the nerve ring, with a small number of synapses directed towards the pm4 pharynx muscle cell. In the two individuals observed, only one of the M2L received input from I1L, but M2R did not receive input.

Although the anatomy of M2 in *P. pacificus* as it projects through the anterior isthmus is quite divergent from that of M2 in *C. elegans*, the pattern of synaptic connectivity along the neuron is highly conserved, including the synapses with I1, pm4 and pm5. The single synapses received from I1 are found in the same location in both species. However, the synapses to pm5 in *C. elegans* are along the entire length of the isthmus, rather than being restricted to the anterior isthmus as in *P. pacificus*. This unusually high degree of conservation in connectivity may indicate some element of conserved function. Although the function of M2 is not well understood in either species, it has been shown to be important for isthmus peristalsis in *Panagrolaimus* (Chiang *et al.*, 2006).

M3

M3 (Fig. 13.9) is a bilaterally symmetrical class of neurons with the most posterior cell bodies in the ventrosublateral metacorpus. A single neurite projects from the medial side of the cell body and passes through the inner side of the pharyngeal nerve ring. Here it forms a subcuticular ending with the cuticle lining the subventral pharyngeal gland, forming junctions with the pharynx muscles pm4 and pm5. From there it projects into the anterior metacorpus commissure and travels along the same side of the body to the dorsal nerve, meeting the M3 neuron from the opposite side. The neurite then orientates posteriorly and projects through the dorsal nerve back to the pharyngeal nerve ring, where it terminates. M3R projects through the left side of the dorsal nerve, and M3L on the right.

Close to the cell body, M3 typically receives extensive synaptic input from I1 and a smaller amount from I6. Most of these synapses originate in the pharynx nerve ring accessory nerves that wrap around the dorsal side of the M3 cell bodies. In one individual, the synapses directed towards M3R from I1 are absent, apparently replaced by a similar number of synapses from I2. In the ventrosublateral and dorsal nerves, M3 receives additional input from I1. Along the neurite that is distal to the subcuticular ending, M3 forms multiple synapses directed towards the pm4 pharynx muscle cells. In the dorsal nerve, a small number of synapses are directed towards the dorsal gland. In both individuals of *P. pacificus* where connectivity data are available, I4 forms a small number of synapses directed towards M3R but not towards M3L. In *C. elegans*, both M3 neurons receive substantial input from I4.

In *C. elegans*, there is an additional neurite originating on the posterior side of the cell body that projects a short distance into the anterior isthmus and receives synaptic input from I4 and NSM. Both the neurite and the synaptic connections are absent in *P. pacificus*. Although there are some differences, synaptic wiring associated with M3 is generally conserved between *C. elegans* and *P. pacificus*.

M4

The cell body of M4 (Fig. 13.9) is located on the dorsal side of the metacorpus. The front of the cell body is adjacent to the pharyngeal nerve ring and it extends to the posterior end of the pm4 pharynx muscle cell. A single neurite originates from the cell body posterior to the pharyngeal nerve ring. It projects to the most medial position of the dorsal nerve and travels within it a short distance to enter the pharyngeal nerve ring. There, it divides into two symmetrical branches that occupy a medial position as they travel through the nerve ring. Each branch projects through the pharyngeal nerve ring to the ventrosublateral nerve, where it forms a short branch that extends to terminate close to the anterior metacorpus commissure. The larger branch continues projecting through the pharyngeal nerve ring, crossing the ventral midline and extending to the ventrosublateral nerve on the opposite side. Here it projects posteriorly into the isthmus. In the anterior isthmus, it briefly leaves the ventrosublateral nerve to wrap around the ventrosublateral pharynx gland cells before returning to the nerve and continuing to travel through the remainder of the isthmus. Unlike *C. elegans*, the neurites from M4 in *P. pacificus* do not enter the terminal bulb commissure. Instead, each branch continues to project posteriorly. The two branches terminate asymmetrically, with the neurite that projects through the right side of the isthmus terminating next to the pm7 cell body. The neurite that projects through the left side of the isthmus extends posteriorly to pm7 where it crosses the ventral midline next to the neurite from M5 and travels to a position on the right side close to the posterior margin of the pharynx, where it terminates.

M4 in *P. pacificus* receives very few synapses, and the presynaptic partners vary between individuals. The only synaptic input found in both individuals was a single synapse from I4 (Table 13.1). Like *C. elegans*, M4 produces numerous synapses along the length of the isthmus directed towards the pm5 muscle cell. *P. pacificus* additionally forms a few synapses directed towards cells in the terminal bulb, including pm6 and the mc3 marginal epithelial cells.

M5

The cell body of M5 (Fig. 13.9) is located peripherally in the right subdorsal sector of the terminal bulb at the level of the most posterior pharynx muscle cell pm8. Two neurites originate from the cell body. Unlike in *C. elegans*, these neurites are not bilaterally symmetrical. A very short neurite projects from the posterior cell body. In one of the two individuals of *P. pacificus* fully reconstructed this short process receives a single synapse from NSM.

From the centre of the cell body, a single and much larger neurite projects ventrally along the right side of the pharynx lumen, meeting the terminus of M4 and travelling with it to cross the ventral midline. From here, it continues to rotate around the pharynx lumen close to M4 as it moves anteriorly to the point where the left ventrosublateral nerve and terminal bulb commissure meet. Rather than entering the commissure and remaining within the terminal bulb as in *C. elegans*, M5 in *P. pacificus* projects into the dorsal nerve alongside I6 and extends through the isthmus. Upon reaching the anterior isthmus, many processes emerge to form a network-like structure in the dorsal isthmus similar to the ones described above for M2. Within this structure, M5 produces a large number of synapses directed towards the dorsal pm5 muscle cell.

MC

MC (Fig. 13.9) is a bilaterally symmetrical class of neurons with cell bodies medial to those of I2 and posterior to I1. An anteriorly projecting neurite exits the cell body to form a subcuticular ending at the transition between the metacorpus and corpus, forming junctions with the pm3 and pm4 pharynx muscle cells. There is no synaptic input or output along this process.

A second neurite projects posteriorly through the medial side of the adjacent ventrosublateral nerve to the pharyngeal nerve ring. It travels on the anterior side of the nerve ring to the dorsal nerve. After crossing over the dorsal midline and passing the MC neuron from the opposite side, it moves to the posterior nerve ring and continues to project to the subdorsal sector. Here, a branch forms that exits on the posterior side of the pharyngeal nerve ring. This branch has a variable morphology, typically branching one more time and forming numerous synapses directed towards the pharynx muscle cell pm4 and the adjacent mc2 marginal epithelial cell. The branch remaining in the pharyngeal nerve ring continues projecting to the ventrosublateral nerve on the opposite side of the animal from the cell body. Here it projects posteriorly, either in the accessory nerves or on the ventral side of the animal. This neurite also forms multiple synapses directed towards the muscle cell pm4 and the marginal epithelial cell mc2.

In *C. elegans*, MC neurons are cholinergic and serve to regulate the rate of pumping in the pharynx. It presumably does so through its synapses onto the mc2 marginal epithelial cells. In *P. pacificus*, MC neurons synapse directly onto the pm4 pharynx muscle cells. As the pumping rate of the pharynx in *P. pacificus* must be regulated during predatory *vs* bacterial feeding, the conservation in synaptic output makes MC the most likely candidate for performing this function. In *C. elegans*, the most important input from chemical synapses is derived from I1 neurons, whereas in *P. pacificus* it is more variable but consistently contains input from I2 neurons and not from I1. These differences in input may help to explain some of the behavioural differences.

MI

The cell body of MI (Fig. 13.9) is located within the metacorpus anterior to the cell body for M4, peripheral to that of M3 and posterior to that of the e3D epithelial cell. A single neurite projects from the medial side of the cell body and enters the dorsal nerve anterior to the anterior metacorpus commissure. It projects posteriorly past the commissure on the medial side and into the pharyngeal nerve ring. From here it projects ventrally down the right side of the pharyngeal nerve ring, crosses ventrally, and continues around the pharyngeal nerve ring to terminate in the left subdorsal sector. In either ventrosublateral nerve, MI forms a large varicosity that tapers into a very short, anteriorly projecting, process.

In *C. elegans*, the connectivity for MI has been described as highly variable. It appears to be less so in *P. pacificus*. In both species, motor output is to the pm4 pharynx muscle cell and in both species MI is also presynaptic to M1 and I5. Chemical synaptic input was highly variable in *C. elegans* but consistently originated from I1, I3 and I4 in *P. pacificus*.

NSM

NSM (Fig. 13.7) is a bilaterally symmetrical class of neurons with cell bodies in the metacorpus posterior to those of I2 neurons and at the level of the anterior metacorpus commissure. A single process originates from the medial side of the cell body and projects to the cuticle lining the pharyngeal lumen. Here it forms a subcuticular ending anterior to where the subventral gland ducts open into the lumen. The neurite turns to project posteriorly through the medial side of the pharyngeal nerve ring without entering or forming synapses. In all other species where it has been observed, including *C. elegans*, NSM forms a branch that projects through the pharyngeal nerve ring to the dorsal side, and then into the isthmus. This makes the anatomy of NSM in *P. pacificus* highly unusual. Once the neurite reaches the cell body of M3, a short branch projects to the posterior tip of the adjacent pharynx nerve ring accessory nerve. The larger branch continues to project through the ventrosublateral nerves in the isthmus. In the anterior isthmus, NSM forms multiple synaptic densities. As in *C. elegans*, most of these synapses are directed towards the body cavity close to the somatic nerve ring. In *C. elegans*, NSM terminates within the isthmus close to the terminal bulb. In *P. pacificus*, however, NSM continues to project into the terminal bulb. They exit the ventrosublateral nerves and project posteriorly along the periphery of the pharynx until they reach the pharyngeal-intestinal valve. Here, they form several irregular branches that wrap around the pharynx with the neuron I6. Within these branches, NSM forms additional synapses directed towards the body cavity, as well as to the pm8 muscle cell. NSM appears to get feedback from I6, forming synapses directed towards I6 in the terminal bulb and receiving input from I6 in the metacorpus.

Conclusions

Microscopy and computing technologies are rapidly reducing the barrier to large-scale and higher-throughput descriptions of neuroanatomy, including maps of synaptic connectivity. Although this comparison of the pharyngeal nervous systems of the nematodes *P. pacificus* and *C. elegans* represents the first of its kind, the near future promises to transform electron microscopy image data into a bioinformatics resource.

Nematodes offer several advantages for comparative and system level studies of synaptic wiring (Jarrel *et al.*, 2012; Bumbarger *et al.*, 2013). Their small size will allow for both a completeness of system description and sample sizes sufficient for a comparative and experimental approach to connectomics. All neurons are identified as neurons with highly specific identities and function, and homologous neurons can be identified for most neurons between individuals of even distantly related species. A number of tools are applicable across species to provide the necessary context to bridge our understanding of network structure and function, including the ability to record simultaneously the activity of large numbers of neurons in response to controlled stimuli (Schrödel *et al.*, 2013).

Therefore, continued studies of nematode connectomics and comparative neuroanatomy focused on completeness, sample size, behavioural context and an emphasis on data quality over data quantity will likely yield generalisable insight into structure-function relationships in nervous systems and play an important role in a modern approach to systems neuroscience.

References

ALBERTSON, D.G. & THOMSON, J.N. (1976). The pharynx of *Caenorhabditis elegans*. *Philosophical Transactions of the Royal Society B: Biological Sciences* 275, 299-325.

BALTZLEY, M.J., GAUDRY, Q. & KRISTAN, W.B.J. (2010). Species-specific behavioral patterns correlate with differences in synaptic connections between homologous mechanosensory neurons. *Journal of Comparative Physiology A: Neuroethology, Sensory, Neural, and Behavioral Physiology* 196, 181-197.

BUMBARGER, D.J., RIEBESELL, M., RODELSPERGER, C. & SOMMER, R.J. (2013). System-wide rewiring underlies behavioral differences in predatory and bacterial-feeding nematodes. *Cell* 152, 109-119.

CHIANG, J.T., STECIUK, M., SHTONDA, B. & AVERY, L. (2006). Evolution of pharyngeal behaviors and neuronal functions in free-living soil nematodes. *Journal of Experimental Biology* 209, 1859-1873.

COOK, S., HALL, D. & EMMONS, S. (2014). New understandings of information flow through the *C. elegans* pharynx. *Poster presented at* C. elegans *topic meeting: neuronal development, synaptic function & behavior*, July 7-10, 2014, Madison, WI, USA.

COWDEN, C., SITHIGORNGUL, P., BRACKLEY, P., GUASTELLA, J. & STRETTON, A.O. (1993). Localization and differential expression of FMRFamide-like immunoreactivity in the nematode *Ascaris suum*. *Journal of Comparative Neurology* 333, 455-468.

DE LEY, P. & BLAXTER, M. (2002). Systematic position and phylogeny. In: Lee, D.L. (Ed.). *The biology of nematodes*. London, UK, Taylor & Francis, pp. 1-30.

HOSCHITZ, M., BRIGHT, M. & OTT, J.A. (2001). Ultrastructure and reconstruction of the pharynx of *Leptonemella juliae* (Nematoda, Adenophorea). *Zoomorphology* 121, 95-107.

JARRELL, T.A., WANG, Y., BLONIARZ, A.E., BRITTIN, C.A., XU, M., THOMSON, J.N., ALBERTSON, D.G., HALL, D.H. & EMMONS, S.W. (2012). The connectome of a decision-making neural network. *Science* 337, 437-444.

JOHNSON, C.D., REINITZ, C.A., SITHIGORNGUL, P. & STRETTON, A.O. (1996). Neuronal localization of serotonin in the nematode *Ascaris suum*. *Journal of Comparative Neurology* 367, 352-360.

KATZ, P.S. & HARRIS-WARRICK, R.M. (1999). The evolution of neuronal circuits underlying species-specific behavior. *Current Opinion in Neurobiology* 9, 628-633.

NEWCOMB, J.M. & KATZ, P.S. (2009). Different functions for homologous serotonergic interneurons and serotonin in species-specific rhythmic behaviours. *Proceedings of the Royal Society B: Biological Sciences* 276, 99-108.

RAGSDALE, E.J., NGO, P.T., CRUM, J., ELLISMAN, M.H. & BALDWIN, J.G. (2011). Reconstruction of the pharyngeal corpus of *Aphelenchus avenae* (Nematoda: Tylenchomorpha), with implications for phylogenetic congruence. *Zoological Journal of the Linnean Society* 161, 1-30.

RAO, V.T., FORRESTER, S.G., KELLER, K. & PRICHARD, R.K. (2010). Localization of serotonin and dopamine in *Haemonchus contortus*. *International Journal for Parasitology* 41, 249-254.

RIVARD, L., SRINIVASAN, J., STONE, A., OCHOA, S., STERNBERG, P.W. & LOER, C.M. (2010). A comparison of experience-dependent locomotory behaviors and biogenic amine neurons in nematode relatives of *Caenorhabditis elegans*. *BMC Neuroscience* 11, 22.

ROBERTSON, W.M. (1975). A possible gustatory organ associated with the odontophore in *Longidorus leptocephalus* and *Xiphinema diversicaudatum*. *Nematologica* 21, 443-448.

ROBERTSON, W.M. (1979). Observations on the oesophageal nerve system of *Longidorus leptocephalus*. *Nematologica* 25, 245-254.

SAKURAI, A., NEWCOMB, J.M., LILLVIS, J.L. & KATZ, P.S. (2011). Different roles for homologous interneurons in species exhibiting similar rhythmic behaviors. *Current Biology* 21, 1036-1043.

SCHRÖDEL, T., PREVEDEL, R., AUMAYR, K., ZIMMER, M. & VAZIRI, A. (2013). Brain-wide 3D imaging of neuronal activity in *Caenorhabditis elegans* with sculpted light. *Nature Methods* 10, 1013-1020.

SHAW, S.R. & MEINERTZHAGEN, I.A. (1986). Evolutionary progression at synaptic connections made by identified homologous neurones. *Proceedings of the National Academy of Science of the United States of America* 83, 7961-7965.

SHAW, S.R. & MOORE, D. (1989). Evolutionary remodeling in a visual system through extensive changes in the synaptic connectivity of homologous neurons. *Visual Neuroscience* 3, 405-410.

SIDDIQI, M.R. (1970). *Oriverutus lobatus* gen. n., sp. n. and *Sicaguttur sartum* gen. n., sp. n. (Nematoda: Dorylaimoidea) from cultivated soils in Africa. *Nematologica* 16, 483-491.

SONG, B.M. & AVERY, L. (2013). The pharynx of the nematode *C. elegans*: a model system for the study of motor control. *Worm* 2. DOI:10.4161/worm.21833.

STEPANYANTS, A., HOF, P.R. & CHKLOVSKII, D.B. (2002). Geometry and structural plasticity of synaptic connectivity. *Neuron* 34, 275-288.

WHITE, J.G., SOUTHGATE, E., THOMSON, J.N. & BRENNER, S. (1986). The structure of the nervous system of the nematode *Caenorhabditis elegans*. *Philosophical Transactions of the Royal Society B: Biological Sciences* 314, 1-340.

ZHANG, Y.C. & BALDWIN, J.G. (2000). Ultrastructure of the post-corpus of *Zeldia punctata* (Cephalobina) for analysis of the evolutionary framework of nematodes related to *Caenorhabditis elegans* (Rhabditina). *Proceedings of the Royal Society of London B: Biological Sciences* 267, 1229-1238.

Nematology Monographs & Perspectives, 2015, Vol. 11, 385-407

Chapter 14

Bacterial interactions and the innate immune system

Amit SINHA [1] and Robbie RAE [2]

[1] *Department of Neurobiology, University of Massachusetts Medical School, 364 Plantation Street, Worcester, MA 01605, USA*
Amit.Sinha@umassmed.edu

[2] *School of Natural Sciences & Psychology, Liverpool John Moores University, Byrom Street, Liverpool, L3 3AF, UK*
r.g.rae@ljmu.ac.uk

Introduction

Nematodes and bacteria are the most numerous organisms on Earth with numbers of nematodes thought to exceed 1 million m^{-2} (Floyd *et al.*, 2002) and bacterial cells in 1 g of soil estimated to be 10^{10} (Faegri *et al.*, 1977). These two groups of organisms have evolved numerous relationships ranging from strict symbiosis, whereby nematodes rely on one bacterial symbiont for food and development, to pathogenicity, whereby bacteria kill nematodes. More specifically, human filarial nematodes, such as *Brugia malayi*, rely on their vertically transmitted endosymbiotic bacterium *Wolbachia* for fertility, development and survival (Taylor *et al.*, 2005). Marine nematodes from the subfamily Stilbonematinae have an ectosymbiotic relationship with sulphur oxidising bacteria that attach to the nematode cuticle, cover its body and provide nutrition (Ott *et al.*, 1991). One of the most striking relationships nematodes have with bacteria is that of the entomopathogenic nematodes from the genera *Steinernema* and *Heterorhabditis*, which use their symbiotic bacteria *Xenorhabdus* and *Photorhabdus*, respectively, to cause mortality to insect hosts (Forst *et al.*, 1997). The free-living genetic model nematodes *Pristionchus pacificus* and *Caenorhabditis elegans* are saprobic nematodes that use bacteria as a food source, which grow in decaying scarab beetles and rotting fruit, respectively. *Pristionchus* nematodes have a necromenic association with several beetle species (Herrmann *et al.*, 2006, 2007). Specifically, *P. pacificus* is attracted to beetle hosts by pheromones (Hong & Sommer, 2006; see Hong, Chapter 12, this volume); the nematode latches onto the passing beetles (Weller *et al.*, 2010; Brown *et al.*, 2011), waits for the host to die, exits the dauer stage and then feeds on associated proliferating fungi and bacteria (Rae *et al.*, 2008). Once the food supply is depleted then the nematodes develop into dauers and search for new beetle hosts. Whilst *C. elegans* and *P. pacificus* are both saprobic nematodes, they differ in one important anatomical feature: *C. elegans* has a grinder in the terminal bulb of the pharynx to break apart bacterial cells, whereas *P. pacificus* and all other species of the Diplogastridae do not have the grinder (Fig. 14.1), an anatomical difference that must affect the physiology and the innate immunity of these nematodes.

Over the past 12 years *C. elegans* has been developed as a model for studying the mechanistic processes governing innate immunity. In order to combat bacterial and fungal pathogens, *C. elegans* uses various signalling pathways, including ERK MAP kinase, p38 MAP

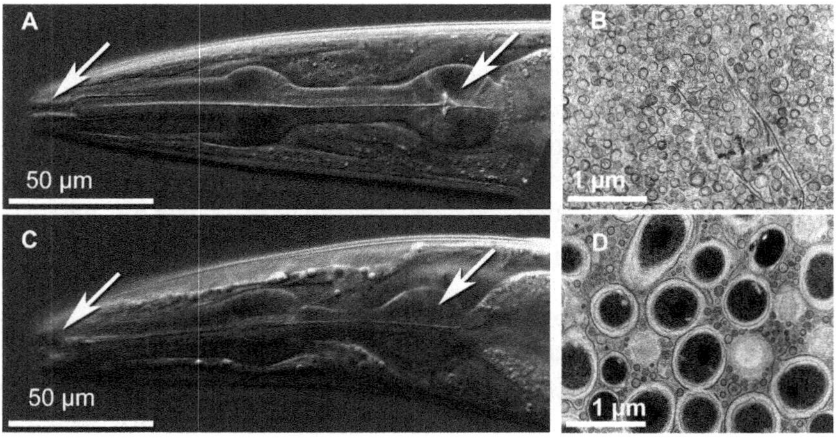

Fig. 14.1. Pristionchus pacificus *and* Caenorhabditis elegans *have marked differences in the morphology of their pharynx which affects the disruption of ingested bacteria. A:* C. elegans *pharynx has a grinder and long, narrow mouth-like suction pump; B: Disruption of the bacterial food* Escherichia coli *OP50 after passage through* C. elegans *pharynx; C:* P. pacificus *pharynx does not have a grinder, and is relatively shorter and broader; D: Incomplete disruption of bacterial food* E. coli *OP50 after passage through the pharynx of* P. pacificus. *Figure from Rae* et al. *(2008).*

kinase, TGF β, JNK-like MAP kinase, the G-protein coupled receptor FSHR-1, bZIP transcription factor *zip-2* and the beta-Catenin/*bar-1* (Ewbank, 2006). Many of these pathways regulate lectins, lysozymes and antimicrobial peptides. The majority of this research has focused on understanding the molecular mechanisms of the *C. elegans* innate immune system when fed opportunistic human pathogens, such as *Staphylococcus aureus*. Building on the results of this body of research and the catalogue of genes and genetic mechanisms that have previously been identified in *C. elegans*, investigations into how *P. pacificus* combats natural ecologically relevant bacterial pathogens have added an evolutionary and a comparative perspective to mechanistic studies of nematode-bacterial interactions.

A survey for naturally associated bacteria of *Pristionchus* nematodes

Pristionchus pacificus can be grown easily under laboratory conditions on *Escherichia coli* OP50 (Sommer *et al.*, 1996), yet before 2008, there was little information on what bacteria these nematodes feed on in nature and what bacterial species are present on the decomposing beetle host. *Pristionchus* species can be readily isolated from a selection of beetles species, *e.g.*, *P. pacificus* from the Oriental beetle (*Anomala orientalis*) (Herrmann *et al.*, 2007), *P. entomophagus* from dung beetles (*Geotrupes* sp.), *P. maupasi* from cockchafers (*Melolontha* sp.) and *P. uniformis* from the Colorado potato beetle (*Leptinotarsa decemlineata*) (Fig. 14.2A) (Herrmann *et al.*, 2006). This system allowed investigation and understanding of the natural bacterial associations these nematodes have in the beetle/host environment. Initial experiments concentrated on several aims: *i*) to understand the bacteria that are associated with several *Pristionchus* species emerging from host beetles; *ii*) to assess the effects these bacteria have on *Pristionchus* species; and *iii*) to assess survival of *P. pacificus* compared to *C. elegans* when fed human- and insect-pathogenic bacteria.

In order to answer these questions, standard microbiological techniques and metagenomics were used to profile the bacteria associated with *Pristionchus* nematodes when emerging from beetle hosts (Rae *et al.*, 2008). This metagenomic approach allowed unculturable bacteria that were present in the intestine of several specimens of *P. lheritieri* and *P. entomophagus* from soil to be profiled. It was shown that at least

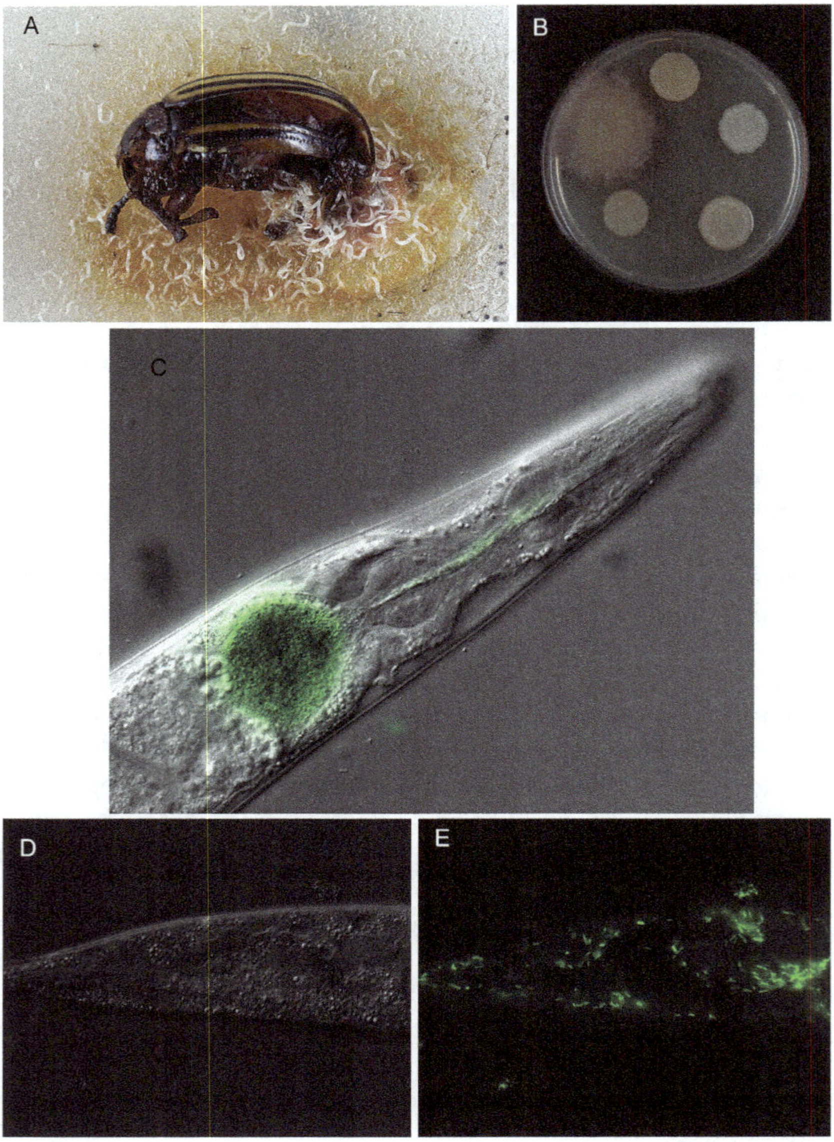

Fig. 14.2. *A:* Pristionchus uniformis *emerging from a dead Colorado beetle and feeding on the associated microorganisms including bacteria and fungi; B: A selection of bacteria isolated from* P. maupasi *emerging from its cockchafer host on LB plate; C:* P. pacificus *feeding on GFP labelled* Serratia marcescens, *which accumulates at the anterior of the intestine behind the pharynx;*

40 different species of bacteria were present in the gut of these nematodes, many of which were animal- and plant-pathogenic bacteria, including *Bordetella* sp., *Burkholderia* sp., *Agrobacterium* sp. and *Microbacterium* sp. However, it must be noted that, by taking this approach, bacteria from the genus *Bacillus* were under-represented because they exist as heat-resistant spores and the cell lysis protocol will not break apart these cells. To remedy this, bacteria from several *Pristionchus* species emerging from beetles were also cultured on standard LB plates (Rae *et al.*, 2008). This allowed quantification of the exact effects these bacteria were having on each *Pristionchus* species when fed individually. Primarily, five bacteria were isolated from *P. entomophagus* emerging from dung beetles (and also five bacteria isolated from soil) (Fig. 14.2B), eight bacteria from *P. maupasi* from cockchafers and four bacteria from *P. pacificus* from Oriental beetles. The most common bacteria isolated were *Bacillus* and *Pseudomonas* species, as well as pathogens such as *Serratia* spp. (Fig. 14.2C, D).

Chemotaxis assays were used to assess the behaviour of each respective *Pristionchus* species when exposed to its associated bacteria (*vs* an *E. coli* OP50 control) (Rae *et al.*, 2008). There was little species specificity, with each *Pristionchus* species responding similarly to all bacteria tested regardless of those isolated from their host beetle. This is in contrast to the *Pristionchus*-scarab beetle association that is largely species-specific (Hong *et al.*, 2008) and to the highly specific association of entomopathogenic nematodes with their respective bacterial symbionts. Nonetheless, one striking result was that, when *Pristionchus* nematodes were exposed to a strain of *Bacillus thuringiensis*, they displayed strong aversion and had dramatically reduced fecundity and development time compared to all the other bacterial strains. In separate experiments, insect pathogens (*Xenorhabdus nematophila*, *Xenorhabdus* spp. and *Photorhabdus luminescens*) and opportunistic human pathogens (*Pseudomonas aeruginosa* and *S. aureus*) were fed to both *C. elegans* and *P. pacificus*. Surprisingly, it was demonstrated that, unlike *C. elegans* that dies when fed *P. aeruginosa* and *S. aureus* (Tan *et*

D: Over time S. marcescens *enters the body of the nematode by lysing the intestinal wall (upper panel) and can be visualised growing throughout the nematode by fluorescence microscopy (lower panel). The bacterium will then go on to reproduce prolifically on the decomposing nematode.*

al., 1999), *P. pacificus* was remarkably resistant (Rae *et al.*, 2008). Similarly, it had also been shown that *P. pacificus* was the only nematode from a diverse collection of species that was resistant to the Cry 5B toxin from *B. thuringiensis* (Wei *et al.*, 2003). How can *P. pacificus* cope with such pathogens? Upon sequencing of the *P. pacificus* genome, it was shown that (compared with *C. elegans*) there were many more genes involved with detoxification of xenobiotic compounds, such as cytochrome P450 enzymes, glucosyl transferases and ABC transporters (Dieterich *et al.*, 2008). Apart from these differences in genome architecture, the absence of the grinder might be another reason for this extreme resistance (Fig. 14.1). Without the grinder, whole bacterial cells might enter and leave the intestine without the toxins inside being released and killing the host. Interestingly, it had already been demonstrated decades ago that *Pristionchus* nematodes could harbour and transport not just bacteria but an array of organisms that were able to survive the *Pristionchus* gut. For example, *P. lheritieri* has been shown to transport beneficial bacteria such as *Rhizobium japonicum*, plant-pathogenic bacteria such as *Agrobacterium tumefaciens*, *Erwinia amylovora*, *E. carotovora* and *Pseudomonas phaseolicola*, and human pathogens such as *Salmonella typhi*, *S. wichita* and *Serratia marcescens* (Chantanao & Jensen, 1969a; Smerda *et al.*, 1971; Jatala *et al.*, 1974). Other organisms that have survived the *Pristionchus* gut include an unnamed phage of *A. tumefaciens* (Chantanao & Jensen, 1969b), four species of green algae (Leake & Jensen, 1970) and fungi that caused potato wilt (*Fusarium oxysporum* and *Verticillium dahliae*) (Jensen & Siemer, 1971). How *P. pacificus* gains energy and nutrition from bacteria and fungi without breaking up cells remains a complete mystery.

Nematode and *Bacillus* interactions

Assessing the pathogenicity of *Bacillus* vegetative cells fed to *P. pacificus* and *C. elegans*

As soil nematodes, *Pristionchus* species are likely to be in contact with *Bacillus* bacteria, which, when ingested, were shown to have extreme effects on brood size and behaviour (Rae *et al.*, 2008). *Bacillus* bacteria are extremely numerous in the soil system with an estimated 10^4-10^6 g^{-1} soil (Martin & Travers, 1989). In general, nematodes such as *C. elegans* and *P. pacificus* avoid *Bacillus* species, *e.g.*, *B. thuringiensis* (Hasshoff *et al.*, 2007), and some species have been shown to be pathogenic to nematodes (Wei *et al.*, 2003). In order to investigate these interactions further, a soil survey of *Bacillus* from around Europe was conducted with the main aim of understanding how pathogenic these bacteria can be to both *P. pacificus* and *C. elegans* (Rae *et al.*, 2010). Soil samples were collected from numerous locations in the UK and around Tübingen, Germany. *Bacillus* can be easily isolated as they are resistant to extreme heat; therefore, soil samples can be mixed with buffer and heated to 80°C to kill the resident bacterial community and then plated out onto standard LB plates. These colonies were then picked individually into 96-well plates filled with LB, grown overnight and then mixed with 50% glycerol and stored at −20°C indefinitely. As well as collecting *Bacillus* from soil, they were also extracted from horse dung and dung beetles (*Geotrupes* sp.), which are hosts for *P. entomophagus*. In total, 768 *Bacillus* strains were fed to both *C. elegans* and *P. pacificus* in a high-throughput agar-based assay and screened for candidates that caused death of nematodes over 5 days. To gain an idea of the diversity of *Bacillus* sampled, the 16S rRNA gene of 768 *Bacillus* strains was sequenced. The most common species isolated were *Bacillus* sp. CM-B72, *Bacillus* sp. RA51, *Bacillus weihenstephanensis*, *B. cereus*, *B. longisporus*, as well as common species such as *B. mycoides*, *B. pumilus*, *B. licheniformis*, *B. subtilis* and *B. simplex*. Twenty *Bacillus* strains that could kill both or either *C. elegans* and *P. pacificus* in less than 5 days were identified. When the nematodes were fed these strains they also had severely affected development time and brood size in both nematodes. The most interesting strains identified were three *B. thuringiensis* strains isolated from dung beetles from around Tübingen, which killed *C. elegans* in less than 12 h (Rae *et al.*, 2010). These strains were designated *B. thuringiensis* DB7, 27 and 73 and were shown to be the most pathogenic bacteria to date that can kill *C. elegans* when grown on standard Nematode Growing Media (NGM) agar. The pathogenicity of bacterial pathogens can be enhanced by growing on specialist agar, for example *P. aeruginosa*, grown on a fast-killing medium, can kill *C. elegans* within 24 h (Tan *et al.*, 1999) but on NGM it takes several days. The most remarkable aspect of this finding was not only that these *B. thuringiensis* strains were incredibly pathogenic to *C. elegans*, but that *P. pacificus* remained resistant. Indeed, *P. pacificus* can live, feed and reproduce on these strains for many days, whereas *C. elegans* can only survive for 8 h. To investigate this further, a selection of previously known bacterial-resistant *C. elegans* mutants were used to discover whether they too would be resistant to *B. thuringiensis* DB27. Principally, *C. elegans daf-2* and *age-1* mutants are resistant to gram-negative and gram-positive pathogenic bacteria, such as *Enterococcus faecalis*, *P. aeruginosa* and *S. aureus* (Garsin *et al.*, 2003). Also, *bre* (bacillus resistant) mutants are resistant to *B. thuringiensis* Cry 5B toxin due to changes in glycolipids in the intestine (Griffitts *et al.*, 2005). When these *C. elegans* mutants were fed *B. thuringiensis* DB27 they showed no significant increase in survival, indicating that these genes were not responsible for aiding the immune response towards *B. thuringiensis* DB27 and that resistance would require a different molecular mechanism.

In order to discover how both *C. elegans* and *P. pacificus* combatted *B. thuringiensis* DB27, large-scale forward genetic screens were carried out to isolate *C. elegans* mutants that were resistant to this bacterium, and *P. pacificus* mutants that were hypersusceptible to *B. thuringiensis* DB27. Several *C. elegans* mutants were isolated that were resistant to *B. thuringiensis* DB27, including the nuclear autoantigenic sperm protein *nasp-1* (Iatsenko *et al.*, 2013). Surprisingly, detailed analysis of *nasp-1* revealed a new role of the endonuclease *dcr-1*/dicer in *C. elegans* innate immunity (Iatsenko *et al.*, 2013). The high specificity of *dcr-1* mutant resistance also suggests that the virulence mechanisms on the bacterial side must be distinct from other *Bacillus* strains. Indeed, genome sequencing of *B. thuringiensis* DB27 revealed the existence of multiple plasmids that are often known to carry toxin genes (Iatsenko *et al.*, 2014a). Detailed molecular studies resulted in the identification of two novel Cry toxins of *B. thuringiensis* DB27 involved in virulence against *C. elegans* (Iatsenko *et al.*, 2014b). By contrast, another nematicidal isolate, *B. thuringiensis* 4A4, also had an effect on *P. pacificus*, and molecular investigations showed a multifactorial nature of virulence based on the Cry toxins Cry21Ha and Cry1Ba and β-exotoxins (Iatsenko *et al.*, 2014c). Consistent with these findings were also the negative findings of genetic screens in *P. pacificus*. When the same genetic screening procedure was repeated with *P. pacificus*, but looking for hypersusceptible mutants that would die on *B. thuringiensis* DB27, no mutants could be isolated. While representing a negative result, this observation provides the first evidence that *P. pacificus* might have evolved a strong systemic response of resistance to soil bacteria, maybe in response to the lack of the grinder.

ASSESSING PATHOGENICITY OF *BACILLUS* SPORES FED TO *P. PACIFICUS* AND *C. ELEGANS*

In nature, *Bacillus* bacteria exist as heat-resistant spores and are the life stage responsible for causing infection to mammals and invertebrates. Therefore, another screen was carried out in order to try to understand the pathogenicity of spore-grown *Bacillus* towards nematodes. *Bacillus* can be easily grown to spore stage after 7-10 days in BT medium (Rae *et al.*, 2010). The spores were then purified using heat treatment and ethanol to remove the remaining vegetative cells and then fed to nematodes on NGM plates without the addition of peptone to inhibit growth of spores. Four hundred *Bacillus* strains were grown to spore stage and fed to both *C. elegans* and *P. pacificus*. Most of the *Bacillus* strains managed to support growth of both nematodes and they could develop and lay eggs, albeit at a lower number than controls. Six *Bacillus* strains grown to spore stage were pathogenic to *C. elegans* but not to *P. pacificus*. Interestingly, when the bacteria were grown to normal vegetative cell stage, *C. elegans* did not die. Thus, higher resistance of *P. pacificus* to bacteria seems to represent a general phenomenon and pathogenic effects are often specifically associated with either the vegetative or the spore stage.

One particularly virulent *Bacillus* strain (*Bacillus* sp. 142) killed *C. elegans* in 3-5 days, whereas *P. pacificus* remained resistant (Rae *et al.*, 2012a). In order to understand the genes responsible for increased resistance of *P. pacificus* towards *Bacillus* sp. 142, hypersusceptible mutants that died after being fed this bacterium were isolated by EMS mutagenesis and two *P. pacificus* mutants that died after 4-5 days of feeding on *Bacillus* sp. 142 were found. These genes were identified as *Ppa-unc-1* and *Ppa-unc-13* (Rae *et al.*, 2012a); *unc* 'uncoordinated' mutant animals have problems with locomotion; they remain stationary and feed in pulses compared to *P. pacificus* wild type. *Ppa-unc-1* encodes the TWITCHIN protein and is essential for myosin function in muscle cells. *Ppa-unc-13* encodes a diacylglycerol binding protein that when mutated reduces neurosecretion (Rae *et al.*, 2012a). *Ppa-unc-1* and *Ppa-unc-13* are homologous to *C. elegans unc-22* and *unc-13*, respectively. During further characterisation of these genes, it was evident that mutant animals have problems digesting food as they retained bacteria for long periods of time. The normal defecation cycle of *P. pacificus* wild type is around 93 s but *Ppa-unc-1* and *Ppa-unc-13* have severely extended defecation times of 129.0 ± 5.69 s and 131.9 ± 8.06 s, respectively. A similar effect was also shown for *C. elegans* defecation mutants, *e.g.*, *egl-8 (md1971)*, *unc-33 (e204)*, *unc-16 (n730)*, *unc-2 (e55)* and *unc-13 (e1091)* that died significantly faster than normal *C. elegans* wild type when fed either *Bacillus* sp. 142 or *S. aureus* (Rae *et al.*, 2012a). Thus, increased residence time in the nematode gut can increase susceptibility to pathogens in two diverse nematode species, highlighting the importance of first line defence mechanisms against bacterial pathogens and underpinning the complexity between immunity and physiology.

Systems biology analysis of *P. pacificus* and *C. elegans* exposed to several pathogens

To search for the genes involved in pathogen response of *C. elegans* and *P. pacificus*, whole-genome gene expression profiling experiments were performed using custom designed *P. pacificus* microarrays and commercially available *C. elegans* microarrays (Sinha *et al.*, 2012a).

Caenorhabditis elegans and *P. pacificus* were fed two gram-negative bacteria (*S. marcescens* and *X. nematophila*) and two gram-positive bacteria (*Staphylococcus aureus* and *B. thuringiensis* DB27) and their transcriptional response was assessed. Both nematodes were susceptible to the gram-negative bacteria tested but *P. pacificus* was resistant to *S. aureus* and *B. thuringiensis* DB27, whereas *C. elegans* was susceptible.

Young adults of *C. elegans* and *P. pacificus* were exposed to each bacterium for 4 h instead of later time-points so that early response genes, instead of genes associated with tissue necrosis and cell death observed at later time-points of exposure to pathogens, could be identified (Wong *et al.*, 2007). Based on these gene expression data, two kinds of analysis could be made: comparison of expression profiles of within a nematode species across different pathogens, and comparison of expression profiles across the two nematode species when exposed to the same bacteria for the same length of time. This comprehensive and systematic analysis revealed many interesting features of pathogen response machinery in both nematodes.

Pathogen exposure resulted in a marked reduction in the total RNA that could be extracted per worm, suggesting that a global transcriptional repression could be potentially one of the earliest effects of pathogenic assault, in addition to the well documented translational repression by bacteria toxins (*e.g.*, Dunbar *et al.*, 2012; Kleino & Silverman, 2012; McEwan *et al.*, 2012; Mohr & Sonenberg, 2012). Further, more genes were found to be downregulated than upregulated in both the nematodes in response to the gram-negative bacteria, while the trend was the opposite for gram-positive bacteria. It will be interesting to see whether this is a general trend or specific to only these four bacteria used in this study. Also, the number of genes differentially expressed in both nematodes varied with the rate of lethality induced by each bacterium, such that more lethal bacteria induced the differential expression of a larger number of genes (Sinha *et al.*, 2012a).

Within-species comparisons of pathogen response genes in each nematode across different pathogens revealed a remarkable specificity in their respective innate immune responses (Fig. 14.3). Within both *C. elegans* and *P. pacificus* there was surprisingly little overlap between genes differentially expressed across different pathogens (Fig. 14.3). Although the gene-by-gene overlap between different expression profiles

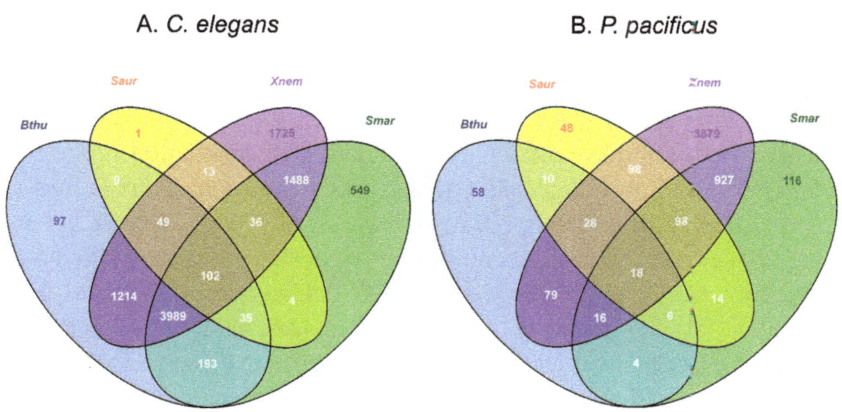

Fig. 14.3. *Induction of a pathogen specific response is evident by the small number of genes found to be common across expression profiles induced in response to four different bacteria in both nematode species. A:* Caenorhabditis elegans*; B:* Pristionchus pacificus. *Abbreviations: Bthu =* Bacillus thuringiensis*; Saur =* Staphylococcus aureus*; Smar =* Serratia marcescens*; and Xnem =* Xenorhabdus nematophila*). Figure from Sinha* et al. *(2012a).*

within each nematode was low, a Pfam protein domain-based analysis revealed enrichment for proteins with common functional domains, *e.g.*, proteasome function in *C. elegans* and proteins involved in lipid metabolism such as lipases and fatty acid desaturases in *P. pacificus*. Both *C. elegans* and *P. pacificus* genomes contain hundreds of genes that encode for C-type lectins and collagen proteins, which have been proposed to have a role in pathogen response (Schulenburg *et al.*, 2008; Iatsenko *et al.*, 2013) and differential expression for many genes encoding proteins of these families was observed. However, there was little commonality between the proteins of either of these families differentially expressed across various pathogens in the same nematode species, indicating that such gene families might have expanded in response to the variety of pathogens encountered in the wild and have acquired specificity during their evolution. It seems that both nematodes are capable of mounting a bacteria-specific response, which is qualitatively different from a generalised stress response.

Evolutionary trends in the innate immune response were revealed by comparing the expression profiles across the two nematodes when exposed to the same pathogen and analysing similarities and differences across the well-defined 1 : 1 orthologous genes (Fig. 14.4). The expression profiles of the two nematodes were more similar to each other in the case of gram-negative bacteria *S. marcescens* and *X. nematophila*, which are equally pathogenic to both *C. elegans* and *P. pacificus* (Fig. 14.4). Most of the genes commonly regulated across the two nematodes were involved in germline function and regulation of translation processes, consistent with the observed translational repression (Dunbar *et al.*, 2012; Kleino & Silverman, 2012; McEwan *et al.*, 2012; Mohr & Sonenberg, 2012) and reduced fecundity upon exposure to various pathogens (*e.g.*, Tan *et al.*, 1999; Marroquin *et al.*, 2000; Garsin *et al.*, 2001; Mylonakis *et al.*, 2002; Tang *et al.*, 2005). Nonetheless, an equally large fraction of differentially expressed genes in response to these gram-negative bacteria did not show any overlap across the two nematodes, underscoring the divergent components of their innate immunity (Fig. 14.4). Moreover, in response to the gram-positive bacteria *B. thuringiensis* and *S. aureus*, which have widely different effects on survival of *C. elegans* and *P. pacificus*, the differentially expressed genes showed a negligible overlap (Fig. 14.4), providing more support to the hypothesis that the innate immune response has widely diverged over the course of evolution in response to the different microbes encountered by each nematode in

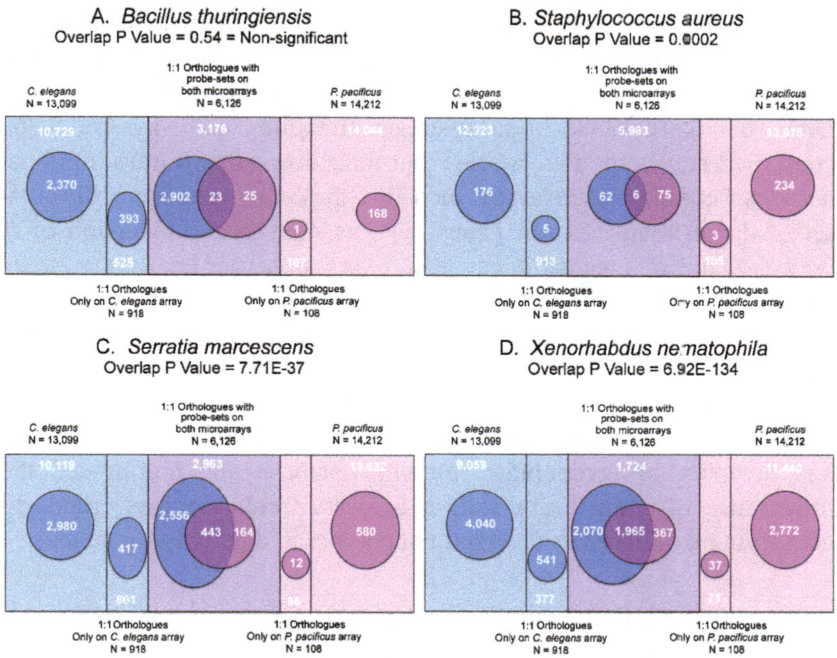

Fig. 14.4. Caenorhabditis elegans *and* Pristionchus pacificus *respond to the same pathogen by regulating very different set of genes. Only a small number of 1 : 1 orthologues were found to be common in the gene expression profiles of* C. elegans *and* P. pacificus *in response to: A:* Bacillus thuringiensis*; B:* Staphylococcus aureus*; C:* Serratia marcescens*; and D:* Xenorhabdus nematophila. *Also, the number of species-specific genes differentially expressed in response to various pathogens is much larger than the number of 1 : 1 orthologous gene pairs. The rectangular boxes represent the entire set of genes represented on microarrays of respective species and their region of overlap indicates the 1 : 1 orthologues across* C. elegans *and* P. pacificus. *The ovals indicate the respective subsets of differentially expressed genes. Figure from Sinha* et al. *(2012a).*

their respective ecological niche. This conclusion was further supported by the observation that about 10-20% of the *P. pacificus* gene expression profiles comprised lineage specific 'pioneer genes' (Borchert *et al.*, 2010), which do not have an orthologue in other organisms.

These analyses so far are based on comparison of downstream effector molecules induced in response to pathogens. To identify the potential upstream regulators of these genes, their overlap with known targets of key innate immunity regulators was also analysed. It was found that DAF-16/FOXO, TGF-beta and p38 MAP Kinase pathways played significant roles on all pathogens, except in the case of *B. thuringiensis*, which did not show differential activation of a significant number of DAF-16 targets. Thus, these upstream signalling pathways might be conserved across the two species, but their downstream transcriptional networks could have diverged such that they now regulate a different set of downstream effector genes, *e.g.*, as seen in a comparison of *P. pacificus* and *C. elegans* dauer gene expression profiles (Sinha *et al.*, 2012b). It will be interesting to test these hypotheses when mutants affecting the genetic components of these pathways become available in *P. pacificus*.

In summary, this comprehensive systems-biology approach between *P. pacificus* and *C. elegans* revealed an under-appreciated complexity and diversity of invertebrate immune response and highlighted the importance of comparative studies across different species crucial to identifying conserved *vs* diverged phenotypes and underlying genes.

Sexual reproductive system signals that increase resistance to bacterial pathogens and lifespan in *P. pacificus*

At the beginning of development in both *C. elegans* and *P. pacificus* (first- and second-stage juveniles) the reproductive system is composed of four cells (Z1, Z2, Z3, Z4). These cells eventually give rise to the gonad (derived from Z1, Z4) and the germ line (derived from Z2, Z3). Interestingly, by removing the germline cells by laser microsurgery, Hsin & Kenyon (1999) showed that the lifespan of *C. elegans* can be extended by 60%. This increase was not due to a trade-off between sterility and longevity as animals that have the gonad removed are sterile but only live as long as the *C. elegans* wild type. Thus, a signal produced from the remaining somatic gonad cells can increase lifespan. There are still numerous questions about this fascinating phenomenon that needed answering. For example, how conserved is this response throughout the Nematoda? What genetic mechanisms are responsible for extending lifespan? Does removal of the germ line also affect any other survival phenotypes?

As with *C. elegans*, the lifespan and resistance to bacterial pathogens (*S. marcescens* and *X. nematophila*) of *P. pacificus* can be extended significantly when germline cells are removed *via* laser microsurgery (Rae *et al.*, 2012b). This was also shown in several *Pristionchus* species and *P. pacificus* strains. In order to understand the genetic processes governing this response, whole genome microarrays comparing non-ablated *P. pacificus* with germline-ablated *P. pacificus* (Z2, Z3 ablated) were performed. This experimental procedure was repeated with *P. pacificus* feeding on *E. coli* OP50 (to understand the genes involved with lifespan) and the natural bacterial pathogen *S. marcescens* (to understand the genes involved with immunity). In total, 3300 genes were found to be differential expressed between germline-intact and germline-ablated *P. pacificus*, and many were involved with processes such as lipid metabolism, proteasomal maintenance and nuclear pore complexes. Also, many of the genes upregulated were known targets of the FOXO-like transcription factor DAF-16 and the nuclear hormone receptor DAF-12, which were shown to be also involved in dauer formation. As numerous mutants have been isolated and identified in *P. pacificus* as being responsible for dauer formation (Ogawa *et al.*, 2009; Bento *et al.*, 2010), the germ line of *Ppa-daf-16* (*tu302* and *tu901*) and *Ppa-daf-12* (*tu390* and *tu389*) was ablated and survival was monitored when these animals were fed *S. marcescens*. Survival of these germline-ablated mutants was remarkably decreased, proving that these components of the dauer pathway were indeed essential for increased resistance to pathogens and that the gonad signal was acting on these pathways. Additionally, it could also be shown that germline ablation-dependent extension in both immunity and lifespan were similarly regulated, as evidenced by very few differences observed in the gene expression profiles across the two scenarios. Therefore, it can be asked if investment in the immune system can increase longevity. For example, it has been shown in nature that longer-lived albatrosses, domestic hens and sheep have higher antibody counts (Graham *et al.*, 2010). Under laboratory conditions, mutants originally isolated as having a longer lifespan were also more resistant to stressors. For example, long-lived *C. elegans age-1* and *daf-2* mutants were resistant to heat, UV light, oxidative stresses such as H_2O_2 and the herbicide Paraquat® (1,1′-Dimethyl-4,4′-bipyridinium dichloride), high oxygen tension, heavy metals and pathogenic bacteria such as *P. aeruginosa* and *E. faecalis* (Garsin *et al.*, 2003). Also, across the *Caenorhabditis* genus, longer-lived species were more resistant to the fungus *Cryptococcus neoformans* and the bacteria *P. aeruginosa* and *S. aureus* (Van den Berg *et al.*, 2006; Amrit *et al.*, 2010). Therefore, investment in a strong immune system could be the reason that lifespan was increased substantially and that the ability to respond to stress is a rate-determining factor leading to ageing and senescence.

Conclusions and questions for the future

It is only recently that researchers have started to understand the evolutionary relationship of nematodes with bacteria. For example, the associated microbiome of the pine wood nematode (*Bursaphelenchus xylophilus*) has been analysed and is responsible for allowing nematodes to cope with pinenes and various compounds produced by secondary metabolism of the pine host (Cheng *et al.*, 2013). Also, the microbiome of the free-living nematode, *Acrobeloides maximus*, has been shown to include three genera, such as *Ochrobactrum*, *Pedobacter* and *Chitinophaga*, and it has been speculated that this is a symbiotic relationship, although the benefit of retaining these bacteria remains unknown (Baquiran *et al.*, 2013). Stilbonematinae marine nematodes have an ectosymbiotic relationship with bacteria, which attach to the nematode cuticle for food (Ott *et al.*, 1991). By contrast, nematodes from the genera *Astomonema*, *Parastomonema* and *Rhabdothyreus* have endosymbiotic relationships with bacteria, which fill the gut of the nematode and are thought to provide nutrition (Musat *et al.*, 2007). Also, bacteria have been shown to be in association with the soybean cyst nematode *Heterodera glycines* (Nour *et al.*, 2003), the burrowing nematode *Radopholus* spp. (Haegeman *et al.*, 2009) and the dagger nematode *Xiphinema americanum* (Vandekerckhove *et al.*, 2000). Although *C. elegans* has provided a remarkable insight into how the innate immune system copes with bacterial pathogens, there is a lack of information on what bacteria these nematodes associate with in nature. The only study so far by Grewal (1991) showed several species of bacteria, including *Acinetobacter* sp., *Bacillus* sp., *Pseudomonas* sp. and *Enterobacter* sp., which altered the growth and fertility of *C. elegans* when used as a food source. With the advent of cheaper sequencing and high throughput genomics, it should be possible to profile the bacteria present in most nematode species and to understand functionally the effects of mutualism, parasitism and symbiosis on both nematode and bacteria and how these remarkable relationships evolved.

Currently, there are several bacteria to which *P. pacificus* is resistant but *C. elegans* is not, namely *P. aeruginosa*, *S. aureus*, *B. thuringiensis*

DB27, *Bacillus* sp. 142 and *B. thuringiensis* Cry 5B toxin (Wei *et al.*, 2003; Rae *et al.*, 2008, 2010, 2012a). How did *P. pacificus* evolve such a robust innate immune system compared to *C. elegans*? What genes are essential for coping with pathogenic bacteria and how do they differ from *C. elegans*? Or perhaps the difference in morphology is more pertinent. As *P. pacificus* does not have a grinder, it swallows its bacterial food whole and does not break open the bacterial cellular wall, which would release the associated toxins. Indeed, it has been known for many years that numerous organisms can survive passage through the gut, including algae, fungi, phages and bacteria. It also remains a complete mystery how *P. pacificus* actually gains any nutrition from these potential food sources when there is no observable lysis of bacterial cells and they are expelled in just 93 s. With the advent of high throughput metabolic analysis perhaps these questions can be addressed.

All research detailed in this chapter was based on one strain of *P. pacificus* isolated from California in 1988. During this time this strain has been cultured under laboratory conditions with a monoxenic diet of *E. coli* OP50 for many years. In order to gain a more realistic ecological perspective of the genes involved with immunity, it is imperative to use recently collected *P. pacificus* strains that have not been grown on *E. coli* OP50 for such a long time. Currently, there are over 600 *P. pacificus* strains collected from around the world that exhibit natural variation in attraction behaviour towards the insect pheromone EDTA (Hong *et al.*, 2008; see Hong, Chapter 12, this volume) and dauer formation (Mayer & Sommer, 2011). By screening through these strains for increased or decreased resistance to naturally associated bacteria, and combining with next generation sequencing approaches such as RAD-seq (Restriction site Associated DNA sequencing) and GWAS (Genome Wide Association Studies), it should now be possible to identify the loci responsible for resistance to bacterial pathogens and to understand the evolutionary history of such alleles.

Acknowledgements

Both authors are extremely grateful to Hanh Witte, who supplied great technical expertise throughout this research and was critical in method development, brute force screening and large scale sequencing of both nematodes and bacteria.

References

AMRIT, F.R.G., BOEHNISCH, C.M.L. & MAY, R.C. (2010). Phenotypic covariance of longevity, immunity and stress resistance in the *Caenorhabditis* nematodes. *PLoS ONE* 5, e9978.

BAQUIRAN, J.-P., THATER, B., SEDKY, S., DE LEY, P., CROWLEY, D. & ORWIN, P.M. (2013). Culture-independent investigation of the microbiome associated with the nematode *Acrobeloides maximus*. *PLoS ONE* 8, e67425.

BENTO, G., OGAWA, A. & SOMMER, R.J. (2010). Co-option of the endocrine signaling module Dafachronic Acid-DAF-12 in nematode evolution. *Nature* 466, 494-497.

BORCHERT, N., DIETERICH, C., KRUG, K., SCHÜTZ, W., JUNG, S., NORDHEIM, A., SOMMER, R.J. & MACEK, B. (2010). Proteogenomics of *Pristionchus pacificus* reveals distinct proteome structure of nematode models. *Genome Research* 20, 837-846.

BROWN, F.D., D'ANNA, I. & SOMMER, R.J. (2011). Host finding behaviour in the nematode *Pristionchus pacificus*. *Proceedings of the Royal Society B: Biological Sciences* 278, 3260-3269.

CHANTANAO, A. & JENSEN, H.J. (1969a). Saprozoic nematodes as carriers and disseminators of plant pathogenic bacteria. *Journal of Nematology* 1, 216-218.

CHANTANAO, A. & JENSEN, H.J. (1969b). Transmission of an *Agrobacterium tumefaciens* phage by *Pristionchus lheritieri*. *Journal of Nematology* 1, 166-168.

CHENG, X.-Y., TIAN, X.-L., WANG, Y.-S., LIN, R.-M., MOA, Z.-C., CHEN, N. & XIE, B.-Y. (2013). Metagenomic analysis of the pinewood nematode microbiome reveals a symbiotic relationship critical for xenobiotics degradation. *Scientific Reports* 3, 1869. DOI:10.1038/srep01869.

DIETERICH, C., CLIFTON, S.W., SCHUSTER, L.N., CHINWALLA, A., DELEHAUNTY, K., DINKELACKER, I., FULTON, L., FULTON, R., GODFREY, J., MINX, P. *ET AL.* (2008). The *Pristionchus pacificus* genome provides a unique perspective on nematode lifestyle and parasitism. *Nature Genetics* 40, 1193-1198.

DUNBAR, T.L., YAN, Z., BALLA, K.M., SMELKINSON, M.G. & TROEMEL, E.R. (2012). *C. elegans* detects pathogen-induced translational inhibition to activate immune signaling. *Cell Host Microbe* 11, 375-386.

EWBANK, J.J. (2006). Signaling in the immune response. In: The *C. elegans* Research Community (Ed.). *Wormbook*. DOI:10.1895/wormbook.1.83.1.

FAEGRI, A., TORSVIK, V.L. & GOKSOYR, J. (1977). Bacterial and fungal activity in soil: separation of bacteria and fungi by a rapid fractionated centrifugation technique. *Soil Biology & Biochemistry* 9, 105-112.

FLOYD, R., ABEBE, E., PAPERT, A. & BLAXTER, M. (2002). Molecular barcodes for soil nematode identification. *Molecular Ecology* 11, 839-850.

FORST, S., DOWDS, B., BOEMARE, N. & STACKEBRANDT, E. (1997). *Xenorhabdus* spp. and *Photorhabdus* spp.: bugs that kill bugs. *Annual Review of Microbiology* 51, 47-72.

GARSIN, D.A., SIFRI, C.D., MYLONAKIS, E., QIN, X., SINGH, K.V., MURRAY, B.E., CALDERWOOD, S.B. & AUSUBEL, F.M. (2001). A simple model host for identifying Gram-positive virulence factors. *Proceedings of the National Academy of Sciences of the United States of America* 98, 10892-10897.

GARSIN, D.A., VILLANUEVA, J.M., BEGUN, J., KIM, D.H., SIFRI, C.D., CALDERWOOD, S.B., RUVKUN, G. & AUSUBEL, F.M. (2003). Long-lived *C. elegans daf-2* mutants are resistant to bacterial pathogens. *Science* 300, 1921.

GRAHAM, A.L., HAYWARD, A.D., WATT, K.A., PILKINGTON, J.G., PEMBERTON, J.M. & NUSSEY, J.H. (2010). Fitness correlates of heritable variation in antibody responsiveness in a wild mammal. *Science* 330, 662-665.

GREWAL, P.S. (1991). Influence of bacteria and temperature on the reproduction of *Caenorhabditis elegans* (Nematoda: Rhabditidae) infesting mushrooms (*Agaricus bisporus*). *Nematologica* 37, 72-82.

GRIFFITTS, J.S., HASLAM, S.M., YANG, T., GARCZYNSKI, S.F., MULLOY, B., MORRIS, H., CREMER, P.S., DELL, A., ADANG, M.J. & AROIAN, R.V. (2005). Glycolipids as receptors for *Bacillus thuringiensis* crystal toxin. *Science* 307, 922-925.

HAEGEMAN, A., VANHOLME, B., JACOB, J., VANDEKERCKHOVE, T.T.M., CLAEYS, M., BORGONIE, G. & GHEYSEN, G. (2009). An endosymbiotic bacterium in a plant-parasitic nematode: member of a new *Wolbachia* super group. *International Journal for Parasitology* 39, 1045-1054.

HASSHOFF, M., BOEHNISCH, C., TONN, D., HASERT, B. & SCHULENBURG, H. (2007). The role of *Caenorhabditis elegans* insulin-like signaling in the behavioural avoidance of pathogenic *Bacillus thuringiensis*. *FASEB Journal* 21, 1808-1812.

HERRMANN, M., MAYER, E.W. & SOMMER, R.J. (2006). Nematodes of the genus *Pristionchus* are closely associated with scarab beetles and the Colorado potato beetle in western Europe. *Zoology* 109, 96-108.

HERRMANN, M., MAYER, W., HONG, R., KIENLE, S., MINASAKI, R. & SOMMER, R.J. (2007). The nematode *Pristionchus pacificus* (Nematoda: Diplogastridae) is associated with the Oriental beetle *Exomala orientalis* (Coleoptera: Scarabaeidae) in Japan. *Zoological Science* 24, 883-889.

HONG, R.L. & SOMMER, R.J. (2006). Chemoattraction in *Pristionchus* nematodes and implication for insect host recognition. *Current Biology* 16, 2359-2365.

Hong, R.L., Witte, H. & Sommer, R.J. (2008). Natural variation in *P. pacificus* insect pheromone attraction involves the protein kinase EGL-4. *Proceedings of the National Academy of Sciences of the United States of America* 105, 7779-7784.

Hsin, H. & Kenyon, C. (1999). Signals from the reproductive system regulate the lifespan of *C. elegans*. *Nature* 27, 362-366.

Iatsenko, I., Sinha, A., Rödelsperger, C. & Sommer, R.J. (2013). A new role of DCR-1/Dicer in *C. elegans* innate immunity against the highly virulent bacterium *Bacillus thuringiensis DB27*. *Infection and Immunity* 81, 3942-3957.

Iatsenko, I., Corton, C., Pickard, D.J., Dougan, G. & Sommer, R.J. (2014a). Draft genome sequence of highly nematicidal *Bacilllus thuringiensis* DB27. *Genome Announcements* 2, e00101-14.

Iatsenko, I., Boichenko, J. & Sommer, R.J. (2014b). *Bacillus thuringiensis* DB27 produces two novel toxins, Cry21Fa1 and Cry21Ha1, which act synergistically against nematodes. *Applied and Environmental Microbiology* 80, 3266-3275.

Iatsenko, I., Nikolov, A. & Sommer, R.J. (2014c). Identification of *Bacillus thuringiensis* 4A4 nematicidal factors using the model nematodes *Pristionchus pacificus* and *Caenorhabditis elegans*. *Toxins* 6, 2050-2063.

Jatala, P., Jensen, H.J. & Russell, S.A. (1974). *Pristionchus lheritieri* as a carrier of *Rhizobium japonicum*. *Journal of Nematology* 6, 130-131.

Jensen, H.J. & Siemer, S.R. (1971). Protection of *Fusarium* and *Verticillium* propagules from selected biocides following ingestion by *Pristionchus lheritieri*. *Journal of Nematology* 3, 23-27.

Kleino, A. & Silverman, N. (2012). UnZIPping mechanisms of effector-triggered immunity in animals. *Cell Host & Microbe* 11, 320-322.

Leake, P.A. & Jensen, H.J. (1970). Survival of chlorophyceae ingested by saprozoic nematodes. *Journal of Nematology* 2, 351-354.

Marroquin, L.D., Elyassnia, D., Griffitts, J.S., Feitelson, J.S. & Aroian, R.V. (2000). *Bacillus thuringiensis* (Bt) toxin susceptibility and isolation of resistance mutants in the nematode *Caenorhabditis elegans*. *Genetics* 155, 1693-1699.

Martin, P.A. & Travers, R.S. (1989). Worldwide abundance and distribution of *Bacillus thuringiensis* isolates. *Applied and Environmental Microbiology* 55, 2437-2442.

Mayer, M.G. & Sommer, R.J. (2011). Natural variation in *Pristionchus pacificus* dauer formation reveals cross-preference rather than self-preference of nematode dauer pheromones. *Proceedings of the Royal Society B: Biological Sciences* 278, 2784-2790.

McEwan, D.L., Kirienko, N.V. & Ausubel, F.M. (2012). Host translational inhibition by *Pseudomonas aeruginosa* exotoxin A triggers an immune response in *Caenorhabditis elegans*. *Cell Host & Microbe* 11, 364-374.

Mohr, I. & Sonenberg, N. (2012). Host translation at the nexus of infection and immunity. *Cell Host & Microbe* 12, 470-483.

Musat, N., Giere, O., Gieseke, A., Thiermann, F., Amann, R. & Dubilier, N. (2007). Molecular and morphological characterization of the association between bacterial endosymbionts and the marine nematode *Astomonema* sp. from the Bahamas. *Environmental Microbiology* 9, 1345-1353.

Mylonakis, E., Ausubel, F.M., Perfect, J.R., Heitman, J. & Calderwood, S.B. (2002). Killing of *Caenorhabditis elegans* by *Cryptococcus neoformans* as a model of yeast pathogenesis. *Proceedings of the National Academy of Sciences of the United States of America* 99, 15675-15680.

Nour, S.M., Lawrence, J.R., Zhu, H., Swerhone, G.D., Welsh, M., Welacky, T.W. & Topp, E. (2003). Bacteria associated with cysts of the soybean cyst nematode (*Heterodera glycines*). *Applied and Environmental Microbiology* 69, 607-615.

Nussbaumer, A.D., Bright, M., Baranyi, C., Beiser, C.J. & Ott, J.A. (2004). Attachment mechanism in a highly specific association between ectosymbiotic bacteria and marine nematodes. *Aquatic Microbial Ecology* 34, 239-246.

Ogawa, A., Streit, A., Antebi, A. & Sommer, R.J. (2009). A conserved endocrine mechanism controls the formation of dauer and infective larvae in nematodes. *Current Biology* 19, 67-71.

Ott, J., Novak, R., Schiemer, F., Hentschel, U., Nebelsick, M. & Polz, M.F. (1991). Tackling the sulfide gradient: a novel strategy involving marine nematodes and chemoautotrophic ectosymbionts. *Marine Ecology* 12, 261-279.

Rae, R., Riebesell, M., Dinkelacker, I., Wang, Q., Herrmann, M., Weller, A.M., Dieterich, C. & Sommer, R.J. (2008). Isolation of naturally associated bacteria of necromenic *Pristionchus* nematodes and fitness consequences. *Journal of Experimental Biology* 211, 1927-1936.

Rae, R., Iatsenko, I., Witte, H. & Sommer, R.J. (2010). Naturally isolated *Bacillus* strains show extreme virulence to the free-living nematodes *Caenorhabditis elegans* and *Pristionchus pacificus*. *Environmental Microbiology* 11, 3007-3021.

Rae, R., Witte, H., Rödelsperger, C. & Sommer, R.J. (2012a). The importance of being regular: *Caenorhabditis elegans* and *Pristionchus pacificus* defecation mutants are hypersusceptible to bacterial pathogens. *International Journal for Parasitology* 42, 747-753.

RAE, R., SINHA, A. & SOMMER, R.J. (2012b). Genome wide analysis of germline signaling genes regulating longevity and innate immunity in the nematode *Pristionchus pacificus*. *PLoS Pathogens* 8, e1002864.

SCHULENBURG, H., HOEPPNER, M.P., WEINER 3RD, J. & BORNBERG-BAUER, E. (2008). Specificity of the innate immune system and diversity of C-type lectin domain (CTLD) proteins in the nematode *Caenorhabditis elegans*. *Immunobiology* 213, 237-250.

SINHA, A., RAE, R., IATSENKO, I. & SOMMER, R.J. (2012a). System wide analysis of the evolution of innate immunity in the nematode model species *Caenorhabditis elegans* and *Pristionchus pacificus*. *PLoS ONE* 7, e44255.

SINHA, A., SOMMER, R.J. & DIETERICH, C. (2012b). Divergent gene expression in the conserved dauer stage of the nematodes *Pristionchus pacificus* and *Caenorhabditis elegans*. *BMC Genomics* 13, 254.

SMERDA, S.M., JENSEN, H.J. & ANDERSON, A.W. (1971). Escape of Salmonellae from chlorination during ingestion by *Pristionchus lheritieri* (Nematoda: Diplogasterinae). *Journal of Nematology* 3, 201-204.

SOMMER, R.J., CARTA, L.K., KIM, S.-Y. & STERNBERG, P.W. (1996). Morphological, genetic and molecular description of *Pristionchus pacificus* sp. n. (Nematoda, Diplogastridae). *Fundamental and Applied Nematology* 19, 511-521.

TAN, M.-W., MAHAJAN-MIKLOS, S. & AUSUBEL, F.M. (1999). Killing of *Caenorhabditis elegans* by *Pseudomonas aeruginosa* used to model mammalian bacterial pathogenesis. *Proceedings of the National Academy of Sciences of the United States of America* 96, 715-720.

TANG, R.J., BREGER, J., IDNURM, A., GERIK, K.J., LODGE, J.K., HEITMAN, J., CALDERWOOD, S.B. & MYLONAKIS, E. (2005). *Cryptococcus neoformans* gene involved in mammalian pathogenesis identified by a *Caenorhabditis elegans* progeny-based approach. *Infection and Immunity* 73, 8219-8225.

TAYLOR, M.J., BANDY, C. & HOERAUF, A. (2005). *Wolbachia* bacterial endosymbionts of filarial nematodes. *Advances in Parasitology* 60, 245-282.

VAN DEN BERG, M.C.W., WOERLEE, J.Z., MA, H. & MAY, R.C. (2006). Sex-dependent resistance to the pathogenic fungus *Cryptococcus neoformans*. *Genetics* 173, 677-683.

VANDEKERCKHOVE, T.T., WILLEMS, A., GILLIS, M. & COOMANS, A. (2000). Occurrence of novel *Verrucomicrobial* species, endosymbiotic and associated with parthenogenesis in *Xiphinema americanum* group species (Nematoda, Longidoridae). *International Journal of Systematic and Evolutionary Microbiology* 50, 2197-2205.

WEI, J.Z., HALE, K., CARTA, L., PLATZER, E., WONG, C., FANG, S.C. & AROIAN, R.V. (2003). *Bacillus thuringiensis* crystal proteins that target nematodes. *Proceedings of the National Academy of Sciences of the United States of America* 100, 2760-2765.

WELLER, A., MAYER, W., RAE, R. & SOMMER, R.J. (2010). Quantitative assessment of the nematode fauna present on *Geotrupes* dung beetles reveals species-rich communities with a heterogenous distribution. *Journal of Parasitology* 96, 525-531.

WONG, D., BAZOPOULOU, D., PUJOL, N., TAVERNARAKIS, N. & EWBANK, J.J. (2007). Genome-wide investigation reveals pathogen-specific and shared signatures in the response of *Caenorhabditis elegans* to infection. *Genome Biology* 8, R194.

Nematology Monographs & Perspectives, 2015, Vol. 11, 409-410

Index of genes and proteins

Genes are listed without the *Cel-* or *Ppa-* prefix.

Index of genera and species

For a full list of species and synonyms of *Pristionchus* see Ragsdale *et al.* (Chapter 4, this volume).

General index

www.ingramcontent.com/pod-product-compliance
Lightning Source LLC
Chambersburg PA
CBHW071231230926
51849CB00001B/71
9789004260290